Timber's Green Revolution: Sustainable Construction from Start to Finish

Timber's Green Revolution: Sustainable Construction from Start to Finish

Guest Editor

Manuel Carpio

MDPI

Basel • Beijing • Wuhan • Barcelona • Belgrade • Novi Sad • Cluj • Manchester

Guest Editor
Manuel Carpio
Department of Construction
Engineering and Project
Management
University of Granada
Granada
Spain

Editorial Office
MDPI AG
Grosspeteranlage 5
4052 Basel, Switzerland

This is a reprint of the Special Issue, published open access by the journal *Sustainability* (ISSN 2071-1050), freely accessible at: https://www.mdpi.com/journal/sustainability/special_issues/SI1L17HL6L.

For citation purposes, cite each article independently as indicated on the article page online and as indicated below:

Lastname, A.A.; Lastname, B.B. Article Title. *Journal Name* **Year**, *Volume Number*, Page Range.

ISBN 978-3-7258-7386-9 (Hbk)
ISBN 978-3-7258-7387-6 (PDF)
https://doi.org/10.3390/books978-3-7258-7387-6

Cover image courtesy of Centro Nacional de Excelencia para la Industria de la Madera (CENAMAD)

Contents

About the Editor

Manuel Carpio

Manuel Carpio is a professor at the Department of Construction Engineering and Project Management, University of Granada, Spain, and a researcher affiliated with the National Center of Excellence for the Timber Industry in Chile. His research focuses on sustainable construction, timber engineering, life cycle assessment, and the environmental performance of buildings, with particular emphasis on mass timber systems and low carbon structural solutions. He is actively involved in research projects addressing circular economy strategies, carbon storage in the built environment, and the integration of timber technologies in urban development. He has contributed to numerous scientific publications in the field of sustainable construction. His work bridges sustainability assessment, and project management, promoting evidence based strategies for reducing embodied emissions in the construction sector.

sustainability

MDPI

Editorial

Timber Construction as a Systemic Pathway Towards Sustainable Built Environments

Manuel Carpio [1,2,3]

[1] Department of Construction Engineering and Project Management, University of Granada, 18071 Granada, Spain; carpio@ugr.es
[2] Department of Construction Engineering and Management, School of Engineering, Pontificia Universidad Católica de Chile, Santiago 7820436, Chile
[3] Centro Nacional de Excelencia para la Industria de la Madera (CENAMAD), Pontificia Universidad Católica de Chile, Santiago 7820436, Chile

The construction sector is at a critical crossroads [1]. Achieving sustainable built environments now requires systemic solutions that integrate materials, processes, and territorial conditions, rather than isolated technological improvements [2,3]. Material choice has therefore become a strategic decision rather than a purely technical one [4,5].

Within this context, timber has re-emerged as a key material for sustainable construction. Its renewability, capacity for biogenic carbon storage, and compatibility with industrialised and prefabricated processes position it as a central component of low-carbon construction strategies [6,7]. However, the sustainability of timber construction cannot be assessed at the level of individual products or buildings alone; it must be examined across the full value chain, from territorial and forest-related conditions to material processing, structural performance, and urban-scale impacts [8,9].

This Special Issue, "Timber Construction as a Systemic Pathway Towards Sustainable Built Environments", was conceived to address timber construction as a continuous and interconnected system. The eleven contributions approach this system from complementary perspectives, covering upstream environmental baselines, material and process innovation, construction-scale performance, and city-level implications.

At the upstream end of the value chain, two contributions focus on the environmental conditions that underpin any credible timber-based construction strategy. One study, on *Pinus radiata* plantations, analyses the relationship between carbon stock and cumulative production at harvesting age, demonstrating how soil conditions influence both productivity and carbon storage potential (1). In parallel, the assessment of agricultural land expansion and groundwater sustainability in north-central Chile highlights how land-use change and water scarcity can constrain long-term territorial sustainability (2). Together, these works emphasise that the environmental performance of timber construction is inseparable from resilient land, forest, and water management.

Material innovation and value recovery are subsequently explored through bio-based and circular approaches. The applicability of pulp and paper industry waste in manufacturing mycelium-based thermoacoustic insulation materials illustrates how industrial by-products can be transformed into functional construction solutions (3). Complementarily, the development of cement pastes incorporating wheat straw ash and silica fume contributes to a reduction in embodied carbon through agricultural residue valorisation and hybrid material systems (4). Although not timber-based, this contribution reinforces the transversal nature of sustainable construction and the importance of low-impact material portfolios.

https://doi.org/10.3390/su18052176

At the product and process scale, engineered wood performance and manufacturing quality are addressed. An investigation into green-glued finger joints examines the influence of wood ring orientation on adhesion performance, providing practical knowledge to improve reliability, material efficiency, and the use of underutilised timber resources (5). Such process-level insights are essential, as connection behaviour often governs the structural performance and robustness of modern timber systems.

Several contributions adopt a broader perspective, analysing how timber construction knowledge evolves and is translated into practice. A bibliometric review of prefabricated and modular timber construction maps research development from 1990 to 2023, identifying dominant themes, emerging trends, and persistent challenges related to industrialisation and circular economy principles (6). A science mapping study focused on mid-rise and tall timber buildings highlights the growing relevance of timber in urban construction while underlining the need to align environmental ambition with structural feasibility and regulatory acceptance (7). In addition, a bibliometric analysis centred on urban coastal development situates timber housing within climate-resilient planning strategies, reinforcing the link between timber construction and adaptation-orientated urban policies (8).

Downstream, whole-building and urban-scale performance are examined. A life cycle assessment and cost analysis comparing mid-rise mass timber and concrete buildings in Australia demonstrate that timber solutions can provide both environmental and economic advantages when assessed across the building life cycle (9). At the city scale, the concept of urban timber buildings as a "second forest" reframes timber structures as long-term biogenic carbon stocks, positioning cities as active components in carbon management strategies (10).

Finally, this Special Issue addresses the transformation of the existing building stock. The proposed use of laminated timber elements to enhance the seismic performance of reinforced concrete buildings illustrates how timber can contribute to retrofit and strengthening strategies, extending its role beyond new construction while supporting sustainability objectives (11).

Together, the contributions to this Special Issue demonstrate that timber construction represents a systemic pathway towards sustainable built environments rather than a singular material substitution. Progress depends on linking robust territorial and environmental baselines, innovative and efficient material processing, reliable engineered products, and credible performance at the building and urban scales. By addressing timber construction as an integrated system, this Special Issue contributes to a deeper understanding of how timber can support climate-resilient, resource-efficient, and scalable construction strategies in the decades ahead.

Funding: This work was funded by the projects ANID BASAL FB210015 CENAMAD and BG23/00134.

Acknowledgments: The Editor would like to thank all the authors who contributed to this Special Issue for their commitment and the high quality of their research.

Conflicts of Interest: The authors declare no conflicts of interest.

List of Contributions:

1. Asmussen, M.V.; Rubilar, R.; Bozo, D.; Alzamora, R.M.; Elissetche, J.P.; Pincheira, M.; Jara, O. Relationship Between Carbon Stock and Stand Cumulative Production at Harvesting Age of *Pinus radiata* Plantations: A Comparison Between Granitic and Metamorphic Soils. *Sustainability* **2025**, *17*, 3614.
2. Pizarro, R.; Borcoski, F.; Ingram, B.; Bustamante-Ortega, R.; Sangüesa, C.; Ibáñez, A.; Toledo, C.; Vidal, C.; Garcia-Chevesich, P.A. Increases in the Amounts of Agricultural Surfaces and Their

Impact on the Sustainability of Groundwater Resources in North-Central Chile. *Sustainability* **2024**, *16*, 7570.
3. Muñoz, H.; Molina, P.; Urzúa-Parra, I.A.; Vasco, D.A.; Walczak, M.; Rodríguez-Grau, G.; Chateau, F.; Sancy, M. Applicability of Paper and Pulp Industry Waste for Manufacturing Mycelium-Based Materials for Thermoacoustic Insulation. *Sustainability* **2024**, *16*, 8034.
4. Bastías, B.; González, M.; Rey-Rey, J.; Valerio, G.; Guindos, P. Sustainable Cement Paste Development Using Wheat Straw Ash and Silica Fume Replacement Model. *Sustainability* **2024**, *16*, 11226.
5. Rodríguez-Grau, G.; Cordonnier, P.-L.; Navarrete, B.; Montero, C.; Alvarado, C.; Pommier, R.; Rosales, V.; Galarce, C. The Adhesion Performance in Green-Glued Finger Joints Using Different Wood Ring Orientations. *Sustainability* **2024**, *16*, 7158.
6. Gutiérrez, N.; Negrão, J.; Dias, A.; Guindos, P. Bibliometric Review of Prefabricated and Modular Timber Construction from 1990 to 2023: Evolution, Trends, and Current Challenges. *Sustainability* **2024**, *16*, 2134.
7. Wenzel, A.; Guindos, P.; Carpio, M. Using Timber in Mid-Rise and Tall Buildings to Construct Our Cities: A Science Mapping Study. *Sustainability* **2025**, *17*, 1928.
8. García-Ruiz, A.; Díez-Minguito, M.; Verichev, K.; Carpio, M.; Bibliometric Analysis of Urban Coastal Development: Strategies for Climate-Resilient Timber Housing. *Sustainability* **2024**, *16*, 1431.
9. Jolly, R.; Fairweather, H.; Rayburg, S.; Rodwell, J. Life Cycle Assessment and Cost Analysis of Mid-Rise Mass Timber vs. Concrete Buildings in Australia. *Sustainability* **2024**, *16*, 6465.
10. Victorero, F.; Bustamante, W. Timber Biogenic Carbon Stock in the Urban Environment: Santiago City as a Second Forest. *Sustainability* **2025**, *17*, 529.
11. Yıldız, Y.; Şermet, F. A New Approach to Improving the Seismic Performance of Existing Reinforced Concrete Buildings Using Laminated Timber. *Sustainability* **2025**, *17*, 7690.

References

1. Röck, M.; Rørvik, P.M.; Steininger, K.W.; Hoxha, E.; Passer, A.; Habert, G.; Saade, M.R.M.; Gustavsen, A. Embodied GHG emissions of buildings—The hidden challenge for effective climate change mitigation. *Appl. Energy* **2020**, *258*, 114107. [CrossRef]
2. United Nations Environment Programme; Yale Center for Ecosystems + Architecture. *Building Materials and the Climate: Constructing a New Future*; UNEP Report; UNEP: Nairobi, Kenya, 2023. [CrossRef]
3. Pomponi, F.; Moncaster, A. Circular economy for the built environment: A research framework. *J. Clean. Prod.* **2017**, *143*, 710–718. [CrossRef]
4. Soust-Verdaguer, B.; García-Martínez, A.; Rey-Álvarez, B.; de Diego, B.; de la Fuente, A.; Röck, M. Data infrastructure for whole-life carbon emissions baselines of buildings in Spain. *Energy Build.* **2025**, *347*, 116307. [CrossRef]
5. Nawarathna, A.; Alwan, Z.; Gledson, B.; Fernando, N. Embodied Carbon as a Material Selection Criterion: Insights from Sri Lankan Construction Sector. *Sustainability* **2021**, *13*, 2202. [CrossRef]
6. Duan, Z.; Huang, Q.; Zhang, Q. Life cycle assessment of mass timber construction: A review. *Build. Environ.* **2022**, *221*, 109320. [CrossRef]
7. Kumar, V.; Lo Ricco, M.; Bergman, R.D.; Nepal, P.; Poudyal, N.C. Environmental impact assessment of mass timber, structural steel, and reinforced concrete buildings based on the 2021 International Building Code provisions. *Build. Environ.* **2024**, *251*, 111195. [CrossRef]
8. Andersen, J.H.; Rasmussen, N.L.; Ryberg, M.W. Comparative life cycle assessment of cross laminated timber building and concrete building with special focus on biogenic carbon. *Energy Build.* **2022**, *254*, 111604. [CrossRef]
9. Hoxha, E.; Pittau, F.; Habert, G.; Passer, A. Biogenic carbon in buildings: A critical overview of LCA methods. *Build. Cities* **2020**, *1*, 504–524. [CrossRef]

https://doi.org/10.3390/su18052176

sustainability

MDPI

Article

Relationship Between Carbon Stock and Stand Cumulative Production at Harvesting Age of *Pinus radiata* Plantations: A Comparison Between Granitic and Metamorphic Soils

Marianne V. Asmussen [1,2], Rafael Rubilar [1,2,*], Daniel Bozo [1,2], Rosa M. Alzamora [2,3], Juan Pedro Elissetche [2,3], Matías Pincheira [4] and Oscar Jara [1,2]

[1] Cooperativa de Productividad Forestal, Departamento de Silvicultura, Facultad de Ciencias Forestales, Universidad de Concepción, Concepción 4030555, Chile; marianneasmussen@gmail.com (M.V.A.); dbozo@udec.cl (D.B.); ojara2016@udec.cl (O.J.)

[2] Centro Nacional de Excelencia para la Industria de la Madera (CENAMAD—ANID BASAL FB210015), Pontificia Universidad Católica de Chile, Santiago 7820436, Chile; ralzamora@udec.cl (R.M.A.); jelisset@udec.cl (J.P.E.)

[3] Departamento de Manejo de Bosques y Medio Ambiente, Facultad de Ciencias Forestales, Universidad de Concepción, Concepción 4030000, Chile

[4] Forestal Mininco S.A., Avenida Alemania 751, Los Ángeles 4440000, Chile; matias.pincheira@cmpc.cl

* Correspondence: rafaelrubilar@udec.cl

Abstract: The relationship between stand cumulative production at harvesting age and carbon stock for different soil types in forest plantations is critical for sustainable forest management and climate change mitigation. This study evaluated carbon stocks in *Pinus radiata* D. Don on granitic and metamorphic soils in central Chile. We selected 10 plantations and established three 1000 m^2 stands per plantation to quantify the carbon stock of total biomass using allometric equations and in situ carbon assessments of the forest floor and mineral soil (up to 1 m deep). A strong positive correlation was observed between stand cumulative production at harvesting age and total carbon stock (r^2 = 0.767), regardless of the soil type. Metamorphic and granitic soils demonstrated a high carbon stock capacity, particularly in deeper soil layers (40–100 cm), with soil contributing over 40% of the total soil carbon stock. Soil bulk density and carbon concentration were strongly correlated (r^2 = 0.74), emphasizing the role of soil physical properties in carbon storage at deep soil horizons. These findings highlight the critical role of subsoils as carbon reservoirs. Predictive linear models developed from this study offer a useful and simple approach for estimating carbon stocks, contributing to national carbon neutrality goals and sustainable forest management.

Keywords: *Pinus radiata*; carbon stock; productivity; metamorphic soils; granitic soils; predictive models

1. Introduction

The increase in greenhouse gas concentrations over the past century, mainly due to anthropogenic activities, has significantly contributed to global warming and changes in global temperature and precipitation patterns [1–3]. Forests play a critical role in mitigating climate change by acting as carbon sinks, offsetting the effects of rising greenhouse gases. Global forest carbon stocks are estimated to be about 861 Pg C., representing 29% of the carbon in major terrestrial and oceanic reservoirs [3]. Of this forest carbon, 44% is stored in soils, 42% in above and below-ground biomass, 8% in deadwood, and 5% in litter [4,5].

 https://doi.org/10.3390/su17083614

This distribution highlights the importance of all components of forest ecosystems in the global carbon cycle. The soil organic matter (SOM) is a key component of soil, which contains more organic carbon than global vegetation and the atmosphere combined [6]. Also, factors such as stand age [7–9], species composition [10–13], climate, site conditions, and soil type [13–15] significantly influence carbon accumulation, with more productive sites having greater carbon storage potential in both biomass and soil [13,16].

Currently, native forests and forestry plantations coexist in the ecosystems, necessitating a balanced approach that leverages the strengths of both systems to maximize carbon storage and enhance ecosystem resilience against climate change [17]. Afforestation and reforestation in managed forests are important strategies for mitigating climate change, especially in degraded or non-forested areas [18,19]. These practices not only enhance CO_2 sequestration but also improve ecosystem services, biodiversity, and soil quality [20–23]. Developing knowledge about the carbon stock capacity of plantations, along with implementing sustainable management practices, is essential to ensure that these plantations continue to provide valuable goods and services while minimizing their environmental impacts [24–28].

In Chile, the forestry sector is a key component of the national economy, with *Pinus radiata* D. Don being the predominant non-native species covering approximately 1,299,451 hectares, which constitutes about 55.9% of the total forest plantation area [29]. This species is recognized for its rapid growth and high biomass production that depends on advanced genotypes, silviculture, and environmental conditions [30–32]. Numerous studies worldwide have explored carbon stock in *Pinus radiata* plantations [13,33–36], highlighting the global interest and efforts in understanding and optimizing carbon stock in these ecosystems. Understanding the carbon stock capacity of *Pinus radiata* plantations is essential; however, most existing studies have predominantly focused on carbon stocks in above-ground biomass, with limited attention given to soils, especially deep horizons, as significant carbon reservoirs [4,33,37]. This gap underscores the importance of assessing carbon stocks at greater soil depths to fully capture the sequestration potential of different soil types [36,38]. Soil types such as sandy, recent volcanic ash, metamorphic, and granitic exhibit distinct physicochemical and biological characteristics that influence their capacity to stock carbon. These differences are primarily driven by variations in organic matter dynamics, nutrient availability, and microbial activity, which collectively govern soil carbon storage and stabilization [39,40].

Sandy soils, characterized by coarse texture and low water and nutrient retention, generally exhibit limited carbon storage potential due to reduced organic matter accumulation and microbial activity [36,41,42]. In contrast, recent volcanic ash soils initially provide high nutrient availability that supports rapid plant growth and carbon input; however, their long-term sequestration potential depends on factors such as soil stabilization and sustained organic matter inputs [36,43,44]. Metamorphic soils, with their complex mineral compositions, are conducive to stabilizing organic matter and reducing decomposition rates, making them effective for long-term carbon storage [45,46]. Granitic soils, while limited by coarse texture and low nutrient-holding capacity, can enhance their carbon sequestration potential through management interventions, such as fertilization, that improve vegetation growth and organic matter accumulation [47,48]. Recent analyses of *Pinus radiata* plantations in Chile reveal significant variability in soil carbon stocks across parent materials. The highest total stocks were observed in soils from marine sediments and recent volcanic ash (445.4 and 473.2 $Mg \cdot ha^{-1}$), while the lowest were in lacustrine soils (144.4 $Mg \cdot ha^{-1}$). Within the 0–30 cm depth, soils from recent volcanic ash over metamorphic rock had the highest carbon values (171.3–172.6 $Mg \cdot ha^{-1}$), followed by old volcanic ash over marine sediments. Lacustrine soils consistently showed the lowest values, with just 37.6 Mg·ha^{-1} in this layer [13,36].

This study evaluated the relationship between total carbon stock in Pinus radiata plantations and stand cumulative production at harvesting age. A model was developed integrating information on carbon stock reservoirs in biomass, forest floor, and soil up to 1 m in depth, based on field data from plantations established on granitic and metamorphic soils—two predominant soil types in Chilean forestry [13]. The objective is to provide a predictive model that enhances the understanding of carbon stocks under different productive conditions. Our findings could serve as a practical tool for landowners to assess how their management practices align with national carbon neutrality goals, contributing to climate change mitigation.

2. Materials and Methods

2.1. Study Area and Experimental Design

The analysis was carried out across ten *Pinus radiata* plantations, selected to represent a stand cumulative production at harvesting age gradient (18–24 years old, the age at which radiata pine plantations are harvested in Chile). The plantations are located in central Chile, spanning the Maule (latitude 35°25′ S) to Araucanía (latitude 38°54′ S) administrative regions. This area features a Mediterranean climate with significant variation in environmental conditions. Mean annual precipitation ranges from 950 mm in the northern areas to 2500 mm in the southern regions, while mean annual temperature decreases progressively from 19 °C in the north to 11 °C in the south (Figure 1). All selected sites correspond to afforestation areas, meaning that plantations were established on previously highly degraded agricultural soils. All plantations followed conventional soil preparation practices, including ripping to a depth of 0.6 cm and disking. Standard operational fertilization was applied, with NPK 15–13.2–12.5 (100–150 g tree^{-1}) and an additional 2 to 3 g of boron per plant, administered after planting. Annual weed control was conducted, followed by banded herbicide applications for two years after plantation establishment. Each plantation was pruned to a minimum height of 5.5 m and underwent one or two commercial thinnings from below, the first to 700 trees per hectare and the second to the final stocking density ranging between 400 and 550 trees per hectare, depending on the specific silvicultural regime of the stand.

Field measurements were conducted between May and November 2023 in plantations under intensive management. The experimental design includes 5 plantations per soil type, metamorphic and granitic [49,50], with 3 independent stands of 1000 m^2 established at each plantation, resulting in a total of 30 independent stands across all evaluated plantations. The selected plantations were planted between 1998 and 2004, each consisting of a single cohort of trees, meaning that all trees within a plantation were of the same age. In each stand, the diameter at breast height (DBH; measured at 1.3 m) and total height (H) were measured for all trees. Additionally, 10 forest floor and 15 soil samples were collected within each stand according to a zig-zag sample design.

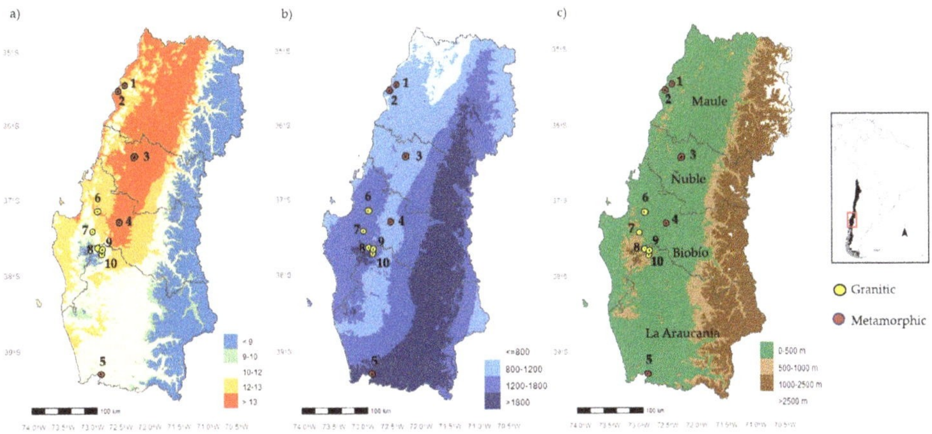

Figure 1. Location of *Pinus radiata* plantations evaluated per soil type and climate variability. (**a**) Mean annual temperature (°C); (**b**) Mean annual precipitation (mm) and (**c**) elevation (m a.s.l.) and administrative regions. Red points represent plantations on metamorphic soils, while yellow points represent plantations on granitic soils. Plantation code: 1–SPC402, 2–EP530, 3–HH252, 4–LQ381, 5–SF690, 6–SM401, 7–HO669, 8–PM437, 9–LD397 and 10–LN574.

2.2. Stand Cumulative Production at Harvesting Age Estimation

Stand cumulative production at harvesting age (S_p, $m^3\ ha^{-1}$) refers to the current volume of the stand at the time of harvest, excluding the volume removed during thinning. To estimate the S_p, the volume for individual trees over bank (V_{tree}, $m^3\ tree^{-1}$) was calculated using the allometric Equation (1) established by CMPC Forestal Mininco (1995) [51,52]:

$$V_{\text{tree}} = -0.00214 + 0.0000295 * \text{DBH2} + 0.001349 * H + 0.00002486 * \text{DBH2} * H \quad (1)$$

where the sum of all V_{tree} per stand was multiplied by 10 to scale the volume on a per-hectare basis, yielding the stand's cumulative production at harvesting age (S_p). Finally, the plantation cumulative production at harvesting age (Pl_p) was then calculated as the mean of the three S_p estimates ($m^3\ ha^{-1}$) in each plantation evaluated.

2.3. Carbon Stock in Above and Belowground Biomass

The aboveground biomass of each tree was calculated using the following published allometric Equations (2)–(5) developed for Pinus radiata plantations [53]:

$$AGB_s = 0.02389 * (D^2 * H)^{0.93216} \quad (2)$$

$$AGB_b = 0.00127 * (D^2 * H)^{0.99646} \quad (3)$$

$$AGB_{br} = 0.00431 * (D^2 * H)^{0.92709} \quad (4)$$

$$AGB_n = 0.19428 * (D^2 * H)^{0.48666} \quad (5)$$

where AGB_s is the steam biomass, AGB_b is the bark biomass, AGB_{br} is the branch, and AGB_n is the needle biomass, all in kg $tree^{-1}$. The total aboveground biomass (AGB_{total}) was then calculated as the sum of all components as Equation (6):

$$AGB_{total} = AGB_s + AGB_b + AGB_{br} + AGB_n \quad (6)$$

The belowground biomass (BGB, kg tree^{-1}). of each tree was calculated using Equation (7), which estimates the total root biomass as follows [53,54]:

$$BGB = \exp\,(0.902 * ln(AGBtotal) - 0.7368) \quad (7)$$

The carbon stock of each individual tree was estimated by multiplying AGB_{total} and BGB by a carbon factor (CF), representing the carbon fraction in each biomass component. For this calculation, a CF of 0.48 was used, as recommended by the IPCC (2013) [55]. The carbon stocks of all trees within each stand were summed, and the mean value across the three stands of each plantation was calculated. This value was then scaled to a per hectare basis to estimate the aboveground biomass carbon stock (CAGB, Mg·ha^{-1}) and belowground biomass carbon stock (CBGB, Mg·ha^{-1}) for each plantation. The sum of both estimates provided the aboveground and belowground carbon stocks (CTGB).

2.4. Forest Floor Sampling and Carbon Estimations

Forest floor samples were collected from all plantations. Ten sample points of litter (L) and coarse woody debris (W) were taken in each stand using a circular cutting frame (25 cm diameter). In total, 600 forest floor samples (10 sample points × 2 sample types (L and W) × 3 stands × 10 plantations) were obtained. Each floor sample was oven-dried at 65 °C for 48 h and weighed. The dried samples for each forest floor type were composited, homogenized, and ground using a 250 µm sieve blade mill for each stand. A 5 g subsample was obtained to analyze the carbon concentration (C) in litter and coarse woody debris of each stand. The values obtained were averaged to provide a value per plantation (FF_L and FF_W). The forest floor carbon stock for each sample type per plantation (FFC_L and FFC_W) was determined by multiplying the dry biomass weight by the carbon concentration (C) and then scaled to a per hectare basis. Finally, the total forest floor carbon stock per plantation (FFC_T) was calculated by summing the FFC_L and FFC_W values. All analyses were conducted at the Soil, Water, and Forest Research Laboratory (LISAB, by its Spanish acronym) Faculty of Forest Sciences, University of Concepción, Chile.

2.5. Soil Sampling and Carbon Estimations

Soil samples were collected from 10 plantations at three depths (0–20, 20–40, and 40–100 cm) using an auger, following a zigzag sampling design. Fifteen samples were taken per depth within each stand and then pooled into a bucket per depth. After completing the sampling for each stand, the contents of each bucket were thoroughly mixed on a tarp to create a composite sample of approximately 250 g per depth. In total, 90 composite soil samples (3 depths × 3 stands × 10 plantations) were obtained. The samples were air-dried and sieved through a 2 mm mesh to remove roots and plant debris. A 10 g subsample of each composite sample was oven-dried at 65 °C for 24 h, and a 5 g subsample was subsequently analyzed for organic carbon (Cd) and total nitrogen (Nd) concentrations using an Infrared Mass Spectrometer (IRMS, SERCON).

Additional soil samples were collected from each stand at three depths (0–20, 20–40, and 40–100 cm) using a metal cylinder with a volume of 100 cm^3. These samples were oven-dried at 65 °C for 24 h to determine their dry mass. Bulk density (BDd, g/cm^3) was calculated as the ratio of the oven-dried mass to the cylinder volume, and soil texture was determined via the Bouyoucos hydrometer method. All analyses were conducted at the Soil, Water, and Forest Research Laboratory (LISAB, by its Spanish acronym) Faculty of Forest Sciences, University of Concepción, Chile.

Soil organic carbon stock for each depth (SOCd, Mg·ha^{-1}) was calculated by using Equation (8) [56]:

$$SOCd = Cd * BD * d * (1 - \delta) \quad (8)$$

where Cd is the soil organic carbon concentration, BD is the soil bulk density, d is the depth (cm), and δ is the proportion of rock fragments (>2 mm).

The mean value for each depth per plantation was calculated, and the total soil organic carbon stock (SOC_T) until 1 m of depth per plantation was determined by summing the mean SOCd estimates at each depth.

2.6. Statistical Analysis

The software R version 4.3.3 [57] was used for all statistical analyses. We tested the data for independence of residuals, equality of variance (homoscedasticity), and normality of residuals using diagnostic plots and Shapiro–Wilk and Levene's tests. Analyses were performed without data transformation. An analysis of variance (ANOVA) was conducted for each carbon reservoir, total carbon stock, and all stand inventory estimates to evaluate mean differences between soil types. Regression models were used to determine the relationship between stand cumulative production at harvesting age and total carbon stock with the function lm () in R. The models were selected based on their relevance, i.e., adjusted coefficient of determination r^2 closer to 1 and lower p-value. Additionally, predictive models were developed to estimate carbon stock based on the factors considered in the analysis. These predictions were generated using the total carbon stock per plantation (C_Total), which was calculated as the sum of carbon stocks estimated in the reservoirs, including soil carbon (SOC_T), forest floor carbon (FFC_T), and biomass carbon (CTGB). The predictive models were developed to estimate C_Total based on S_p and soil type (SoilType, categorical: granitic or metamorphic). The selected model included an interaction term to account for potential differences in the relationship between S_p and C_Total across soil types:

$$C\_Total = \beta0 + \beta1\ (S_p) + \beta\ (SoilType) + \beta3\ (S_p \times SoilType) + \epsilon \quad (9)$$

Predictions were generated by creating a sequence of S_p values and combining them with the two soil type categories. Using the fitted model, we applied the predict() function in R to estimate C_Total under different combinations of predictor variables.

3. Results

3.1. Plantation Cumulative Production at Harvesting Age

No significant differences were observed on average for diameter at breast height (DBH, p = 0.96), total height (H, p = 0.68), number of trees (N_{trees}, p = 0.33), and plantation cumulative production at harvesting age (Pl_p, p = 0.7) estimates between metamorphic and granitic soils (Table 1). Results showed the existence of a gradient of cumulative production at harvesting age among the plantations for both soil types.

Table 1. Mean stand characteristics for each plantation.

Administrative Region	Plantation Code	Soil Type	Soil Order [a]	Age (Years)	N_{trees} (Tree ha^{-1})	DBH (Tree ha^{-1})	H (m)	Pl_p (m^3 ha^{-1})
Biobío	SM-362	Granitic	Inceptisol	18.8	550.0	29.1 (+1.5)	27.3 (+0.6)	362.9 (+29.0)
La Araucanía	LD-397	Granitic	Inceptisol	19.8	473.3	30.4 (+0.8)	30.6 (+0.4)	397.9 (+6.8)
La Araucanía	PM-437	Granitic	Ultisol	21.8	503.3	34.2 (+1.0)	25.8 (+0.1)	437.4 (+28.1)
La Araucanía	LN-574	Granitic	Ultisol	20.8	510.0	36.6 (+1.6)	30.6 (+0.3)	574.6 (+35.0)
Biobío	HO-669	Granitic	Inceptisol	24.8	526.7	36.0 (+0.7)	34.2 (+1.2)	669.8 (+15.3)
Ñuble	HH-252	Metamorphic	Entisol	19.8	496.7	27.7 (+0.9)	22.7 (+1.6)	252.7 (+28.9)
Biobío	LQ-381	Metamorphic	Inceptisol	20.8	550.0	29.7 (+1.0)	27.1 (+1.2)	381.5 (+64.8)
Maule	SPC-402	Metamorphic	Alfisol	19.8	473.3	32.0 (+0.8)	30.0 (+1.3)	402.8 (+19.0)
Maule	EP-530	Metamorphic	Alfisol	20.8	490.0	35.8 (+2.2)	30.2 (+1.2)	530.1 (+50.8)
La Araucanía	SF-690	Metamorphic	Ultisoles	24.8	436.7	40.6 (+0.7)	33.5 (+0.6)	690.4 (+6.0)

Table 1. *Cont.*

Administrative Region	Plantation Code	Soil Type	Soil Order [a]	Age (Years)	N_{trees} (Tree ha^{-1})	DBH (Tree ha^{-1})	H (m)	Pl_p (m^3 ha^{-1})
	Granitic Soils (Mean)				512.7 ns	33.3 ns	29.7 ns	488.5 ns
	Metamorphic (Mean)				489.3 ns	33.2 ns	28.7 ns	451.5 ns
	p-value				0.33	0.96	0.68	0.7

[a] [58], Standard errors of the mean (SEM) are given in parentheses. Tree number (N_{trees}), diameter at breast height (DBH), tree height (H), and plantation cumulative production at harvesting age (Pl_p). (ns) indicate no-significant differences between the means of soil types considering an ANOVA analysis ($p < 0.05$).

3.2. Soil Horizon Characteristics

We evaluated bulk density, C/N ratio, carbon concentration, and the percentage of clay, sand, and silt in three soil horizons from forest plantations with varying Pl_p levels using soil pits (Table 2). Linear regression analyses for each soil type showed a significant relationship between carbon concentration and bulk density, in which higher carbon concentrations were associated with lower bulk density values. High Pl_p, such as SF-690 (690.4 m^3·ha^{-1}) and HO-669 (669.8 m^3·ha^{-1}), are associated with surface horizons (0–20 cm) exhibiting higher total carbon content (4.33% and 5.78%, respectively) and lower bulk density (1.16 g·cm^3 and 0.71 g·cm^3, respectively). The relationships were stronger in metamorphic soils ($r^2 = 0.74$, p-value < 0.0001) compared to granitic soil ($r^2 = 0.5$, p-value < 0.0002).

Table 2. Soil horizon characteristics based on pit descriptions.

Soil Type	Plantation Code	Depth (cm)	B.D. (g cm^3)	C total	C/N.	Clay %	Silt	Sand
Granitic soil	SM-362	0–20	1.49	1.61	24	24.0	26.4	49.7
		20–40	1.38	0.91	10	27.9	26.7	45.4
		40–100	1.44	0.6	5	38.1	43.0	18.9
	LD-397	0–20	1.35	4.12	31	11.7	33.2	55.1
		20–40	1.28	0.63	25	24.7	30.2	45.2
		40–100	1.46	0.23	2	29.8	29.4	40.8
	PM-437	0–20	1.23	1.76	23	27.6	25.8	46.6
		20–40	0.89	1.99	26	29.7	22.8	47.5
		40–100	1.36	0.45	14	35.2	31.1	33.6
	LN-574	0–20	0.71	5.78	21	26.2	24.2	49.6
		20–40	1.15	2.18	23	45.9	19.3	34.8
		40–100	1.21	1.01	16	27	11	61.8
	HO-669	0–20	1.23	2.56	29	24.0	38.4	37.6
		20–40	1.45	1.21	22	15.9	38.2	45.9
		40–100	1.50	0.6	13	50.9	22.4	26.7
Metamorphic	HH-252	0–20	1.58	0.75	15	15.7	54.0	30.3
		20–40	1.56	0.81	16	20.4	39.7	39.8
		40–100	1.71	0.25	9	27.1	25.6	47.3
	LQ-381	0–20	1.52	1.78	17	38.3	17.1	44.6
		20–40	1.42	1.32	14	30.1	54.0	15.8
		40–100	1.48	0.68	9	41.4	46.1	12.4
	SPC-402	0–20	1.20	2.97	31	31.7	52.0	16.3
		20–40	1.39	1.77	29	11.3	38.3	50.4
		40–100	1.43	0.39	10	23.1	38.9	38.0
	EP-530	0–20	1.16	4.33	21	33.7	41.4	24.9
		20–40	1.30	2.35	16	23.8	47.1	29.2
		40–100	1.39	2.17	14	49.4	30.1	20.9
	SF-690	0–20	0.92	11.97	14	64.1	28.2	7.7
		20–40	1.13	9.06	17	15.2	38.0	46.8
		40–100	1.29	3.81	12	13.9	23.1	63.0

B.D. = Soil Bulk density; C total = Total carbon; C/N = carbon-to-nitrogen ratio.

3.3. Carbon Stock

No significant difference was observed for aboveground and belowground carbon stock estimates between soil types (p-values of 0.46 and 0.41, respectively) (Table 3). The linear regression model analyzing CTGB as a function of stand cumulative production at harvesting age (S_P), combining data from both soils, showed an overall good fit (r^2 = 0.99, p-value = 1.92×10^{-41}). When analyzed separately by soil type, the regression lines showed almost identical intercepts (13.06) and slopes (0.318). This indicates that soil type does not significantly alter the relationship between S_P and CTGB, emphasizing the dominant role of cumulative production at harvesting age in determining biomass carbon storage (Figure 2a).

Table 3. Mean carbon stock across soil types and plantations.

Plantation Code	Soil Type	Pl_P ($m^3\ ha^{-1}$)	CAGB ($Mg{\cdot}ha^{-1}$)	CBGB ($Mg{\cdot}ha^{-1}$)	FFC_T ($Mg{\cdot}ha^{-1}$)	SOC_T ($Mg{\cdot}ha^{-1}$)	C Total ($Mg{\cdot}ha^{-1}$)
SM-362	Granitic soil	362.9 (±29.0)	102	27	19.8	143	291.8
LD-397	Granitic soil	397.9 (±6.8)	110	28	14.64	220	372.64
PM-437	Granitic soil	437.4 (±28.1)	121	31	9.24	181	342.24
LN-574	Granitic soil	574.6 (±35.0)	157	40	24.06	353	574.06
HO-669	Granitic soil	669.8 (±15.3)	181	45	31.2	213	470.2
HH-252	Metamorphic	252.7 (±28.9)	72	20	15.8	138	245.8
LQ-381	Metamorphic	381.5 (±64.8)	107	28	11.2	250	396.2
SPC-402	Metamorphic	402.8 (±19.0)	112	29	5.6	367	513.6
EP-530	Metamorphic	530.1 (±50.8)	145	37	13.7	329	524.7
SF-690	Metamorphic	690.4 (±6.0)	185	45	26.8	829	1085.8
Granitic Soils (Mean)		488.5 ns	134.2 ns	34.37 ns	16.89 ns	221.97 a	407.47 ns
Metamorphic (Mean)		451.5 ns	124.05 ns	31.77 ns	12.01 ns	382.44 b	550.29 ns
p-value		0.7	0.46	0.41	0.056	0.032	0.1

Standard errors of the mean (SEM) are given in parentheses. Plantation cumulative production at harvesting age (P_P), carbon stock in aboveground (CAGB) and belowground biomass (CBGB), forest floor carbon stock (FFC_T), soil organic carbon stock (SOC_T), and total carbon stock (CTotal). (ns) indicates no significant differences between soil types based on an ANOVA analysis ($p < 0.05$), while different letters indicate significant differences ($p < 0.05$).

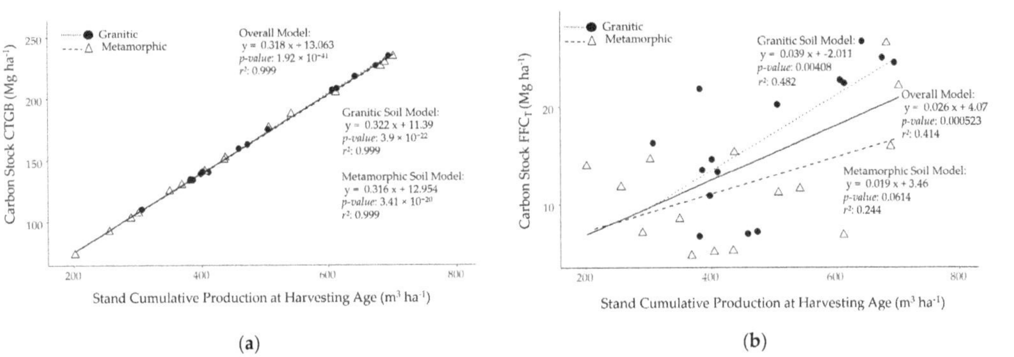

Figure 2. (**a**) Relationship between stand cumulative production at harvesting age (S_P) and biomass carbon stock (CTGB), and (**b**) relationship between stand cumulative production at harvesting age (S_P) and forest floor carbon stock (FFC_T). The black circles represent data points (mean stand observations) from granitic soils, while the white triangles correspond to metamorphic soils. The solid black line represents the overall linear regression model, which includes both soil types. The dashed line represents the regression for metamorphic soils, and the dotted line represents the regression for granitic soils. The p-value, r^2, and equation of the regression line for all three models are shown on the graph to provide insight into the statistical significance and fit of the models.

For forest floor carbon stock (FFC_T), the mean values per plantation ranged from 5.6 Mg·ha^{-1} in SPC-402 to 26.8 Mg·ha^{-1} in SF-690 (Table 3). There was no significant difference in average FFC_T between soil types (p-value = 0.056). The overall linear regression model for FFC_T explained 41.4% of the variance (r^2 = 0.41; p-value = 0.000523) with a slope of 0.026 (Figure 2b). The ANCOVA analysis for the linear regressions by soil type showed that slopes (p-value = 0.19) and intercepts (p-value = 0.06) were statistically similar for both soil types, despite granitic soils having a slightly steeper slope than metamorphic soils.

For soil organic carbon stock (SOC_T), the mean values per plantation ranged from 138 Mg·ha^{-1} in HH-252 to 829 Mg·ha^{-1} in SF-690 (Table 3). There was a significant difference in average SOC_T between soil types (p-value = 0.032). The overall linear regression model explained 69% of the variance (r^2 = 0.69) and showed a significant positive effect of S_p (p-value = 7.63 × 10^6), with a slope of 0.376. The ANCOVA showed statistical differences in intercepts (p-value = 0.000187) and slopes (p-value = 0.00899) between soil types. The linear regression analysis for the relationship between S_p and SOC_T showed a higher determination coefficient for metamorphic soils than for granitic soils (0.68 vs. 0.27). This suggests that metamorphic soils have a greater capacity to accumulate carbon in response to increased S_p. Notably, SF-690, a metamorphic soil site with the highest S_p (690.4 m^3·ha^{-1}), recorded the highest SOC_T value (829 Mg·ha^{-1}), significantly exceeding the overall model predictions. This very high SOC_T value may be attributed to higher soil organic matter concentrations, which enhance carbon accumulation. Such deviations highlight the potential influence of local factors, such as organic matter enrichment or site-specific conditions, on the relationship between S_p and soil carbon stocks (Figure 3a).

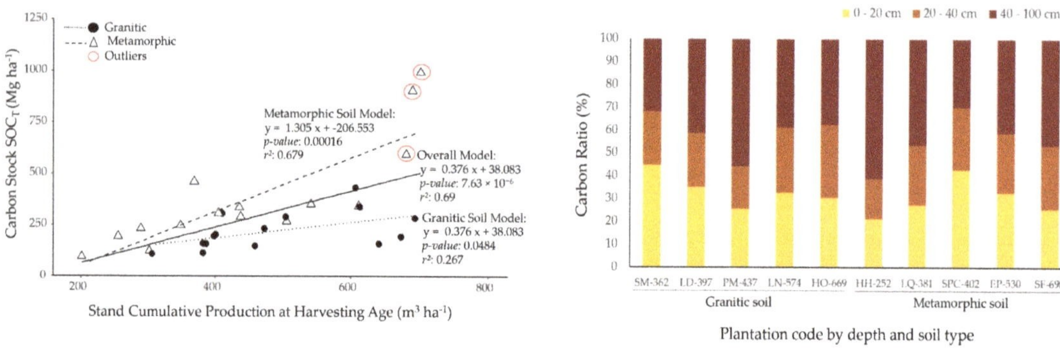

(a) (b)

Figure 3. (**a**) Relationship between stand cumulative production at harvesting age (S_p) and soil carbon stock (SOC_T), and (**b**) average plantation carbon ratio by depth and soil type. The black circles represent data points (mean stand observations) from granitic soils, while the white triangles correspond to metamorphic soils. The solid black line represents the overall linear regression model, which includes both soil types. The dashed line represents the regression for metamorphic soils, and the dotted line represents the regression for granitic soils. The red circles highlight outlier values that deviate significantly from the predicted trends, indicating sites with unusual soil carbon stocks relative to their S_p. The p-value, r^2, and equation of the regression line for all three models are shown on the graph to provide insight into the statistical significance and fit of the models.

There was no effect of sampling depth on the distribution of soil organic carbon stock (SOC_T) (p-value = 0.28), with values ranging from 25% to 44% (Figure 3b).

For total carbon stock (CTotal), the mean values per plantation ranged from 245.8 Mg·ha^{-1} in HH-252 to 1085.8 Mg·ha^{-1} in SF-690 (Table 3). There was no significant difference in average CTotal between soil types (p-value = 0.1). The overall linear regression model explained 76.7% of the variance ($r^2 = 0.767$; p-value = 3.14×10^{-8}) with a slope of 0.737. The ANCOVA for the linear regressions detected significant differences in intercepts (p-value = 0.000255) and slopes (p-value = 0.010914) between the two soil types. The regression model showed a stronger relationship between S_p and carbon stock for the metamorphic soil ($r^2 = 0.768$, p-value = 1.83×10^{-5}) than the granitic soil ($r^2 = 0.586$, p-value = 0.0008). For metamorphic soils, three outliers were observed at one site where high stand cumulative production at harvest age was associated with unusually high carbon stock values (Figure 4a).

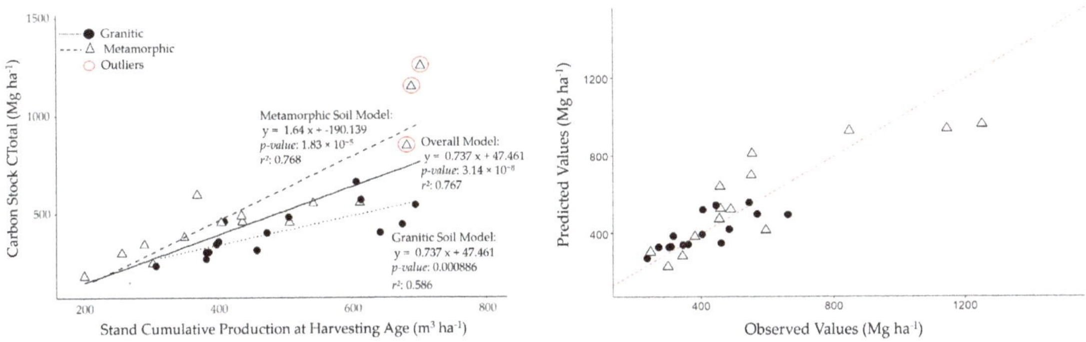

(a) (b)

Figure 4. (**a**) Relationship between stand cumulative production at harvesting age (S_p) and total carbon stock (CTotal), and (**b**) Relationship between observed and predicted total carbon stocks. The black circles represent data points (mean stand observations) from granitic soils, while the white triangles correspond to metamorphic soils. The solid black line represents the overall linear regression model, which includes both soil types. The dashed black line represents the regression for metamorphic soils, and the dotted line represents the regression for granitic soils. The dashed red line represents the 1:1 line, indicating perfect agreement between observed and predicted values. The red circles highlight outlier values that deviate significantly from the predicted trends, indicating sites with unusual soil carbon stocks relative to their S_p. The p-value, r^2, and equation of the regression line for all three models are shown on the graph to provide insight into the statistical significance and fit of the models.

The relationship between observed and predicted total carbon stocks was evaluated to assess the prediction accuracy of the model (Figure 4b). The predicted values are generally in line with the observed values, with most data points clustering around the 1:1 line. The Root Mean Square Error (RMSE) of the model was calculated to be 114.26 Mg·ha^{-1}.

4. Discussion

This study highlights the close and significant relationship of stand cumulative production at harvesting age (S_p) on carbon stock in *Pinus radiata* plantations. Our findings demonstrated a strong positive linear relationship between S_p and total carbon stocks ($r^2 = 0.767$, p-value < 0.001). This relationship is useful for sustainable forest management and climate change mitigation, as it suggests that plantation productivity can be used as a practical tool to estimate carbon stock across various sites [36]. These insights are relevant for the carbon market, planning, and monitoring of forest resources [18]. Additionally, they provide an indicator for evaluating how the forestry sector can meet the national carbon neutrality goals set by international agreements for 2050.

The positive and strong relationship ($r^2 = 0.74$, p-value < 0.0001) between soil bulk density and soil carbon concentration indicates that variations in soil physical characteristics, such as compaction or porosity, may play a more critical role in determining carbon availability at deeper soil layers than soil particle size variations. Since bulk density affects the movement of air and water, root growth, microbial activity, and the storage and distribution of organic matter, this property is crucial in determining soil carbon content (Ray, 2019 [59]). This supports the hypothesis that soil structural characteristics significantly influence carbon stabilization and storage at depth.

The importance of soils as a major carbon reservoir in Pinus radiata plantations agrees with Olmedo et al. (2020) [13], who reported the distribution of carbon stock across four pools in Pinus radiata plantations: 196.8 Mg·ha^{-1} in aboveground biomass, 36.8 Mg·ha^{-1} in belowground biomass, 10.5 Mg·ha^{-1} in the forest floor, and 100 Mg·ha^{-1} in 0–30 cm soil organic carbon. While our results were consistent with their overall observations, the deeper evaluation of soils in our study reveals a significant shift in the relative importance of these pools. In Olmedo et al. (2020) [13], soil accounted for approximately 29% of the total carbon stock. In contrast, by assessing soil carbon stocks down to 1 m depth, we observed that soil contributes over 40% of the total carbon stock. These results agree with Bozo et al. (2024) [36], in which the stock of sequestered carbon is 42.3%. This depth-specific insight aligns with global assessments, such as those by Pan et al. (2011) [4], which underscore the importance of subsoils in long-term carbon storage. Our findings highlight that while aboveground and belowground biomass remain as significant contributors to carbon sequestration, the deeper soil layers serve as a substantial and previously underestimated carbon pool, emphasizing the need to measure carbon sequestration at deep soil layers.

Carbon stocks were not significantly different among sampling depths, confirming that deep soil layers are important contributors to total soil carbon stocks. The high values for carbon stocks below 40 cm depth may be associated with the developmental pattern of roots in commercial plantations of *Pinus radiata*, as the majority of the roots are typically found within the top 1.5 m of the soil. Therefore, belowground carbon characterization should not be restricted to shallow soil horizons.

Biomass carbon stocks were predominantly driven by cumulative production at harvesting age, with minimal variation attributed to soil type. This suggests that genetic improvements, silvicultural practices, and climatic conditions are the primary determinants of aboveground and belowground biomass carbon. While differences in forest floor carbon stocks were not statistically significant, the higher decomposition rates and nutrient cycling efficiency in metamorphic soils may explain a tendency to show lower forest floor carbon accumulation [44].

For plantations on granitic soils, interventions such as enhanced fertilization and organic matter inputs are crucial to improving carbon stock potential. Conversely, plantations on metamorphic soils may benefit from practices that protect soil integrity and minimize disturbances to improve carbon stock capacity. The observed relationship between productivity and total carbon stock underscores the importance of balancing growth optimization with sustainable management to achieve long-term carbon sequestration goals.

While this study provides valuable insights, some limitations warrant consideration. The relatively short temporal scope may overlook long-term dynamics of carbon stock influenced by soil aging, climatic variability, or changes in management practices. Additionally, local factors such as historical land use or genotypic differences in *Pinus radiata* were not accounted for and may have contributed to the observed variability. Future research should focus on integrating landscape-level analyses and extending soil assessments to other parent materials to refine predictive models.

The predictive linear models developed in this study offer a practical tool for estimating carbon stocks in forest plantations, providing a scientific basis for decision-making in forest management. Incorporating these models into national strategies for carbon neutrality can help optimize carbon stock while supporting economic and ecological objectives.

The predicted values were generally in line with the observed values, with most data points clustering around the 1:1 line. This suggests that the model is effectively capturing the general trends. However, some deviations were observed, particularly in plantations with higher productivity, indicating potential model limitations or the influence of unaccounted for local factors. The Root Mean Square Error (RMSE) of the model was calculated to be 114.26 Mg·ha^{-1}, reflecting the average magnitude of the prediction error. This low level of error suggests that the model performs reasonably well but could benefit from additional predictors or refinement to account for unexplained variance.

5. Conclusions

This study underscores the strong relationship between stand cumulative production at harvesting age and total carbon stock in Pinus radiata plantations, highlighting its potential as a practical proxy for carbon stock estimates. Our findings demonstrate that both granitic and metamorphic soils exhibit similar carbon stock capacity. Notably, deeper soil layers play a critical role in long-term carbon stock, emphasizing the need for comprehensive assessments that extend beyond surface horizons. The predictive models developed provide a scalable tool for estimating carbon stocks, offering a practical application for forest management and climate change mitigation strategies. These insights contribute to optimizing plantation management practices in alignment with national and international carbon neutrality goals. Future research should explore the influence of long-term soil dynamics, land-use history, and genetic variability in Pinus radiata to refine carbon stock estimations and enhance sustainable forestry practices in different soil types.

Author Contributions: Conceptualization, R.R., R.M.A. and J.P.E.; data curation, M.V.A., O.J. and D.B.; formal analysis, M.V.A. and R.R.; funding acquisition, R.R., R.M.A. and J.P.E.; investigation, D.B., R.R., O.J., R.M.A. and J.P.E.; methodology, R.R., R.M.A. and J.P.E.; project administration, R.R., R.M.A. and J.P.E.; resources, R.R., R.M.A., J.P.E. and M.P.; software, M.V.A. and R.R.; supervision, R.R. and R.M.A.; visualization, M.V.A.; writing—original draft, M.V.A.; writing—review and editing, R.R. and D.B. All authors have read and agreed to the published version of the manuscript.

Funding: This research was funded by the Forest Productivity Cooperative at Universidad de Concepción and the Chilean National Commission for Scientific and Technological Research (Project Grant ANID BASAL FB210015 "CENAMAD" and Project Grant ANID FOVI220029).

Institutional Review Board Statement: Not applicable.

Informed Consent Statement: Not applicable.

Data Availability Statement: Restrictions apply to the datasets. The datasets presented in this article are not readily available because the data are part of an ongoing study.

Acknowledgments: We acknowledge funding and scientific support from the Forest Productivity Cooperative at Universidad de Concepción and the Chilean National Commission for Scientific and Technological Research (Project Grant ANID BASAL FB210015 "CENAMAD", Project Grant ANID FOVI220029, ANID-Chile, project ANID-ANILLO ACT210060: FiRING:), and Scholarship Program, DOCTORADO BECAS CHILE/2023-21240118; and we gratefully acknowledge CMPC S.A. (Forestal Mininco) for support to access to forest farms.

Conflicts of Interest: The author M.P. was employed by the Forestal Mininco SpA. company. The remaining authors declare that the research was conducted in the absence of any commercial or financial relationships that could be construed as potential conflicts of interest. The author M.P. had the following involvement with the study: conclusions and support to access to forest farms.

References

1. Friedlingstein, P.; O'Sullivan, M.; Jones, M.W.; Andrew, R.M.; Hauck, J.; Olsen, A.; Peters, G.P.; Peters, W.; Pongratz, J.; Sitch, S.; et al. Global Carbon Budget 2020. *Earth Syst. Sci. Data* **2020**, *12*, 3269–3340. [CrossRef]
2. Dellsén, F. Consensus Versus Unanimity: Which Carries More Weight? *Br. J. Philos. Sci.* **2021**. [CrossRef]
3. IPCC. Global Carbon and Other Biogeochemical Cycles and Feedbacks. In *Climate Change 2021—The Physical Science Basis: Working Group I Contribution to the Sixth Assessment Report of the Intergovernmental Panel on Climate Change, Intergovernmental Panel on Climate;* Cambridge University Press: Cambridge, UK, 2023; pp. 673–816.
4. Pan, Y.; Birdsey, R.A.; Fang, J.; Houghton, R.; Kauppi, P.E.; Kurz, W.A.; Phillips, O.L.; Shvidenko, A.; Lewis, S.L.; Canadell, J.G.; et al. A Large and Persistent Carbon Sink in the World's Forests. *Science* **2011**, *333*, 988–993. [CrossRef] [PubMed]
5. Kipping, L.; Goßner, M.M.; Koschorreck, M.; Muszynski, S.; Maurer, F.P.; Weiser, W.W.; Jehmlich, N.; Noll, M. Emission of CO_2 and CH_4 From 13 Deadwood Tree Species Is Linked to Tree Species Identity and Management Intensity in Forest and Grassland Habitats. *Glob. Biogeochem. Cycles* **2022**, *36*, e2021GB007143. [CrossRef]
6. Lehmann, J.; Kleber, M. The contentious nature of soil organic matter. *Nature* **2015**, *528*, 60–68. [CrossRef]
7. Zhang, X.; Zhang, X.; Han, H.; Shi, Z.; Yang, X. Biomass accumulation and carbon sequestration in an age-sequence of Mongolian pine plantations in Horqin sandy land, China. *Forests* **2019**, *10*, 197. [CrossRef]
8. Diao, J.; Liu, J.; Zhu, Z.; Wei, X.; Li, M. Active forest management accelerates carbon storage in plantation forests in Lishui, southern China. *For. Ecosyst.* **2022**, *9*, 100004. [CrossRef]
9. Zha, T.S.; Barr, A.G.; Bernier, P.Y.; Lavigne, M.B.; Trofymow, J.A.; Amiro, B.D.; Arain, M.A.; Bhatti, J.S.; Black, T.A.; Margolis, H.A.; et al. Gross and aboveground net primary production at Canadian forest carbon flux sites. *Agric. For. Meteorol.* **2013**, *174–175*, 54–64. [CrossRef]
10. Díaz Villa, M.V.E.; Cristiano, P.M.; De Diego, M.S.; Rodríguez, S.A.; Efron, S.T.; Bucci, S.J.; Scholz, F.; Goldstein, G. Do selective logging and pine plantations in humid subtropical forests affect aboveground primary productivity as well as carbon and nutrients transfer to soil? *For. Ecol. Manag.* **2022**, *503*, 119736. [CrossRef]
11. López-Senespleda, E.; Calama, R.; Ruiz-Peinado, R. Estimating forest floor carbon stocks in woodland formations in Spain. *Sci. Total Environ.* **2021**, *788*, 147734. [CrossRef]
12. Watt, M.S.; Kimberley, M.O. Spatial comparisons of carbon sequestration for redwood and radiata pine within New Zealand. *For. Ecol. Manag.* **2022**, *513*, 120190. [CrossRef]
13. Olmedo, G.F.; Guevara, M.; Gilabert, H.; Montes, C.R.; Arellano, E.C.; Barría-Knopf, B.; Gárate, F.; Mena-Quijada, P.; Acuña, E.; Bown, H.E.; et al. Baseline of Carbon Stocks in Pinus radiata and *Eucalyptus* spp. Plantations of Chile. *Forests* **2020**, *11*, 1063. [CrossRef]
14. Mizuta, K.; Grunwald, S.; Bacon, A.R.; Cropper, W.P.; Phillips, M.A.; Moss, C.B.; Gonzalez-Benecke, C.A.; Markewitz, D.; Clingensmith, C.M.; Xiong, X. Holistic aboveground ecological productivity efficiency modeling using data envelopment analysis in the southeastern U.S. *Sci. Total Environ.* **2022**, *824*, 153802. [CrossRef] [PubMed]
15. Bellassen, V.; Luyssaert, S. Carbon sequestration: Managing forests in uncertain times. *Nature* **2014**, *506*, 153–155. [CrossRef] [PubMed]
16. Kranabetter, J.M. Site carbon storage along productivity gradients of a late-seral southern boreal forest. *Can. J. For. Res.* **2009**, *39*, 1053–1060. [CrossRef]
17. Waring, B.; Neumann, M.; Prentice, I.C.; Adams, M.; Smith, P.; Siegert, M. Forests and Decarbonization—Roles of Natural and Planted Forests. *Front. For. Glob. Change* **2020**, *3*, 534891. [CrossRef]
18. Smith, P.; Bustamante, M.; Ahammad, H.; Clark, H.; Dong, H.; Elsiddig, E.; Haberl, H.; Harper, R.; House, J.; Jafari, M.; et al. Agriculture, forestry and other land use (AFOLU). In *Climate Change 2014: Mitigation of Climate Change. Contribution of Working Group III to the Fifth Assessment Report of the Intergovernmental Panel on Climate Change;* Cambridge University Press: Cambridge, UK, 2014; pp. 811–922.
19. McKinley, D.C.; Ryan, M.G.; Birdsey, R.A.; Giardina, C.P.; Harmon, M.E.; Heath, L.S.; Houghton, R.A.; Jackson, R.B.; Morrison, J.F.; Murray, B.C.; et al. A synthesis of current knowledge on forests and carbon storage in the United States. *Ecol. Appl.* **2011**, *21*, 1902–1924. [CrossRef] [PubMed]
20. Štraus, H.; Podvinšek, S.; Klopčić, M. Identifying Optimal Forest Management Maximizing Carbon Sequestration in Mountain Forests Impacted by Natural Disturbances: A Case Study in the Alps. *Forests* **2023**, *14*, 947. [CrossRef]
21. Begotti, R.A.; Pacífico, E.d.S.; Ferraz, S.; Galetti, M. Landscape Context of Plantation Forests in the Conservation of Tropical Mammals. *J. Nat. Conserv.* **2018**, *41*, 97–105. [CrossRef]
22. López-Bedoya, P.A.; Magura, T.; Edwards, F.A.; Edwards, D.P.; Rey-Benayas, J.M.; Löveï, G.L.; Noriega, J.A. What Level of Native Beetle Diversity Can Be Supported by Forestry Plantations? A Global Synthesis. *Insect Conserv. Divers.* **2021**, *14*, 736–747. [CrossRef]
23. Heinrichs, S.; Pauchard, A.; Schall, P. Native Plant Diversity and Composition Across a Pinus radiata D. Don Plantation Landscape in South-Central Chile—The Impact of Plantation Age, Logging Roads and Alien Species. *Forests* **2018**, *9*, 567. [CrossRef]

24. Becerra, P.I.; Simonetti, J.A. Native and exotic plant species diversity in forest fragments and forestry plantations of a coastal landscape of central Chile. *Bosque* **2020**, *41*, 125–136. [CrossRef]
25. Kaipainen, T.; Liski, J.; Pussinen, A.; Karjalainen, T. Managing carbon sinks by changing rotation length in European forests. *Environ. Sci. Policy* **2004**, *7*, 205–219. [CrossRef]
26. Nghiem, N. Optimal rotation age for carbon sequestration and biodiversity conservation in Vietnam. *For. Policy Econ.* **2014**, *38*, 56–64. [CrossRef]
27. López-Bedoya, P.A.; Cardona-Galvis, E.A.; Urbina-Cardona, J.N.; Edwards, F.A.; Edwards, D.P. Impacts of Pastures and Forestry Plantations on Herpetofauna: A Global Meta-analysis. *J. Appl. Ecol.* **2022**, *59*, 3038–3048. [CrossRef]
28. Heilmayr, R.; Echeverría, C.; Lambin, E.F. Impacts of Chilean forest subsidies on forest cover, carbon and biodiversity. *Nat. Sustain.* **2020**, *3*, 701–709. [CrossRef]
29. Escobar-Avello, D.; Ferrer, V.; Bravo-Arrepol, G.; Reyes-Contreras, P.; Elissetche, J.P.; Santos, J.; Fuentealba, C.; Cabrera-Barjas, G. Pretreated *Eucalyptus globulus* and *Pinus radiata* Barks: Potential Substrates to Improve Seed Germination for a Sustainable Horticulture. *Forests* **2023**, *14*, 991. [CrossRef]
30. Rubilar, R.; Bozo, D.; Albaugh, T.; Cook, R.; Campoe, O.; Carter, D.; Allen, H.; Lvarez, J.; Pincheira, M.; Zapata, L. Rotation-age effects of subsoiling, fertilization, and weed control on radiata pine growth at sites with contrasting soil physical, nutrient, and water limitations. *For. Ecol. Manag.* **2023**, *544*, 121213. [CrossRef]
31. Ojeda, H.; Rubilar, R.A.; Montes, C.; Cancino, J.; Espinosa, M. Leaf area and growth of Chilean radiata pine plantations after thinning across a water stress gradient. *N. Z. J. For. Sci.* **2018**, *48*, 10. [CrossRef]
32. Alvarez, J.; Allen, H.; Albaugh, T.; Stape, J.; Bullock, B.; Song, C. Factors influencing the growth of radiata pine plantations in Chile. *Forestry* **2012**, *86*, 13–26. [CrossRef]
33. Guo, L.B.; Cowie, A.L.; Montagu, K.D.; Gifford, R.M. Carbon and nitrogen stocks in a native pasture and an adjacent 16-year-old Pinus radiata D. Don. plantation in Australia. *Agric. Ecosyst. Environ.* **2008**, *124*, 205–218. [CrossRef]
34. Guedes, B.S.; Olsson, B.A.; Egnell, G.; Sitoe, A.A.; Karltun, E. Plantations of Pinus and Eucalyptus replacing degraded mountain miombo woodlands in Mozambique significantly increase carbon sequestration. *Glob. Ecol. Conserv.* **2018**, *14*, e00401. [CrossRef]
35. Balboa-Murias, M.Á.; Rodríguez-Soalleiro, R.; Merino, A.; Álvarez-González, J.G. Temporal variations and distribution of carbon stocks in aboveground biomass of radiata pine and maritime pine pure stands under different silvicultural alternatives. *For. Ecol. Manag.* **2006**, *237*, 29–38. [CrossRef]
36. Bozo, D.; Rubilar, R.; Campoe, O.C.; Alzamora, R.M.; Elissetche, J.P.; Valverde, J.C.; Pizarro, R.; Pincheira, M.; Valencia, J.C.; Sanhueza, C. Soil and Site Productivity Effects on Above- and Belowground Radiata Pine Carbon Pools at Harvesting Age. *Plants* **2024**, *13*, 3482. [CrossRef] [PubMed]
37. Temesgen, H.; Affleck, D.; Poudel, K.; Gray, A.; Sessions, J. A review of the challenges and opportunities in estimating above ground forest biomass using tree-level models. *Scand. J. For. Res.* **2015**, *30*, 326–335. [CrossRef]
38. Pfeiffer, M.; Padarian, J.; Osorio, R.; Bustamante, N.; Olmedo, G.F.; Guevara, M.; Aburto, F.; Albornoz, F.; Antilén, M.; Araya, E.; et al. CHLSOC: The Chilean Soil Organic Carbon database, a multi-institutional collaborative effort. *Earth Syst. Sci. Data* **2020**, *12*, 457–468. [CrossRef]
39. Post, W.M.; Kwon, K.C. Soil carbon sequestration and land-use change: Processes and potential. *Glob. Change Biol.* **2000**, *6*, 317–327. [CrossRef]
40. Six, J.; Conant, R.T.; Paul, E.A.; Paustian, K. Stabilization mechanisms of soil organic matter: Implications for C-saturation of soils. *Plant Soil* **2002**, *241*, 155–176. [CrossRef]
41. Zhang, Y.; Ma, W.; Sun, X.; Jiang, J.; Li, D.; Tang, G.; Xu, W.; Jia, H. Biochar Aged for Five Years Altered Carbon Fractions and Enzyme Activities of Sandy Soil. *Land* **2023**, *12*, 1645. [CrossRef]
42. Su, Y.Z.; Wang, X.F.; Yang, R.; Lee, J. Effects of Sandy Desertified Land Rehabilitation on Soil Carbon Sequestration and Aggregation in an Arid Region in China. *J. Environ. Manag.* **2010**, *91*, 2109–2116. [CrossRef]
43. Amin, M.N.; Shil, S.C.; Ghosh, R.C.; Shamsuzzoha, M. Influence of Conservation Tillage on Carbon Sequestration Mechanism Related to Aggregation. *J. Environ. Sci. Nat. Resour.* **2016**, *9*, 23–27. [CrossRef]
44. Katayama, S.; Omori, T.; Tateno, M. Fresh Litter Acts as a Substantial Phosphorus Source of Plant Species Appearing in Primary Succession on Volcanic Ash Soil. *Sci. Rep.* **2021**, *11*, 11497. [CrossRef] [PubMed]
45. Hernández, C.M.; Rodrigo-Comino, J.; Díaz, A. Impact of Lithology and Soil Properties on Abandoned Dryland Terraces During the Early Stages of Soil Erosion by Water in South-East Spain. *Hydrol. Process.* **2017**, *31*, 3095–3109. [CrossRef]
46. Setia, R.; Smith, P.; Marschner, P.; Baldock, J.A.; Chittleborough, D.J.; Smith, J.U. Introducing a Decomposition Rate Modifier in the Rothamsted Carbon Model to Predict Soil Organic Carbon Stocks in Saline Soils. *Environ. Sci. Technol.* **2011**, *45*, 6396–6403. [CrossRef]
47. Zhou, D.; Mao, D.; Ye, M.; Li, S.; Ma, X.-D.; Liu, S. Fractal Features of Soil Grain-Size Distribution in a Typical Tamarix Cones in the Taklimakan Desert, China. *Sci. Rep.* **2022**, *12*, 16461.

48. Vadeboncoeur, M.A.; Hamburg, S.P.; Yanai, R.D.; Blum, J.D. Rates of Sustainable Forest Harvest Depend on Rotation Length and Weathering of Soil Minerals. *For. Ecol. Manag.* **2014**, *318*, 194–205. [CrossRef]
49. CIREN. *Descripción de Suelos, Materiales y Símbolos: Estudio Agrológico VIII Región*; CIREN: Santiago, Chile, 1999.
50. Stolpe, N. *Descripciones de los Principales Suelos de la VIII Región de Chile*; Universidad de Concepción: Concepción, Chile, 2006.
51. Mininco, F. *Compendium of Functions for Species of Interest to Forestal Minin S.A.*; Forestal Mininco: Concepción, Chile, 1995. (In Spanish)
52. Albaugh, T.J.; Alvarez, J.; Rubilar, R.A.; Fox, T.R.; Allen, H.L.; Stape, J.L.; Mardones, O. Long-Term Pinus radiata Productivity Gains from Tillage, Vegetation Control, and Fertilization. *For. Sci.* **2015**, *61*, 800–808. [CrossRef]
53. Sandoval, S.; Montes, C.R.; Olmedo, G.F.; Acuña, E.; Mena-Quijada, P. Modelling above-ground biomass of Pinus radiata trees with explicit multivariate uncertainty. *For. Int. J. For. Res.* **2022**, *95*, 380–390.
54. Zerihun, A.; Montagu, K.D. Belowground to aboveground biomass ratio and vertical root distribution responses of mature Pinus radiata stands to phosphorus fertilization at planting. *Can. J. For. Res.* **2004**, *34*, 1883–1894. [CrossRef]
55. IPCC. *Climate Change 2013: The Physical Science Basis. Contribution of Working Group I to the Fifth Assessment Report of the Intergovernmental Panel on Climate Change*; Cambridge University Press: Cambridge, UK; New York, NY, USA, 2013; p. 1535.
56. Poeplau, C.; Vos, C.; Don, A. Soil organic carbon stocks are systematically overestimated by misuse of the parameters bulk density and rock fragment content. *Soil* **2017**, *3*, 61–66. [CrossRef]
57. R Core Team, R. *A Language and Environment for Statistical Computing*; R Foundation for Statistical Computing: Vienna, Austria, 2024.
58. Staff, S.S. *Keys to Soil Taxonomy*, 13th ed.; USDA-Natural Resources Conservation Service: Washington, DC, USA, 2022.
59. Ray, M. Relation of Soil Bulk Density with the Soil Carbon in the Tropical Dry Deciduous Forest of Jharkhand, India. *Int. J. Sci. Eng. Res.* **2019**, *3*, 1439–1443.

sustainability

MDPI

Article

Increases in the Amounts of Agricultural Surfaces and Their Impact on the Sustainability of Groundwater Resources in North-Central Chile

Roberto Pizarro [1,2,3,4], Francisca Borcoski [1], Ben Ingram [1], Ramón Bustamante-Ortega [1], Claudia Sangüesa [1,2], Alfredo Ibáñez [1,2], Cristóbal Toledo [1], Cristian Vidal [5] and Pablo A. Garcia-Chevesich [6,7,*]

1 UNESCO Chair Surface Hydrology, University of Talca, Talca 3467769, Chile; rpizarro@utalca.cl (R.P.); francisca.borcoski@ug.uchile.cl (F.B.); b.ingram@cranfield.ac.uk (B.I.); bustamante.ortega@gmail.com (R.B.-O.); claudiasanguesa@gmail.com (C.S.); alfredoibacor@gmail.com (A.I.); cristobal.toledo@utalca.cl (C.T.)

2 Centro Nacional de Excelencia para la Industria de la Madera (CENAMAD)-ANID BASAL FB210015, Pontificia Universidad Católica de Chile, Santiago 7810128, Chile

3 Faculty of Forest Sciences and Nature Conservancy, University of Chile, Santiago 8820808, Chile

4 Dirección de Innovación, Universidad de Talca, Talca 3467769, Chile

5 Faculty of Engineering, School of Videogame Development and Virtual Reality Engineering, University of Talca, Campus Talca, Talca 3480260, Chile; cvidal@utalca.cl

6 Department of Civil and Environmental Engineering, Colorado School of Mines, Golden, CO 80401, USA

7 Intergovernmental Hydrological Program, United Nations Educational, Scientific, and Cultural Organization (UNESCO), Montevideo 11200, Uruguay

* Correspondence: pablogarcia@mines.edu

Abstract: Water is a fundamental resource for Chile's productive structure, which is more important in arid areas, and especially with agricultural uses. This study was based on two basins (Cogotí and Illapel) located in the Coquimbo Region of north-central Chile. In this region, surface water rights were closed in 2002 and the only current option is the use of groundwater. These basins have high water demands due to the use of surface and groundwater for agricultural purposes, a fact that should influence the sustainability of groundwater reserves over time. The objective of this study was to determine how much agricultural use has affected the availability of groundwater in two basins. Under the previous context, the evolution of agricultural irrigation surfaces was evaluated using Landsat images and forest classifications. Similarly, groundwater reserves were evaluated using the recessive curves of hydrographs associated with the beginning of each hydrological year. The results show an increase in the agricultural area between 1996 and 2016, with a subsequent decrease, while groundwater reserves denoted significant decreases over time. In conclusion, a significant decrease in the volumes of groundwater reserves in both basins was observed, a decrease that is consistent with the increase in irrigated areas.

Keywords: land use; Copiapó Region; overconsumption; Cogotí; Illapel

1. Introduction

Water resources are a fundamental component for life, which are involved in a host of biological and physical processes [1]. The presence of groundwater constitutes a key factor for most physiological and biochemical activities [2–4], as well as the functioning of various ecosystems [5].

In Chile, a generalized decrease is projected for water resources due to the increase in demands from water users, changes in land use, and climatic variability [6,7]. Likewise, the availability of water is manifested with limitations in the national territory due to climate change [8], a phenomenon that is manifested as a lower supply of water resources in Chile.

Water demands in the north and central zones of Chile permanently exceed the existing availability, with agriculture, mining, and drinking needs being the main demanders of the resource [9,10]. There is clear empirical evidence in Chile that indicates that where mining and large-scale agriculture coexist in the same region, these areas are affected by water scarcity [11].

In this scenario, the evaluation of hydrological variables that account for the general water balance of basins is important. Thus, the flow variable, as an expression of the water production capacity of a hydrographic system, is a determining factor in the objective of knowing the state of a given watershed, in the face of scenarios of climate change and the use of water resources [12].

In the above context, changes in land use for intensive agricultural purposes is a variable that would have a very strong impact on the water production of various basins, as well as on groundwater reserves [13]. In this sense, various authors (e.g., [14–19]) have pointed out that current agricultural practices are not sustainable over time, given that this activity represents 85% of global water consumption and it is likely that this will double by 2050 [17]. Another aspect to consider, especially in areas with low amounts of rainfall, is the use of groundwater to supply the differential demand required for crop irrigation [13,20–22]. However, this practice reduces the volume of water stored in aquifers and, if maintained, it is possible to alter groundwater availability or even dry terminate it [22–25]. This imbalance between water supply and demand is also a factor that will impact local ecosystems and populations, and its effects will be accentuated by climate variability and change [26].

Though the decrease in surface water has been widely studied in Chile (e.g., [13,27–31]), the evolution of groundwater availability is not well understood because its analysis is more complex [32]. However, it is possible to analyze the variation in reserves using baseflows as a proxy unit [32,33], thus facilitating the estimation of recharge. Thus, baseflow and its impact based on overuse is an interesting factor to analyze, which can shed light on the impact on aquifer reserves. One way to estimate baseflows is through exponential distributions [34], distributions that have been used to model surface and groundwater flows at a basin level [35,36].

Based on the above, it is proposed to carry out an analysis of the effects of land use changes for large-scale agricultural purposes in two basins in north-central Chile, establishing whether the trends in the values of water reserves have remained stable over time or have varied significantly to determine whether or not those land use changes have affected the availability of groundwater in the last two decades.

2. Materials and Methods

2.1. Study Area

The Cogotí and Illapel basins are located in the Coquimbo administrative Region, between latitudes 29° S and 32° S (Figure 1). The local climate is arid to semi-arid with a precipitation pattern defined by 2-to-3 wet winter months and a territorial gradient where precipitation increases from north to south and from the coast (west) to the Andes (east) [13,37].

Qgis 3.28 [38] and Rstudio 2023 [39] software were used to carry out this study. The research aims to determine how the surface area of intensively irrigated agricultural areas has varied in four time periods (1996, 2006, 2016, and 2022). In addition, it aims to establish the behavior of trends in groundwater storage volumes in each basin studied from 1996 to 2022.

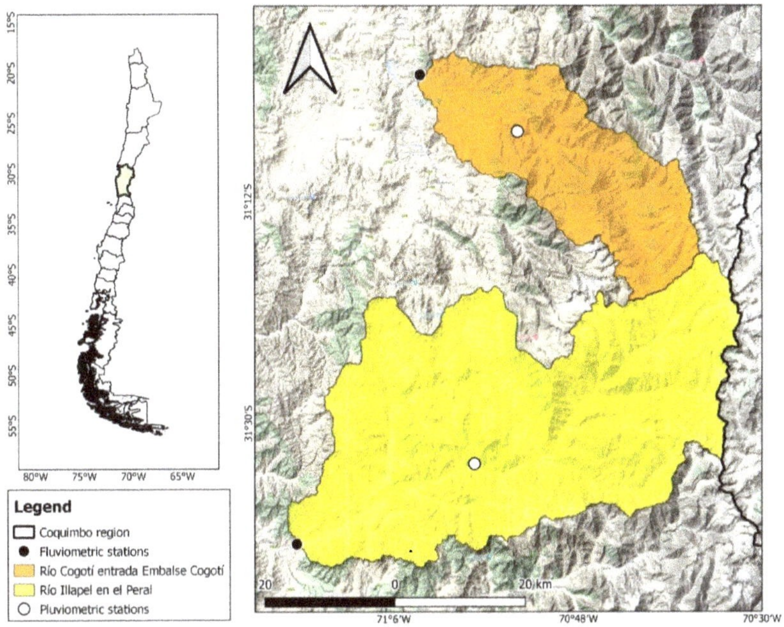

Figure 1. Location map of fluviometric and pluviometric stations within the two selected basins under study. Source: Own work.

2.2. Quantification of Land Use Changes

2.2.1. Selection and Delimitation of Basins

The selection of the basins was carried out between the regions of Coquimbo and Valparaíso, in the north-central zone of Chile, which were subjected to a sorting process based on the hydrological information available at fluviometric stations. The procedure began with the selection of fifteen basins, which were identified using the "CR2 Search" (CR2, 2023), to end with a selection of two basins (Cogotí and Illapel) with fluviometric information between 1996 and 2022, as previously pointed out. Both basins were delimited using the QGIS program, based on the methodology proposed by Dixon and Uddameri [40] (Figure 1 and Table 1).

Table 1. Summary of the fluviometric stations used in this study.

Name of Station	ID	Drainage Area (km^2)	Recording Period
Río Cogotí Entrada Embalse Cogotí	4531002	741	1996–2022
Rio Illapel en el Peral	4726001	2000	1996–2022

2.2.2. Collection and Processing of Landsat Images across Different Time Frames

The effective irrigation areas were determined for four years (1996, 2006, 2016, and 2022) using QGIS and Landsat images. Landsat images (Table 2) from the United States Geological Survey (USGS) were used for the established periods, thereby allowing the analysis of the variation of the irrigation surfaces, especially for intensive uses within each basin.

The satellite codes used were "001 081" and "233 082". After downloading these USGS images, the bands from each year were merged using the "Build Virtual Raster" tool. Subsequently, both areas were combined into a single "Tiff" layer using the "Mosaic" tool of the Orfeo Toolbox extension (OTB, [41]). The result was clipped to the area of the two selected basins using the "Clip Raster by Mask Layer" tool.

Table 2. List of images used for classification.

Year	Sensor	Cloud Cover (%)	Satellite Area *	Image Date
1996	Landsat 5-TM	<10	001 081 233 082	15 January 1996 28 March 1996
2006	Landsat 7-ETM+	<10	001 081 233 082	31 March 2006 16 March 2006
2016	Landsat 8-OLI	<10	001 081 233 082	23 February 2016 4 April 2016
2022	Landsat 9	<10	001 081 233 082	4 April 2022 28 March 2022

* Images sourced from United States Geological Survey (USGS).

2.2.3. Training Polygons

Once the Landsat images were obtained for the areas of interest (basins) and the four periods, the bands were configured for agricultural areas. Training polygons were identified, defining two categories: "Agriculture" and "Others". Intensively irrigated agriculture was defined as that visible in satellite images, including grazing areas. "Other" encompasses non-agricultural areas such as urban areas, bodies of water, native vegetation, and bare soils, among others. Photointerpretation of satellite images (1996–2022) was carried out due to the lack of records documenting agricultural growth within the study area.

2.2.4. Land Use Classification

Land cover monitoring using remote sensing data requires robust and accurate classification methods that enable the mapping of complex land cover and land use categories [42]. The estimation of the agricultural area was carried out using "Random Forest (RF)", a machine learning classifier applied to the Landsat images described in Table 2. RF is a technique that collects several base models and combines them to obtain improved results [43]. This technique consists of the combination of the concept of "Decision trees" [44] and Breiman's "Bagging" [45]. Thus, the concept of "Decision trees" represents an algorithm to classify information from a set of training data (previously selected) with the aim of automating and simplifying this process [46]. On the other hand, "Bagging" or "bootstrap aggregation" would be a type of joint learning, which improves the precision of a weak classifier by creating a set of classifiers [47]. That is, "Bagging" helps with the generation of multiple decision trees through the combination of these methods, improving the precision and generalization of the model [43]. A simplified version of its operation is represented in Figure 2.

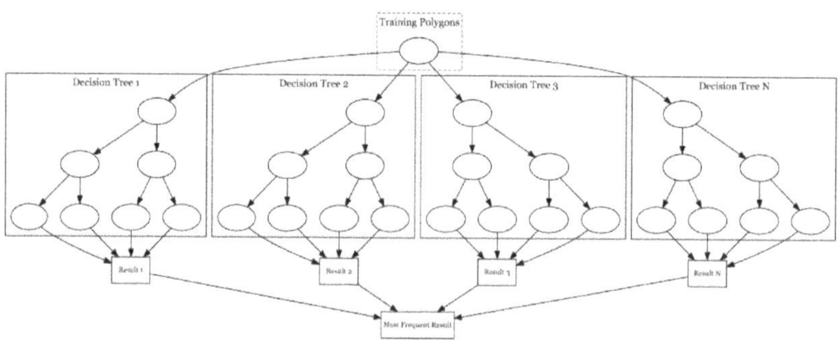

Figure 2. Simplified model of the operation of Random Forest.

This technique has been previously used in other Chilean studies, providing good results [48,49]. Furthermore, training polygons were provided to RF, allowing unsupervised classification data analysis by estimating missing values through the algorithm. The use of the Google Earth Engine cloud geoprocessing platform allowed the implementation of RF-supervised classification, pixel by pixel, for the four proposed years (1996, 2006, 2016, and 2022) of the Landsat images.

2.2.5. Estimation of Irrigation Surface

Once classifications were obtained, the results of the RF algorithm were transformed into vector algorithms using QGIS to calculate the amount of intensive irrigation surfaces in hectares for the four periods. Its result gave rise to the generation of intensively irrigated agricultural areas, for each basin and for the four defined years.

2.3. *Groundwater Volume Estimation Methodology*

Unfortunately, information regarding the size and shape of aquifers in Chile is limited (non-existent for the studied basins), making it difficult to determine whether surface basins coincide with them. However, the methodology used in this study allows for estimating the volume of recharge that the surface basin contributes to the aquifer, as detailed further down in this paper.

2.3.1. Preprocessing of Data for Calculating Baseflows

The calculation of baseflows was carried out in the two selected basins. This calculation was based on the information provided by the CR2 flow database [50], which includes daily flows recorded at its gauging stations (see Figure 1).

The flow analysis period considered data from 1996 to 2022, with particular interest in those data that were recorded between the beginning of the hydrological year and the end of the previous one, that is, between 30 March and 2 April for each year of analysis. This was carried out to determine the recessive behavior of floods when there were lower water reserves in the basin.

2.3.2. Hydrograph Separation and Model Fitting for the 1996–2022 Period

The focus of this research is on baseflow recession, observable in the typical recession hydrograph illustrated by Jain [51]. Similarly, Balocchi et al. [52] shows that the recession curve used includes the second break point of the descending curve of the hydrograph. According to Pizarro et al. [53], the exclusive inflow of groundwater begins from the third breakpoint, especially in arid and semi-arid areas. Keeping this in mind, for this research, the mathematical model of Remenieras, Potential, and Exponential Model 2, here and after the study by Balocchi et al. [52] (Table 3), were used.

Table 3. Equations for baseflow estimation.

Model	Equation	Source
Remenieras	$Q(t) = Q_0 e^{-\alpha t}$	Balocchi et al. [52]
Potential	$Q(t) = Q_0(1+\alpha t)^{-2}$	Cirugeda [54]
Balocchi et al.	$Q(t) = Q_0 e^{(-2\alpha\sqrt{t})}$	Balocchi et al. [52]

Where Q(t): Recessive flow at time t in $m^3\ s^{-1}$, Q_0: Initial recessive flow at time t_0. When the flow starts to be fed exclusively by groundwater reserves in $m^3\ s^{-1}$, e: Neper's constant, t: time, α: Recession constant.

To estimate the groundwater volumes stored in each basin, it was necessary to adjust each predefined model to the data available for each year and each basin at the beginning of the hydrological period (end of March or beginning of April). The first step was to define the value of Q_0, that is, the maximum flow rate of the recessive or exhaustion curve and the moment at which this occurs. To do this, on 31 March of each year and in each basin, the last major flood produced in the area was selected, establishing the last maximum flow or peak that defined the flood. This could be close to the date under study or could be delayed by a few months, especially if these are basins located in arid and semi-arid areas, as was the case. Thus, once the maximum flow is established, the hydrograph separation methodology [52] establishes that the logarithms of the flows must be graphed as a function of time, establishing a semi-logarithmic relationship, lnQ versus t to graphically establish the second break point of the descending curve, which defines the beginning of the recessive curve at the coordinate (t_0, Q_0). To clarify how to obtain the initial flow rate Q_0, Figure 3 shows an example of a last flood for the Illapel basin, before the start of the hydrological year, which occurred in January 1998. An average daily maximum flow can be seen to be slightly higher than 47 m^3s^{-1}. The previous result (Figure 3) is transformed into a semi-logarithmic graph to estimate the break points in the curve (Figure 3). Thus, for the second break point, a flow value is obtained on the "Y" axis (applying an antilogarithm), which is called the baseflow, hereinafter "Q_0", and from the "X" axis a time, hereinafter "t_0". Once the initial baseflow Q_0 is defined, another value is taken corresponding to subsequent weeks or days, that is, a Q(t) at time $t_0 + \Delta t$. With this new value, the value of the parameter α for each model considered is obtained (Table 3).

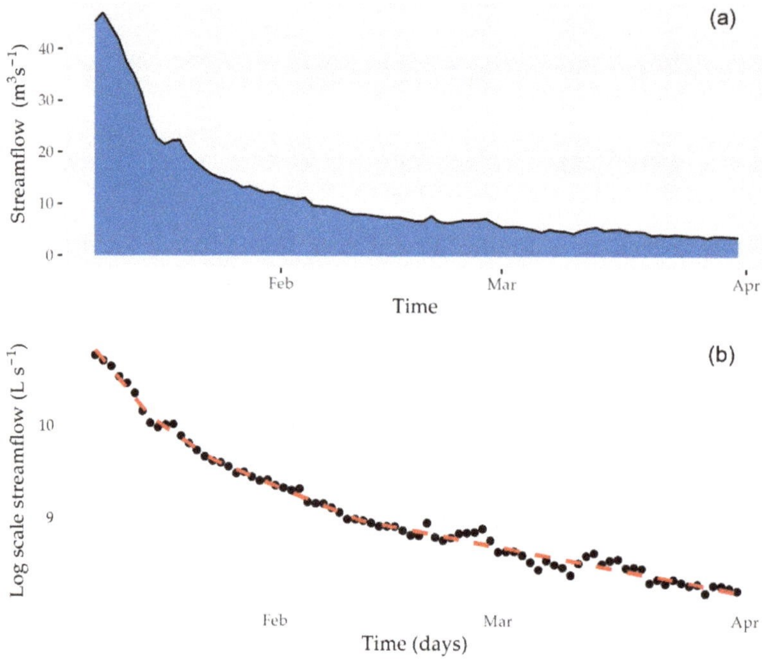

Figure 3. (**a**) The last flood of the hydrological year; (**b**) semi-logarithmic representation of the flood. Black dots indicate daily Flow while red dashed line is the trend. Source: Own work.

2.3.3. Goodness of Fit (GOF)

In the field of hydrology, error metrics play a fundamental role in evaluating the quality of fit or in quantifying disparities between predicted and observed values in a time series [55]. They provide a measure for evaluating and interpreting hydrological models in terms of their predictive effectiveness based on real data.

The most popular system-scale performance metrics in hydrological modeling are Nash–Sutcliffe Efficiency (NSE) [56] and Kling–Gupta Efficiency (KGE) [57]. To validate the authenticity of the data, the use of NSE is ruled out as its value may not be as significant for highly variable data sets [58]. Instead, KGE is used, which is a more robust metric for outliers or that considers the synchronization of peaks and valleys [59]. The KGE equation is defined by Gupta et al. [57] as expressed in Equation (1). For this purpose, Rstudio was used to establish the KGE values corresponding to each basin and year, using the "Metric" package [60].

$$KGE = 1 - \sqrt{(r-1)^2 + (\alpha-1)^2 + (\beta-1)^2} \quad (1)$$

where r is the correlation between the observed and simulated series, α is the relative standard deviation of the simulated series compared to the observed ones, and β is the bias (the difference between the averages of the simulated and observed series).

2.3.4. Estimation of Base Volume

With the fitted models, a $Q(t)$ or baseflow (Equation (2)) function was obtained. Subsequently, to calculate the annual volume of stored groundwater (m^3), the expression of each model adjusted with its corresponding parameters was used. In this context, the objective was to estimate the volume stored as of 31 March of each year under analysis, or dates close to them. To this end, it was assumed that there will be no new contributions of water to the channel, whether from precipitation, snowmelt, or other sources. Therefore, the area under the curve of the equation that represents each model defined the total volume stored in the basin after each flood. Since it is very likely that the coordinate (t_0, Q_0) is on a date before 31 March, the volume flowing between t_0 and 31 March must be subtracted to find the volume stored as of 31 March, which is represented in the expressions shown in Equations (2) and (3).

$$\int_{t_0}^{\infty} Q(t)dt - \int_{t_0}^{31\text{March}} Q(t)dt = \int_{31\text{March}}^{\infty} Q(t)dt \quad (2)$$

$$\text{Volume}\left(\text{Hm}^3\right) = \frac{24\ (\text{h}) * 3600\ (\text{s})}{1{,}000{,}000} \int_{31\text{March}}^{\infty} Q(t)dt \quad (3)$$

2.4. *Estimation of Precipitation*

The precipitation calculation was carried out at two rainfall stations within the selected basins using daily data from CR2 [61], from 1996 to 2022 (Figure 1). The selected stations were Río Cogotí en Fraguita (04531003) and Río Illapel en Huintil (04726003).

The analysis was carried out at a monthly level (i.e., the monthly variation of precipitation per year was evaluated) to determine the behavior of precipitation and its possible impact on the water reserves of the basin.

2.5. *Trends*

Trend analysis focuses on the long-term trend component. However, it is unfeasible to achieve infinitely high sampling rates or perfect measurements [62], hence the need for statistical trend analysis methods. Therefore, in this investigation, the non-parametric Mann–Kendall (MK) test was applied, which has been frequently used to calculate the significance of trends in hydrometeorological time series [63].

The MK test examines each possible pair of data points, (x_i, y_i) and (x_j, y_j), to determine whether the pairs have the same relationship with each other, that is, whether there is a monotonic relationship or not [64]. This makes it possible to determine whether trends in water reserves show a significant decline over time. According to Pizarro et al. [13], the MK statistic for a time series of data $\{Z_k, k = 1, 2, \ldots, n\}$ is developed through the following process: (a) The values of the variables are listed in order, that is, from x_1 to x_n (x_1, x_2, …, x_n). (b) The sign of the difference in each pair of values is sought by comparing their magnitudes ($x_j - x_k$), with ($j > k$), following Equation (4). (c) The Mann–Kendall statistical S is obtained, where if the value of S is positive, it is inferred that the trend is increasing, while if it is negative, it is inferred that there is a decreasing trend, in which j and k represent two consecutive years (Equation (5)). (d) Then, based on the indicators, a variance is estimated for the MK S statistic, which considers the case of ties (Sign $x_j - x_k = 0$) obtained in step "b", using Equation (6). And, (e) the statistical calculation of Z is carried out using Equation (7).

$$Sgn\ (X_j - X_k) = \begin{cases} 1\ if\ (X_j - X_k) > 0 \\ 0\ if\ (X_j - X_k) = 0 \\ -1\ if\ (X_j - X_k) < 0 \end{cases} \tag{4}$$

$$S = \sum_{k=1}^{n-1} \sum_{j=k+1}^{n} Sgn\ (X_j - X_k) \tag{5}$$

$$VAR = \frac{1}{18} * \left[n\ (n-1)\ (2\ n+5) - \sum_{q=1}^{g} t_q\ (t_q\ -1)\ (2\ t_q+5) \right] \tag{6}$$

where n is the sample size, t_q is the frequency of the ties in group q, and q is the number of groups with ties.

$$Z = \begin{cases} \frac{S-1}{[VAR(S)]^{0.5}}\ if\ S > 0 \\ 0\ if\ S = 0 \\ \frac{S+1}{[VAR(S)]^{0.5}}\ if\ S < 0 \end{cases} \tag{7}$$

Once the Z statistic is calculated, it is compared with the p value (0.05), allowing the null hypothesis to be accepted or rejected. Thus, H_0 is rejected if the p value returned by the MK test is less than 0.05 [65].

The MK calculation was performed using the Rstudio program with the "rkt" package [66] for the volume series of both basins. Additionally, the rate of change in the time series was estimated using the Theil–Sen formula (TS), a derivative that is robust to outliers compared to traditional linear regression [67]. From Equation (8), it is considered that $i < j$; $i = 1, 2, \ldots, (n-1)$; $j = 2, 3, \ldots, n$. Therefore, by combining them, it is possible to analyze trends in time series, and their use together offers a more comprehensive assessment of the presence and significance of these trends [67].

$$Theil - Sen\ slope = median\ \frac{(y_j - y_i)}{(x_j - x_i)} \tag{8}$$

The application of the non-parametric MK test to the data series of groundwater volumes stored in each basin and each year was carried out in two periods. First, the 1998–2010 period was analyzed, excluding the megadrought that has affected the country since that year, and the period between 1998 and 2020. The objective was to visualize the trend that stored volumes had before the incidence of a prolonged drought. The simplified process is illustrated in Figure 4.

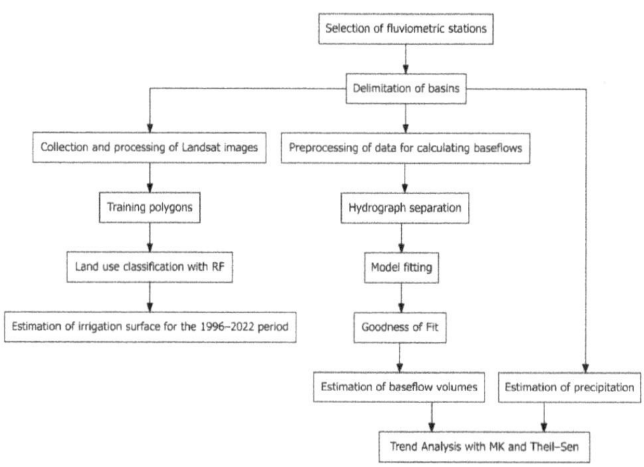

Figure 4. Flowchart illustrating the applied methodology. Source: Own work.

3. Results

3.1. Land Use Changes

The results of the classification of agricultural land use are presented in Figure 5 for the four years under evaluation. This surface was estimated by Random Forest and Landsat images, using the basins delimited in previous steps ("Cogotí" and "Illapel"). Moreover, each year presented an accuracy of 0.9; 0.8; 0.8, and 0.8, respectively, through Random Forest classification and using the training polygons (i.e., the estimate is appropriate). In general, an incremental trend in the agricultural irrigation area amount is observed over time for each basin until 2016, with a strong decrease in 2022 (Figure 6), most likely due to the lack of water availability.

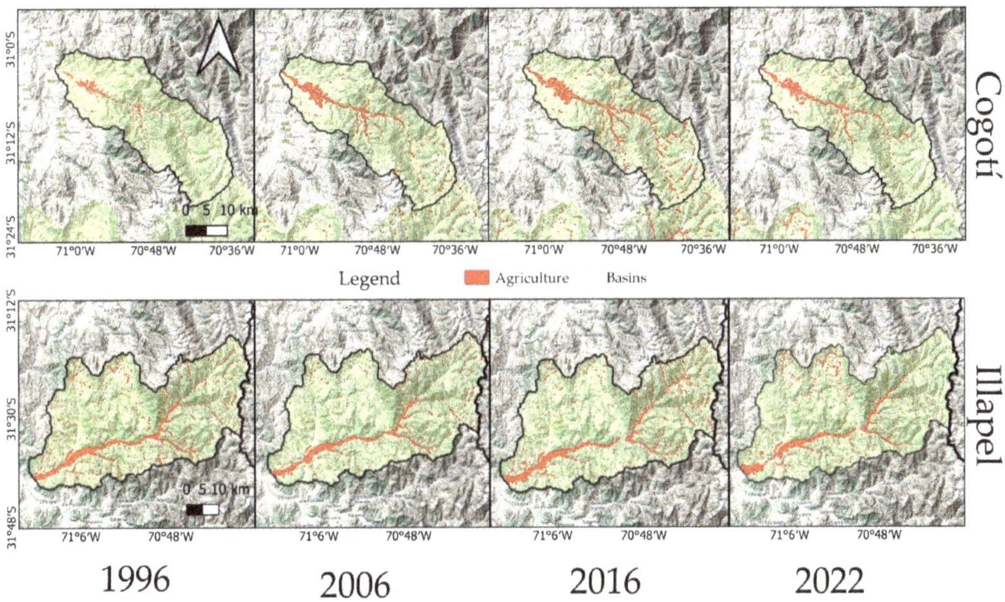

Figure 5. Locations of irrigated agriculture (red) for the Cogotí and Illapel basins (green) in the four years of analysis. Source: Own work.

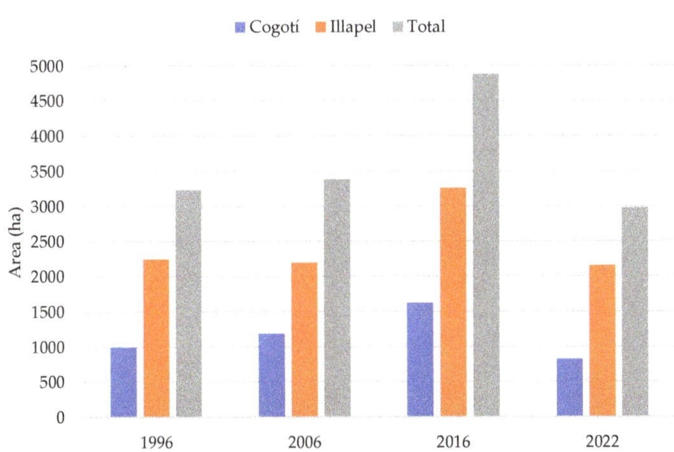

Figure 6. Temporal evolution of irrigated areas. Source: Own work.

3.2. Estimation of Underground Volumes

During the investigation, it was confirmed that the daily flow data obtained from CR2 showed a lack of information for different years in the two basins. As a result, the absence of certain years in the analysis was observed. However, this lack did not prevent an adequate analysis of the data.

In terms of the determination of the start times of the selected recessive flows as of 31 March, the first aspect that needed to be addressed was the determination of the pairs of values (t_0, Q_0) for each year and for each basin, in order to model the baseflow in a given period. This determined the need to estimate the second break point of the descending curve of the flood hydrograph closest to and prior to 31 March. Examples of results associated with the breakpoints identified in the data for the 1996–2020 period for the two basins are detailed in Figures A1 and A2, which provide a visual representation of the results segment, allowing for a more detailed understanding of the trends and patterns identified. An important detail to keep in mind is that the flows, having small values, which is typical of arid and semi-arid areas, had to be converted to liters per second (l s^{-1}), so that the logarithm could have value.

Once the second break points were determined, it was necessary to adjust the mathematical models of Remenieras, Potential, and Balocchi et al. [52] to the flow curve. Goodness of fit was assessed with the KGE test and the results of which are detailed in Tables A1 and A2. In the evaluation of the three models, the average for each basin was considered. When observing these averages, it stands out that both basins show better performance in terms of KGE with the Remenieras model, which is the reason it was finally chosen for further analyses (Figure 7).

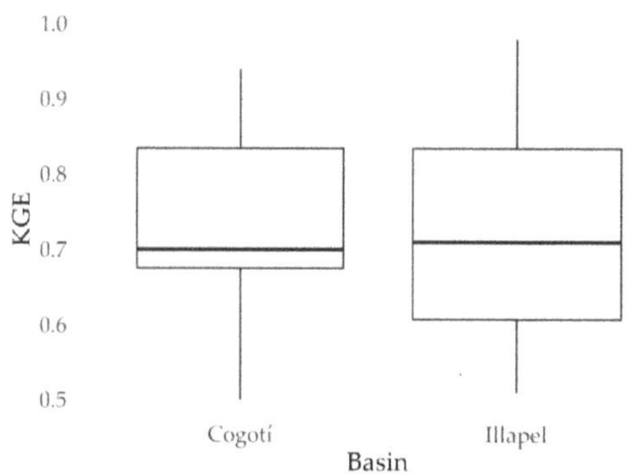

Figure 7. KGE results for the Remenieras model. Source: Own work.

An analysis of the results of the estimation of underground volumes as of 31 March, for each year and each basin, based on the Remenieras Model, was performed. This volume estimate is a default because part of the underground volumes in a basin may be evacuated to another basin, may be evacuated to the sea, or may evaporate from areas upstream of the control station. However, it is a first approximation to estimating water reserves at a given time. With this, it was possible to obtain an estimate of the volumes stored in various temporal scenarios. Based on the integral of the Remenieras's expression (Equation (9)), storage volumes were obtained for each basin and times considered. Results are illustrated in Figure 8. The figure visually denotes a decrease in the volumes stored underground for the two basins. In fact, the values for some years are so low that they are not seen in the graph, as exemplified in Illapel for 2013 or in Cogotí for 2020. Cogotí exhibits the lowest average volume (Table A3), a situation that can be attributed to its smaller area compared to the other basin under study.

$$\int_0^{\infty} Q_0\, e^{-\alpha t} dt = \frac{Q_0}{\alpha} \quad (9)$$

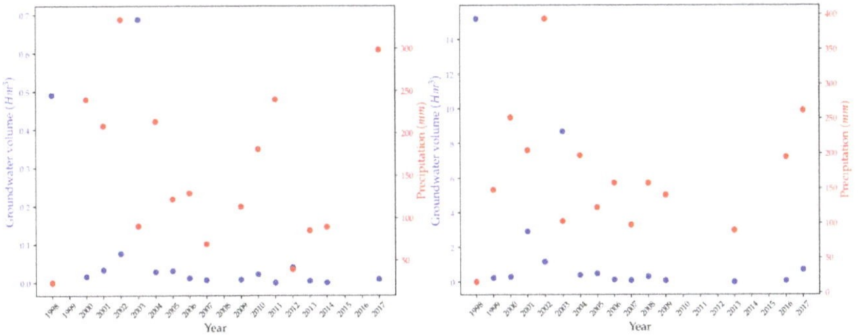

Figure 8. Estimation result of underground reserves for the Cogotí (**left**) and Illapel (**right**) station with the Remenieras Model.

At the macro level, the trends found are negative and significant in both periods. The results obtained are presented in Table 4, where it is possible to observe a significant negative trend of 0.0056 and 0.00178 Hm^3 for both the first and second periods of analysis, respectively, at the Cogotí basin. Therefore, it is inferred that an important part of the underground volumes has decreased considerably towards the end of March or the beginning of April, which is the beginning of the hydrological year.

Table 4. Mann–Kendall trend results for two periods in the two basins.

Basin	Period	Theil-Sen Slope (Hm^3)	*p*-Value	Z
Cogotí	1998–2009	−0.0056	0.049	−1.97
	1998–2020	−0.00178	0.014	−2.45
Illapel	1998–2009	−0.1389	0.0467	−1.99
	1998–2020	−0.0296	0.006	−2.75

For the Illapel basin, on the other hand, a significant negative trend is observed for both periods (Table 4), with 0.1389 and 0.0296 Hm^3, respectively. In the 1998–2009 period, the volume is almost five times greater than in the 1998–2020 period, indicating an accelerated decline in Illapel even before 2010, when the megadrought began, attributing such a depletion to the overconsumption of the resource. This is reinforced when considering that the trends in precipitation, despite being negative, are not significant (Table 5). In other words, the results of the MK test and the TS slope for the Cogotí and Illapel basins do not verify trends that are significantly different from zero in both basins.

Table 5. Mann–Kendall trend results for rainfall in two periods in the two stations.

	Cogotí			Illapel		
Month	Theil-Sen Slope (mm)	*p*-Value	Z	Theil-Sen Slope (mm)	*p*-Value	Z
January *						
February	0	1	−1.247	0	0.66	−0.633
March	0	0.21	−1.247	0	0.53	−0.633
April	0	0.52	−0.641	0	0.09	−1.72
May	0	0.88	0.2	−0.003	0.58	−0.547
June	−0.0012	0.98	−0.026	0.008	0.84	0.248
July	−0.0204	0.51	−0.653	−0.018	0.47	−0.72
August	−0.0054	0.8	−0.255	−0.018	0.26	−1.118
September	0	0.5	−0.681	0	0.53	−0.622
October	0	0.6	−0.531	0	0.73	−0.352
November	0	0.41	0.909	0	0.44	−0.765
December	0	0.16	1.576	0	0.16	1.576

* The test was not carried out for January due to the fact that there was no variation in the data.

3.3. Analysis of the Relationship between Intensive Agricultural Use of Irrigated Land and the Trend in Stored Underground Volumes over Time

Figure 9 shows a comparison between the variables of intensive agricultural use under irrigation and the trend in groundwater reserves for the Cogotí and Illapel basins.

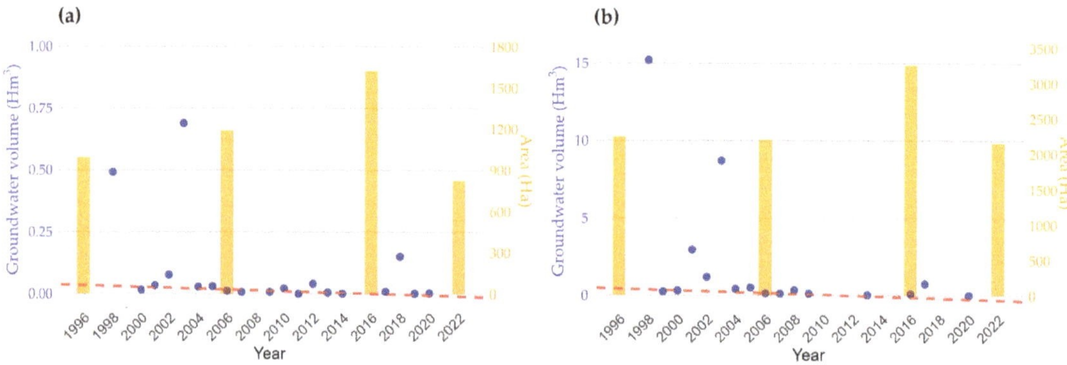

Figure 9. Comparative graph between the trend in stored underground volumes and the use of agricultural land for irrigation over time at the Cogotí (**a**) and Illapel (**b**) basins. The red line is the Theil–Sen robust trend.

4. Discussion

The results achieved show an increase in the amounts of agricultural areas between 1996 and 2016 at the two study areas; such an increase was 64% and 45% for the Cogotí and Illapel basins, respectively. However, in 2022, there was a notable drop in the amount of irrigation surfaces, returning to the 1996 amount. These results coincide with the agricultural area and production reported by the inventories of the Chilean Natural Resources Research Center (CIREN) for the Coquimbo Region, where there was a higher value in 2018, but a significant decrease in 2021 [68]. The above could be explained by the presence of a megadrought that affected the central area of Chile between 2010 and 2020 [69–71].

On the other hand, the drip irrigation method has exceeded 89% in both basins, highlighting the production of high-value crops such as table grapes, avocados, and citrus. In this scenario, it is relevant to note that the evapotranspiration of grapes can vary between 30% and 70% [72], while citrus trees exhibit transpiration rates that fluctuate between 0.15 and 2.30 mm/day [73] and agro-industrial avocado production consumes up to 120% of the surface and groundwater volumes allocated to agricultural use during dry years [74]. Based on the above, the intense use of agricultural land, with the incorporation of highly water-demanding crops, is an evident example of the overuse of resources in both basins. Thus, the most evident proof is that from 2016 onwards, the ecosystem was not able to withstand the pressure and in a scenario of restricted supplies, it was necessary to abandon a significant proportion of the land because the amounts of water resources were not sufficient.

The estimation of underground stored water volumes was made through the integration of baseflows and the application of various mathematical models, where the Remenieras model presented the best fit to the existing data. As noted, this analysis was carried out at the beginning of the hydrological year in the southern hemisphere, that is, 31 March to 1 April of each year, with results showing relevant situations in both basins under study. Thus, it is important to highlight that the estimated volumes present low values in general, which is to be expected in basins located in arid areas with average annual rainfall that is less than 200 mm. In this way, the maximum reserve estimated for Cogotí in the period studied was 0.69 Hm^3, while for Illapel, it was 15.2 Hm^3. Likewise, the lowest value for the same period for Cogotí was 0.3 Hm^3 and for Illapel, it was only a few cubic meters, values that are generally very low and reflect the situation of arid areas.

A similar study was conducted by Balocchi et al. [75], where the influence of native and exotic vegetation (Pine and Eucalyptus) on baseflows in eight basins in central Chile was evaluated. The results of this study show no significant differences between native and exotic vegetation, nor an explicit relationship between cover and baseflow. That is, no significant effect of cover on baseflow was detected. It should be noted that the exotic cover used in by Balocchi et al. [75] does not require irrigation and, therefore, the effect on flows is lower. In this study, an inverse relationship was found between the cover of fruit plantations and water reserves, mainly explained by a higher water consumption given that this type of monoculture needs constant irrigation. Another factor to consider is the prolonged drought affecting the country [69,70], derived from the fact that authors such as Lee and Ajami [76], when analyzing 358 basins in the United States, found that prolonged droughts (9–104 months) can impact baseflow amounts, and that these effects can last up to 41 months after the drought ends. This is important, since the megadrought in Chile has lasted more than 180 months and, therefore, can have a significant impact on baseflows. Parra et al. [77] studied water reserves in central-southern Chile, concluding that the depletion coefficient of the baseflow model is sensitive to dry periods, decreasing the value of the parameter during these periods.

When analyzing the trends obtained through the MK test, these were separated into two periods (1998–2009 and 1998–2020). The objective was to be able to see if the situation prior to the megadrought (i.e., the first period) was similar to the entire period integrating the megadrought. The results achieved indicate that for both the first and second periods in the Cogotí basin, a significant negative trend of 0.0056 and 0.00178, respectively, was observed. The existence of a negative and significant trend, even before the megadrought, suggests that this situation could be attributed to a variable other than this event. In this context and as previously mentioned, the excessive use of water resources emerges as a much more influential factor.

In the case of the Illapel basin, a significant negative trend (TS test) is also seen for both periods (0.1389 and 0.0296, respectively). The 1998–2009 period presents a volume almost five times greater than the 1998–2020 period, that is, an accelerated decrease is observed in Illapel even before 2010, which again reaffirms the overuse of the resource before the megadrought occurred.

The above is relevant because it indicates that before the existence of the mega-drought, stored volumes already showed a significant decrease. Likewise, carrying out flow analysis at the beginning of the hydrological year, that is, when water reserves reach their lowest point and do not receive new water inflows, provides crucial information on the retention, storage, and depletion of groundwater [51]. In other words, performing the analysis in the month of March reduces the relevance of precipitation, since more than 85% of it occurs between the months of June and September [78].

Having carried out the analysis on the recession flow is relevant because although it represents the gradual depletion of water storage during periods of little or no precipitation [79], it could still be slightly dependent on the distribution of precipitation [80].

Previous studies [13,81,82] analyzed precipitation trends in the Coquimbo Region, finding no significant variations. The results from precipitation analysis for the Illapel and Cogotí basins show a similar behavior, where precipitation trends are negative, but not significant. Regarding surface flows, negative and significant trends are observed in the average and maximum flows, which has determined an increase in the use of groundwater [13]. Similarly, Pizarro et al. [13] investigated the decline of groundwater in the region and found that the main driver of this decline is overuse (i.e., an anthropogenic factor). Therefore, this study focuses on crop production and its impact on baseflows. Moreover, future investigations should re-evaluate the behavior of baseflows and how climate change has influenced them.

Furthermore, Pizarro et al. [13] point out that in 2004, the basins of the region were closed for the granting of surface water rights, increasing since that date due to the use of groundwater. Thus, the apparent growth trend in this type of crop then focused on a very drastic use of groundwater, which only decreased in the second half of the 2010s, when extensive surfaces had to be left unused due to the non-availability of water in a period affected by the megadrought that hit the country and from which the Coquimbo Region has not yet recovered. Therefore, it follows that water resources were overexploited beyond their real capacities. This overexploitation had a significant impact on groundwater availability, especially considering that this water source supported the increase in agricultural activity from 2004 onwards.

It is important to mention that rainfall in the central-northern area of the country is greatly affected by climatic phenomena such as the El Niño Southern Oscillation (ENSO) or the Pacific Decadal Oscillation (PDO) [83], which determine that the recharge of aquifers occurs sporadically [84]. In this framework, the fraction of precipitation that eventually converts to recharge is therefore comparable in magnitude to the uncertainty range in recharge quantification methods [85]. However, despite these episodic events, there are studies that have not considered this variable for the estimation of recharges [86,87].

An important aspect of water scarcity is reflected in the quality of life of nearby populations. Reduced water availability limits crop and livestock yields, thereby decreasing family income, a situation that has deteriorated the mental health of these communities [85]. This deterioration has led to migration from rural to urban areas in search of better job opportunities and quality of life [85,86]. Furthermore, the overuse detected by Pizarro et al. [13] has the potential to accelerate water scarcity in rural areas, which are usually used (through deep wells) to address water scarcity in the middle term [87].

On the other hand, it is important to note that in Chile, there is a clear lack of water resources management at the basin level [9,88]. This deficiency makes it difficult to implement sustainable water resource use policies that integrate productive, social, and environmental aspects in their management and thus mitigate the impact of water scarcity on the population.

Finally, though cadastral the amount of information on agricultural land in Chile is limited, extending the timeframe or examining different time scales could provide a more comprehensive understanding of dynamic changes in groundwater reserves. Additionally, the studied basins were selected because they are in an area with high amounts of agricultural production and low amounts of annual rainfall. However, further studies should expand the geographic area under analysis to identify patterns in the behavior of baseflows. Similarly, factors such as urbanization and mining, as well as climate variation models, should be integrated in further analyses.

5. Conclusions

Based on the results from this investigation, it is possible to conclude that there was an increase in the areas allocated to crops with high water demand, in a proportion that was above the ecosystem's capabilities, which determined that, in the midst of a megadrought, large areas of crops were abandoned.

It is also possible to conclude that the use of groundwater has increased and has determined a significant reduction in existing reserves, as demonstrated by the negative and significant trends obtained in this study.

Finally, it is concluded that overuse is a problem in these basins, above and beyond climatic factors, since the negative trends in groundwater availability manifest prior to the existence of the 2010–2021 megadrought and that determines the need to rethink the use of water resources under sustainability schemes.

Author Contributions: Conceptualization, R.P. and F.B.; methodology, F.B., A.I. and R.B.-O.; software, R.B.-O., B.I., F.B. and C.V.; validation, R.P., A.I., P.A.G.-C. and F.B.; formal analysis, R.P., A.I., F.B. and C.S.; investigation, R.P., A.I. and F.B.; data curation, F.B., R.B.-O., B.I. and A.I.; writing—original draft preparation, R.P., A.I., F.B., P.A.G.-C. and B.I.; writing—review and editing, R.P., A.I., F.B., C.T., C.S. and P.A.G.-C.; visualization, F.B. and A.I.; supervision, R.P., P.A.G.-C. and C.S.; project administration, R.P. and F.B.; funding acquisition, R.P. All authors have read and agreed to the published version of the manuscript.

Funding: This research received no external funding.

Institutional Review Board Statement: Not applicable.

Informed Consent Statement: Not applicable.

Data Availability Statement: Data are contained within the article.

Acknowledgments: The authors thank the Cenamad ANID FB 210015 project, which provided adequate methodologies and information to develop this investigation.

Conflicts of Interest: The authors declare no conflicts of interest.

Appendix A

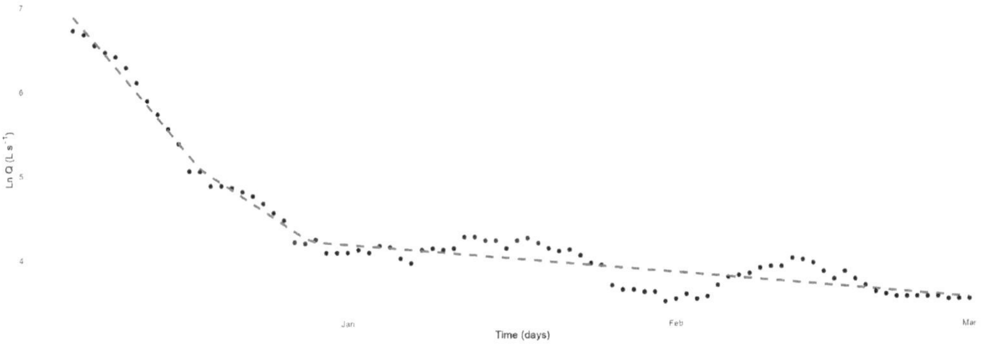

Figure A1. Graph of Ln Q (l s^{-1}) vs. Time (Days), with the three trend lines in Cogotí for the year of 2010.

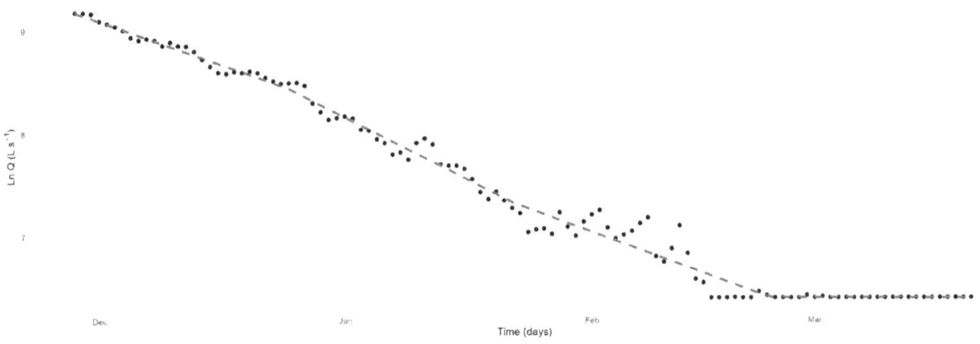

Figure A2. Graph of Ln Q (l s^{-1}) vs. Time (Days), with the three trend lines in Illapel for the year of 2001.

Table A1. Values obtained for the Kling–Gupta Efficiency (KGE) Test for the Río Cogotí Station, the Cogotí Reservoir Entrance (04531002), for the estimation of recessive flows.

Year	Remenieras	Potential	Balocchi et al. [52]
1998	0.92	0.87	0.79
2000	0.80	0.81	0.76
2001	0.56	0.58	0.60
2002	0.70	0.71	0.65
2003	0.73	0.88	0.73
2004	0.68	0.85	0.77
2005	0.58	0.58	0.53
2006	0.89	0.89	0.81
2007	0.74	0.74	0.62
2009	0.87	0.87	0.95
2010	0.69	0.69	0.57
2011	0.59	0.46	0.35
2012	0.67	0.33	0.34
2013	0.50	0.83	0.69
2014	0.69	0.60	0.45
2017	0.69	0.70	0.58
2018	0.94	−0.04	0.13
2019	0.91	0.78	0.62
2020	0.74	0.70	0.53
$\overline{x}$	0.73	0.67	0.61

Table A2. Values obtained for the Kling–Gupta Efficiency (KGE) Test for the Rio Illapel Station, in El Peral (04726001), for the estimation of recessive flows.

Year	Remenieras	Potential	Balocchi et al. [52]
1998	0.85	0.95	0.84
1999	0.76	0.81	0.67
2000	0.83	0.84	0.69
2001	0.66	0.68	0.62
2002	0.63	0.64	0.59
2003	0.83	0.92	0.73
2004	0.79	0.79	0.68
2005	0.63	0.64	0.55
2006	0.98	0.62	0.53
2007	0.86	0.49	0.43
2008	0.55	0.56	0.50
2009	0.86	0.64	0.54
2013	0.53	0.55	0.54
2016	0.62	0.77	0.64
2017	0.51	0.51	0.41
2020	0.57	0.59	0.64
$\overline{x}$	0.72	0.69	0.60

Table A3. Estimated underground volumes for each basin (Hm^3).

Year	Cogotí	Illapel
1996	No data	No data
1997	No data	No data
1998	0.4918	15.211
1999	No data	0.254
2000	0.0168	0.314
2001	0.0340	2.953
2002	0.0767	1.181
2003	0.6887	8.704
2004	0.0289	0.420

Table A3. *Cont.*

Year	Cogotí	Illapel
2005	0.0311	0.504
2006	0.0124	0.145
2007	0.0075	0.116
2008	No data	0.338
2009	0.0085	0.110
2010	0.0226	No data
2011	0.0007	No data
2012	0.0406	No data
2013	0.0054	0.02655
2014	0.0011	No data
2015	No data	No data
2016	No data	0.10089
2017	0.00972	0.73790
2018	0.15065	No data
2019	0.00175	No data
2020	0.00297	1.3×10^{-7}
2021	No data	No data
2022	No data	No data

References

1. Rodríguez-Pérez, L. Implicaciones fisiológicas de la osmorregulación en plantas. *Agron. Colomb.* **2006**, *24*, 28–37.
2. Chavarria, G.; Dos Santos, H.P. Plant Water Relations: Absorption, Transport and Control Mechanisms. In *Advances in Selected Plant Physiology Aspects*; Montanaro, G., Ed.; InTech: London, UK, 2012; ISBN 978-953-51-0557-2.
3. Blatt, M.R.; Chaumont, F.; Farquhar, G. Focus on Water. *Plant Physiol.* **2014**, *164*, 1553–1555. [CrossRef]
4. Gul, M.U.; Paul, A.; S, M.; Chehri, A. Hydrotropism: Understanding the Impact of Water on Plant Movement and Adaptation. *Water* **2023**, *15*, 567. [CrossRef]
5. Vargas-Pineda, O.I.; Trujillo-González, J.M.; Torres-Mora, M.A. Huella Hídrica: Una Herramienta Eficaz Para El Desafío de La Sostenibilidad Del Agua. *Ing. Compet.* **2020**, *22*, 8429. [CrossRef]
6. DGA. *Actualización Del Balance Hídrico Nacional*; SIT; Dirección General de Aguas: Santiago, Chile, 2017; p. 348.
7. Fuentes, I.; Fuster, R.; Avilés, D.; Vervoort, W. Water Scarcity in Central Chile: The Effect of Climate and Land Cover Changes on Hydrologic Resources. *Hydrol. Sci. J.* **2021**, *66*, 1028–1044. [CrossRef]
8. Medrano, H.; Bota, J.; Cifre, J.; Flexas, J.; Ribas-Carbó, M.; Gulías, J. Eficiencia en el uso del agua por las plantas. *Investig. Geográficas* **2007**, *43*, 63–84. [CrossRef]
9. Valdés-Pineda, R.; Pizarro, R.; García-Chevesich, P.; Valdés, J.B.; Olivares, C.; Vera, M.; Balocchi, F.; Pérez, F.; Vallejos, C.; Fuentes, R.; et al. Water Governance in Chile: Availability, Management and Climate Change. *J. Hydrol.* **2014**, *519*, 2538–2567. [CrossRef]
10. Donoso, G. Integrated Water Management in Chile. In *Integrated Water Resources Management in the 21st Century: Revisiting the paradigm*; Martinez-Santos, P., Aldaya, M., Llamas, M., Thuy, L., Eds.; CRC Press: Boca Raton, FL, USA, 2014; pp. 217–233. ISBN 978-1-138-00143-5.
11. Roco, L.; Poblete, D.; Meza, F.; Kerrigan, G. Farmers' Options to Address Water Scarcity in a Changing Climate: Case Studies from Two Basins in Mediterranean Chile. *Environ. Manag.* **2016**, *58*, 958–971. [CrossRef]
12. Morgan Uliana, E.; David Da Silva, D.; Morgan Uliana, E.; Silveira Rodrigues, B.; De Paula Corrêdo, L. Análise de Tendência Em Séries Históricas de Vazão e Precipitação: Uso de Teste Estatístico Não Paramétrico. *Ambient. Agua-Interdiscip. J. Appl. Sci.* **2015**, *10*, 82–88. [CrossRef]
13. Pizarro, R.; Garcia-Chevesich, P.A.; McCray, J.E.; Sharp, J.O.; Valdés-Pineda, R.; Sangüesa, C.; Jaque-Becerra, D.; Álvarez, P.; Norambuena, S.; Ibáñez, A.; et al. Climate Change and Overuse: Water Resource Challenges during Economic Growth in Coquimbo, Chile. *Sustainability* **2022**, *14*, 3440. [CrossRef]
14. Pimentel, D.; Berger, B.; Filiberto, D.; Newton, M.; Wolfe, B.; Karabinakis, E.; Clark, S.; Poon, E.; Abbett, E.; Nandagopal, S. Water Resources: Agricultural and Environmental Issues. *BioScience* **2004**, *54*, 909–918. [CrossRef]
15. Rijsberman, F.R. Water Scarcity: Fact or Fiction? *Agric. Water Manag.* **2006**, *80*, 5–22. [CrossRef]
16. Gordon, L.J.; Finlayson, C.M.; Falkenmark, M. Managing Water in Agriculture for Food Production and Other Ecosystem Services. *Agric. Water Manag.* **2010**, *97*, 512–519. [CrossRef]
17. Pfister, S.; Bayer, P.; Koehler, A.; Hellweg, S. Projected Water Consumption in Future Global Agriculture: Scenarios and Related Impacts. *Sci. Total Environ.* **2011**, *409*, 4206–4216. [CrossRef] [PubMed]
18. Richter, B.D.; Brown, J.D.; DiBenedetto, R.; Gorsky, A.; Keenan, E.; Madray, C.; Morris, M.; Rowell, D.; Ryu, S. Opportunities for Saving and Reallocating Agricultural Water to Alleviate Water Scarcity. *Water Policy* **2017**, *19*, 886–907. [CrossRef]

19. Tuninetti, M.; Tamea, S.; Dalin, C. Water Debt Indicator Reveals Where Agricultural Water Use Exceeds Sustainable Levels. *Water Resour. Res.* **2019**, *55*, 2464–2477. [CrossRef]
20. De Graaf, I.E.M.; Van Beek, L.P.H.; Wada, Y.; Bierkens, M.F.P. Dynamic Attribution of Global Water Demand to Surface Water and Groundwater Resources: Effects of Abstractions and Return Flows on River Discharges. *Adv. Water Resour.* **2014**, *64*, 21–33. [CrossRef]
21. Peña-Arancibia, J.L.; Mainuddin, M.; Kirby, J.M.; Chiew, F.H.S.; McVicar, T.R.; Vaze, J. Assessing Irrigated Agriculture's Surface Water and Groundwater Consumption by Combining Satellite Remote Sensing and Hydrologic Modelling. *Sci. Total Environ.* **2016**, *542*, 372–382. [CrossRef]
22. Niazi, H.; Wild, T.B.; Turner, S.W.D.; Graham, N.T.; Hejazi, M.; Msangi, S.; Kim, S.; Lamontagne, J.R.; Zhao, M. Global Peak Water Limit of Future Groundwater Withdrawals. *Nat. Sustain.* **2024**, *7*, 413–422. [CrossRef]
23. Mehta, P.; Siebert, S.; Kummu, M.; Deng, Q.; Ali, T.; Marston, L.; Xie, W.; Davis, K.F. Half of Twenty-First Century Global Irrigation Expansion Has Been in Water-Stressed Regions. *Nat. Water* **2024**, *2*, 254–261. [CrossRef]
24. Feng, G.; Jin, W.; Ouyang, Y.; Huang, Y. The Role of Changing Land Use and Irrigation Scheduling in Groundwater Depletion Mitigation in a Humid Region. *Agric. Water Manag.* **2024**, *291*, 108606. [CrossRef]
25. Perez, N.; Singh, V.; Ringler, C.; Xie, H.; Zhu, T.; Sutanudjaja, E.H.; Villholth, K.G. Ending Groundwater Overdraft without Affecting Food Security. *Nat. Sustain.* **2024**, *7*, 1007–1017. [CrossRef]
26. Garcia, X.; Estrada, L.; Saló, J.; Acuña, V. Blueing Green Water from Forests as Strategy to Cope with Climate Change in Water Scarce Regions: The Case of the Catalan River Basin District. *J. Environ. Manag.* **2024**, *353*, 120249. [CrossRef] [PubMed]
27. Sangüesa, C.; Pizarro, R.; Ingram, B.; Balocchi, F.; García-Chevesich, P.; Pino, J.; Ibáñez, A.; Vallejos, C.; Mendoza, R.; Bernal, A.; et al. Streamflow Trends in Central Chile. *Hydrology* **2023**, *10*, 144. [CrossRef]
28. Álamos, N.; Alvarez-Garreton, C.; Muñoz, A.; González-Reyes, Á. The Influence of Human Activities on Streamflow Reductions during the Megadrought in Central Chile. *Hydrol. Earth Syst. Sci.* **2024**, *28*, 2483–2503. [CrossRef]
29. Barría, P.; Rojas, M.; Moraga, P.; Muñoz, A.; Bozkurt, D.; Alvarez-Garreton, C. Anthropocene and Streamflow: Long-Term Perspective of Streamflow Variability and Water Rights. *Elem. Sci. Anthr.* **2019**, *7*, 2. [CrossRef]
30. Pizarro, R.; Araya, S.; Jordán, C.; Farías, C.; Flores, J.P.; Bro, P.B. The Effects of Changes in Vegetative Cover on River Flows in the Purapel River Basin of Central Chile. *J. Hydrol.* **2006**, *327*, 249–257. [CrossRef]
31. Little, C.; Lara, A.; McPhee, J.; Urrutia, R. Revealing the Impact of Forest Exotic Plantations on Water Yield in Large Scale Watersheds in South-Central Chile. *J. Hydrol.* **2009**, *374*, 162–170. [CrossRef]
32. Schilling, K.E.; Langel, R.J.; Wolter, C.F.; Arenas-Amado, A. Using Baseflow to Quantify Diffuse Groundwater Recharge and Drought at a Regional Scale. *J. Hydrol.* **2021**, *602*, 126765. [CrossRef]
33. Zomlot, Z.; Verbeiren, B.; Huysmans, M.; Batelaan, O. Spatial Distribution of Groundwater Recharge and Base Flow: Assessment of Controlling Factors. *J. Hydrol. Reg. Stud.* **2015**, *4*, 349–368. [CrossRef]
34. Gomez Isidro, S.; Gómez-Ríos, V.L. Baseflow Analysis Using Master Recession Curves and Numerical Algorithms in Mountain Basins: Suratá's River and Oro's River (Santander, Colombia). *DYNA* **2016**, *83*, 213–222. [CrossRef]
35. Troutman, B.M.; Karlinger, M.R. Gibbs' Distribution on Drainage Networks. *Water Resour. Res.* **1992**, *28*, 563–577. [CrossRef]
36. Schiavo, M. The Role of Different Sources of Uncertainty on the Stochastic Quantification of Subsurface Discharges in Heterogeneous Aquifers. *J. Hydrol.* **2023**, *617*, 128930. [CrossRef]
37. Sarricolea, P.; Herrera-Ossandon, M.; Meseguer-Ruiz, Ó. Climatic Regionalisation of Continental Chile. *J. Maps* **2017**, *13*, 66–73. [CrossRef]
38. QGIS Development Team. QGIS Geographic Information System 2024. Available online: https://www.qgis.org/ (accessed on 13 June 2023).
39. RStudio Team. *RStudio: Integrated Development for R*; RStudio Team: Boston, MA, USA, 2020.
40. Dixon, B.; Uddameri, V. *GIS and Geocomputation for Water Resource Science and Engineering*, 1st ed.; Wiley: Hoboken, NJ, USA, 2015; ISBN 978-1-118-35414-8.
41. Grizonnet, M.; Michel, J.; Poughon, V.; Inglada, J.; Savinaud, M.; Cresson, R. Orfeo ToolBox: Open Source Processing of Remote Sensing Images. *Open Geospatial Data Softw. Stand.* **2017**, *2*, 15. [CrossRef]
42. Rodriguez-Galiano, V.F.; Ghimire, B.; Rogan, J.; Chica-Olmo, M.; Rigol-Sanchez, J.P. An Assessment of the Effectiveness of a Random Forest Classifier for Land-Cover Classification. *ISPRS J. Photogramm. Remote Sens.* **2012**, *67*, 93–104. [CrossRef]
43. Breiman, L. Random Forests. *Mach. Learn.* **2001**, *45*, 5–32. [CrossRef]
44. Breiman, L. *Classification and Regression Trees*; Chapman and Hall/CRC: Boca Raton, FL, USA, 1984.
45. Breiman, L. *Bagging Predictors*; Technical Report 421; Chapman and Hall/CRC: Boca Raton, FL, USA, 1994.
46. Zhang, L.; Huettmann, F.; Liu, S.; Sun, P.; Yu, Z.; Zhang, X.; Mi, C. Classification and Regression with Random Forests as a Standard Method for Presence-Only Data SDMs: A Future Conservation Example Using China Tree Species. *Ecol. Inform.* **2019**, *52*, 46–56. [CrossRef]
47. Azar, A.T.; Elshazly, H.I.; Hassanien, A.E.; Elkorany, A.M. A Random Forest Classifier for Lymph Diseases. *Comput. Methods Programs Biomed.* **2014**, *113*, 465–473. [CrossRef]
48. Reyes Rojas, L.A.; Adhikari, K.; Ventura, S.J. Projecting Soil Organic Carbon Distribution in Central Chile under Future Climate Scenarios. *J. Environ. Qual.* **2018**, *47*, 735–745. [CrossRef]

49. Mashalaba, L.; Galleguillos, M.; Seguel, O.; Poblete-Olivares, J. Predicting Spatial Variability of Selected Soil Properties Using Digital Soil Mapping in a Rainfed Vineyard of Central Chile. *Geoderma Reg.* **2020**, *22*, e00289. [CrossRef]
50. Centro de Ciencia Del Clima y la Resiliencia (CR2). Base de Datos de Caudales 2020. Available online: https://www.cr2.cl/datos-de-caudales (accessed on 5 June 2023).
51. Jain, M.K. Recession of Discharge. In *Encyclopedia of Snow, Ice and Glaciers*; Singh, V.P., Singh, P., Haritashya, U.K., Eds.; Springer: Dordrecht, The Netherlands, 2011; pp. 922–924. ISBN 978-90-481-2641-5.
52. Balocchi, F.; Pizarro, R.; Morales, C.; Olivares, C. Modelamiento matemático de caudales recesivos en la región mediterránea andina del Maule: El caso del estero Upeo, Chile. *Tecnol. Cienc. Agua* **2014**, *5*, 179–188.
53. Pizarro-Tapia, R.; Balocchi-Contreras, F.; Garcia-Chevesich, P.; Macaya-Perez, K.; Bro, P.; León-Gutiérrez, L.; Helwig, B.; Valdés-Pineda, R. On Redefining the Onset of Baseflow Recession on Storm Hydrographs. *Open J. Mod. Hydrol.* **2013**, *3*, 269–277. [CrossRef]
54. Cirugeda, J. *Curso Internacional de Hidrología General y Aplicada*; Centro de Estudios y Experimentación de Obras Públicas, Gabinete de Formación y Documentación: Madrid, Spain, 1985; pp. 66–77.
55. Jackson, E.K.; Roberts, W.; Nelsen, B.; Williams, G.P.; Nelson, E.J.; Ames, D.P. Introductory Overview: Error Metrics for Hydrologic Modelling—A Review of Common Practices and an Open Source Library to Facilitate Use and Adoption. *Environ. Model. Softw.* **2019**, *119*, 32–48. [CrossRef]
56. Nash, J.E.; Sutcliffe, J.V. River Flow Forecasting through Conceptual Models Part I—A Discussion of Principles. *J. Hydrol.* **1970**, *10*, 282–290. [CrossRef]
57. Gupta, H.V.; Kling, H.; Yilmaz, K.K.; Martinez, G.F. Decomposition of the Mean Squared Error and NSE Performance Criteria: Implications for Improving Hydrological Modelling. *J. Hydrol.* **2009**, *377*, 80–91. [CrossRef]
58. Lamontagne, J.R.; Barber, C.A.; Vogel, R.M. Improved Estimators of Model Performance Efficiency for Skewed Hydrologic Data. *Water Resour. Res.* **2020**, *56*, e2020WR027101. [CrossRef]
59. Knoben, W.J.M.; Freer, J.E.; Woods, R.A. Technical Note: Inherent Benchmark or Not? Comparing Nash–Sutcliffe and Kling–Gupta Efficiency Scores. *Hydrol. Earth Syst. Sci.* **2019**, *23*, 4323–4331. [CrossRef]
60. Correndo, A.A.; Rosso, L.H.M.; Hernandez, C.H.; Bastos, L.M.; Nieto, L.; Holzworth, D.; Ciampitti, I.A. Metrica: An R Package to Evaluate Predictionperformance of Regression and Classification Point-Forecastmodels. *J. Open Source Softw.* **2022**, *7*, 4655. [CrossRef]
61. Centro de Ciencia Del Clima y la Resiliencia (CR2). Base de Datos de Precipitaciones 2020. Available online: https://www.cr2.cl/datos-de-precipitacion/ (accessed on 5 June 2023).
62. McBride, G.; Cole, R.G.; Westbrooke, I.; Jowett, I. Assessing Environmentally Significant Effects: A Better Strength-of-Evidence than a Single P Value? *Environ. Monit. Assess.* **2014**, *186*, 2729–2740. [CrossRef]
63. Hamed, K.H. Trend Detection in Hydrologic Data: The Mann–Kendall Trend Test under the Scaling Hypothesis. *J. Hydrol.* **2008**, *349*, 350–363. [CrossRef]
64. Helsel, D.R.; Mueller, D.K.; Slack, J.R. *Computer Program for the Kendall Family of Trend Tests*; Scientific Investigations Report; Version 1.0.; U.S. Geological Survey: Reston, VA, USA, 2006; p. 7.
65. Muñoz, C. Análisis de La Variabilidad En Series Hidrometeorológicas En Una Cuenca de Cabecera Del Río Segura: Cuenca Del Río Mundo. Bachelor's Thesis, Universidad Politécnica de Cartagena, Cartagena, España, 2008; 146p.
66. Marchetto, A. Rkt: Mann-Kendall Test, Seasonal and Regional Kendall Tests 2021. R Package V1.6. Available online: https://cran.r-project.org/web/packages/rkt/rkt.pdf (accessed on 5 June 2023).
67. Helsel, D.R.; Hirsch, R.M.; Ryberg, K.R.; Archfield, S.A.; Gilroy, E.J. *Statistical Methods in Water Resources*; Techniques and Methods; U.S. Geological Survey: Reston, VA, USA, 2020; p. 484.
68. Centro de Información de Recursos Naturales (Ciren); Oficina de Estudios y Políticas Agrarias (ODEPA). *Catastro Frutícola: Principales Resultados Región de Coquimbo*; Catastro Frutícola; Centro de Información de Recursos Naturales (Ciren): Santiago, Chile, 2021.
69. Garreaud, R.D.; Alvarez-Garreton, C.; Barichivich, J.; Boisier, J.P.; Christie, D.; Galleguillos, M.; LeQuesne, C.; McPhee, J.; Zambrano-Bigiarini, M. The 2010–2015 Megadrought in Central Chile: Impacts on Regional Hydroclimate and Vegetation. *Hydrol. Earth Syst. Sci.* **2017**, *21*, 6307–6327. [CrossRef]
70. Garreaud, R.D.; Boisier, J.P.; Rondanelli, R.; Montecinos, A.; Sepúlveda, H.H.; Veloso-Aguila, D. The Central Chile Mega Drought (2010–2018): A Climate Dynamics Perspective. *Int. J. Climatol.* **2020**, *40*, 421–439. [CrossRef]
71. Kim, Y.; Kim, S.; Jeong, H.; An, H. Quantitatively Defining Megadrought Based on Drought Events in Central Chile. *Geomat. Nat. Hazards Risk* **2022**, *13*, 975–992. [CrossRef]
72. Vijayakumar, P.; Rao, S.; Chandran, S. *Agrometeorological Data Collection, Analysis and Management*; Central Research Institute for Dryland Agriculture: Hyderabad, India, 2016; 152p.
73. Teunis Vahrmeijer, J.; Taylor, N.J. Citrus Water Use. In *Citrus—Health Benefits and Production Technology*; Sajid, M., Amanullah, Eds.; IntechOpen: London, UK, 2019; ISBN 978-1-78985-427-5.
74. Gómez-Tagle, A.F.; Gómez-Tagle, A.; Fuerte-Velázquez, D.J.; Barajas-Alcalá, A.G.; Quiroz-Rivera, F.; Alarcón-Chaires, P.E.; Guerrero-García-Rojas, H. Blue and Green Water Footprint of Agro-Industrial Avocado Production in Central Mexico. *Sustainability* **2022**, *14*, 9664. [CrossRef]

75. Balocchi, F.; Flores, N.; Arumí, J.L.; Iroumé, A.; White, D.A.; Silberstein, R.P.; Ramírez De Arellano, P. Comparison of Streamflow Recession between Plantations and Native Forests in Small Catchments in Central-Southern Chile. *Hydrol. Process.* **2021**, *35*, e14182. [CrossRef]
76. Lee, S.; Ajami, H. Comprehensive Assessment of Baseflow Responses to Long-Term Meteorological Droughts across the United States. *J. Hydrol.* **2023**, *626*, 130256. [CrossRef]
77. Parra, V.; Muñoz, E.; Arumí, J.L.; Medina, Y. Analysis of the Behavior of Groundwater Storage Systems at Different Time Scales in Basins of South Central Chile: A Study Based on Flow Recession Records. *Water* **2023**, *15*, 2503. [CrossRef]
78. Valois, R.; MacDonell, S.; Núñez Cobo, J.H.; Maureira-Cortés, H. Groundwater Level Trends and Recharge Event Characterization Using Historical Observed Data in Semi-Arid Chile. *Hydrol. Sci. J.* **2020**, *65*, 597–609. [CrossRef]
79. Singh, P.; Huebl, H.; Weinmeister, H.W. Use of the Recession Characteristics of Snowmelt Hydrographs in the Assessment of Snow Water Storage in a Basin. *Hydrol. Process.* **2000**, *14*, 91–101. [CrossRef]
80. Chow, V.; Maidment, D.; Mays, L. *Hidrología Aplicada*; McGraw-Hill: New York, NY, USA, 1994; 584p.
81. Souvignet, M.; Oyarzún, R.; Verbist, K.M.J.; Gaese, H.; Heinrich, J. Hydro-Meteorological Trends in Semi-Arid North-Central Chile (29–32°S): Water Resources Implications for a Fragile Andean Region. *Hydrol. Sci. J.* **2012**, *57*, 479–495. [CrossRef]
82. Favier, V.; Falvey, M.; Rabatel, A.; Praderio, E.; López, D. Interpreting Discrepancies between Discharge and Precipitation in High-altitude Area of Chile's Norte Chico Region (26–32°S). *Water Resour. Res.* **2009**, *45*, W02424. [CrossRef]
83. Núñez, J.; Rivera, D.; Oyarzún, R.; Arumí, J.L. Influence of Pacific Ocean Multidecadal Variability on the Distributional Properties of Hydrological Variables in North-Central Chile. *J. Hydrol.* **2013**, *501*, 227–240. [CrossRef]
84. Kinzelbach, W.; Aeschbach, W. *A Survey of Methods for Analysing Groundwater Recharge in Arid and Semi-Arid Regions*; Division of Early Warning and Assessment, United Nations Environment Programme: Nairobi, Kenya, 2002; ISBN 978-92-807-2131-7.
85. Scanlon, B.R.; Healy, R.W.; Cook, P.G. Choosing Appropriate Techniques for Quantifying Groundwater Recharge. *Hydrogeol. J.* **2002**, *10*, 18–39. [CrossRef]
86. DGA. *Evaluación de Los Recursos Hídricos Subterráneos de La Cuenca Del Río Limarí*; S.D.T; Dirección General de Aguas: Santiago, Chile, 2008; p. 30.
87. Sandoval, E.; Baldo, G.; Núñez, J.; Oyarzún, J.; Fairley, J.P.; Ajami, H.; Arumí, J.L.; Aguirre, E.; Maturana, H.; Oyarzún, R. Groundwater Recharge Assessment in a Rural, Arid, Mid-Mountain Basin in North-Central Chile. *Hydrol. Sci. J.* **2018**, *63*, 1873–1889. [CrossRef]
88. Donoso, G. Management of Water Resources in Agriculture in Chile and Its Challenges. *Int. J. Agric. Nat. Resour.* **2021**, *48*, 171–185. [CrossRef]

sustainability

MDPI

Article

Applicability of Paper and Pulp Industry Waste for Manufacturing Mycelium-Based Materials for Thermoacoustic Insulation

Hugo Muñoz [1], Paulo Molina [2,*], Ignacio A. Urzúa-Parra [3], Diego A. Vasco [3], Magdalena Walczak [1], Gonzalo Rodríguez-Grau [4,5], Francisco Chateau [5,6] and Mamié Sancy [2,4,5,*]

1 Department of Ingeniería Mecánica y Metalúrgica, Facultad de Ingeniería, Pontificia Universidad Católica de Chile, Santiago 6904411, Chile; hamunoz2@uc.cl (H.M.); mwalczak@uc.cl (M.W.)
2 Millennium Institute on Green Ammonia as Energy Vector, Pontificia Universidad Católica de Chile, Santiago 6904411, Chile
3 Department of Ingeniería Mecánica, Universidad de Santiago de Chile, Avenida Libertador Bernardo O'Higgins N 3363, Estación Central, Santiago 9170022, Chile; ignacio.urzuap@usach.cl (I.A.U.-P.); diego.vascoc@usach.cl (D.A.V.)
4 Escuela de Construcción Civil, Pontificia Universidad Católica de Chile, Santiago 6904411, Chile; grodriguezg@uc.cl
5 Centro Nacional de Excelencia para la Industria de la Madera (CENAMAD), Santiago 9170022, Chile; fchateau@uc.cl
6 Faculty of Arquitectura, Diseño y Estudios Urbanos, Pontificia Universidad Católica de Chile, Santiago 9170022, Chile
* Correspondence: paulo.molina@uc.cl (P.M.); mamiesancy@uc.cl (M.S.)

Abstract: Cellulose and paper produce significant waste such as ash, activated sludge, and sludge from the pulp and paper industry. Depending on the raw material, legislation, and subprocesses, these sludges contain around 30–50% organic matter, mainly composed of less than 0.02 mm cellulose fibers and hemicellulose and lignin. This work used sludge from the pulp and paper industry as a substrate for manufacturing mycelium-based biomaterials using the white rot fungus *Trametes versicolor*. Chemical and surface analyses revealed the formation of new materials. Acoustic impedance analyses revealed that these materials have a noise reduction coefficient and sound absorption average comparable to extruded polystyrene and polyurethane. In addition, the material's thermal conductivity was near that of sheep wool. Therefore, the biomaterials fabricated using sludge and *Trametes versicolor* have the potential to be a game-changer in the industry as promising thermoacoustic insulators.

Keywords: pulp and paper industry; industry waste; mycelium; thermal conductivity; acoustic conductivity; construction; insulation

1. Introduction

Expanded polystyrene (EPS) is one of the most widely used thermal insulators in the construction and packaging industry. EPS-based materials are mainly preferred due to their low cost and versatility, for almost half of thermal. These materials represent 30% of insulator materials for the construction market in the European Union [1]. According to the life cycle assessment developed by Kulakovskaya in Switzerland, the economic and environmental impacts of EPS boards in 2020 were 662 million CHFs and 134 kilotons of CO_2-eq, respectively [2]; however, they have a low recycling rate. A 2018 study revealed that in 2014, of the 2.2 million tons of polystyrene produced in the United States, only 20,000 tons were recycled, representing a rate of just 0.9% [3]. This is why the pollution associated with this type of material is a growing environmental concern, not only due to the large amount of waste but also due to the adverse effects of its waste on ecosystems, such as the marine ecosystem and human health [4].

Developing new biodegradable materials that can replace petroleum-based materials while mitigating their environmental footprint represents a tremendous scientific and technological challenge. In this context, mycelium-based materials that use white rot fungi are a cutting-edge alternative to replace EPS in the construction industry. These materials are manufactured using specialized white rot fungi, which decompose wood or lignocellulosic waste [5], forming a white and spongy substance that can be used to produce affordable and environmentally sustainable biomaterials, as reported previously. For instance, Al-Qahtani et al. [6] reported a review on mycelium-based thermal insulation for domestic cooling footprint reduction. Volk et al. [7] studied the life cycle assessment of mycelium-based materials. Charpentier-Alfaro et al. [8] reported on decaying fungi for mycelium-based materials. Ghazvinian et al. [9] investigated the mycelium-based approach for architectural applications. Attias et al. [10] studied nanocellulose–mycelium materials. Teixeira et al. [11] analyzed the production of biodegradable mycelium-based materials and their mechanical properties using white rot fungi. Huang et al. reported that white rot fungi can degrade lignin and its derived structures, colonizing wood from living or dead trees through their intra and extracellular systems of oxidative enzymes [12]. Meanwhile, Harshvardhan et al. [13] proposed that these biomaterials can be promising tools for the biotransformation of endocrine-disrupting compounds.

Therefore, white rot fungi can colonize a substrate based on the hyphae bind. The particles that form the substrate act as an adhesive that holds a mycelial network. This characteristic and its ability to degrade lignin and cellulose make white rot fungi ideal for creating materials from lignocellulosic waste, such as those from the pulp industry. These biomaterials, known as mycelium-based materials (MBMs), possess properties that would allow them to be positioned as a replacement for traditional insulating and packaging materials due to their thermal conductivity, sound absorption, and fire safety properties, being similar to materials such as EPS, which are frequently used in this kind of application [14–16]. In addition, a 2022 study showed that thermal inertia allows the manufacturing of bricks that reduce temperature fluctuations in buildings compared with other materials, such as expanded vermiculite and lightweight expanded clay [17].

The addition of nano- and microcellulose has been widely researched to enhance the mechanical properties of fungal biomaterials. For instance, incorporating 30% cellulose nanofibers improves tensile strength by 30% and elasticity by 44% [10]. Bhagavathi et al. [18] reported that adding microcrystalline cellulose accelerated the curing process of polyurethane. Marcuello et al. [19] proposed that hemicellulose improves the interfacial adhesion properties of lignocellulosic polymers. Studies also show that mycelium foams grown on cellulose from wastewater treatment have compressive strengths comparable to expanded polystyrene (EPS) but with double the density [20]. Sayfutdinova and colleagues found that adding cellulose results in cohesive biomaterials, with microcellulose degrading in two weeks and nanocellulose in five days, promoting mycelium growth [21]. Bacterial cellulose improves particle cohesion and increases density and tensile strength, thus enhancing compressive strength [14,22]. Additionally, Sun et al. showed that adding 2.5% cellulose nanofibers to a wood particle substrate yields a biomaterial with improved flexural strength and elastic modulus [23]. These findings highlight the potential of cellulose additives to significantly enhance fungal biomaterials, paving the way for sustainable and high-performance materials.

The sludge generated as waste by the pulp industry, called pulp and paper mill sludge (PPMS), represents a valuable alternative for incorporating cellulose from these residues. Including micrometric cellulose could significantly improve the mechanical properties of the resulting biomaterials, especially in terms of compressive strength. This is partly because for each ton of paper, around 40 to 50 kg of dry solid waste is generated, of which 70% is primary sludge. It is worth noting that world paper production exceeded 400 million tons in 2020, and the generation of primary sludge far exceeds 12 million tons annually [24]. These residues, generated during the cellulose fiber liberation process, are primarily composed of short fibers of less than 0.2 mm, which are unsuitable for manufacturing cellulose products, as well as hemicellulose, lignin, and peptides hydrolyzed during the process (see Table 1) [25–27].

Table 1. Physicochemical parameters of pulp and paper mill sludge. The range is a compilation from refs. [25–29].

Apparent Density (g/cm^3)	Humidity (%)	pH	Cellulose (%)	Hemicellulose (%)	Lignin (%)
0.419–0.598	50–75	6.8–8.2	57–72	8.0–12.4	19.7–32.5

Building on these premises, this study verified the hypothesis that applying PPMS to fabric mycelium-based materials is suitable for thermoacoustic insulation in construction applications. In particular, MBM was manufactured using sludge from the pulp and paper industry. For this purpose, the white rot fungus *Trametes versicolor* was used, which has a high potential for producing this type of biomaterial [14,16,30,31]. The materials produced were characterized using physicochemical, morphological, microstructural, and acoustic analyses, compared with commercially available materials currently available as thermal and acoustic insulators. This way, its potential use in a sustainable construction industry was evaluated.

2. Materials and Methods

2.1. Materials

The PPMS was provided by a local pulp- and paper-processing mill as original waste. In particular, it was kept humid and stored in a plastic bag at room temperature for about two weeks. The solid agar medium was malt extract Agar (Merck, Rahway, NJ, USA) and agar-agar (Winkler Ltd., Santiago, Chile).

2.2. Physicochemical and Microstructural Characterization of Sludge

The pH of the received sludge was evaluated by suspending 5.0 g of dry sludge in 100 mL of distilled water. The humidity was determined by drying 100.0 g of the sludge at 105 °C for 24 h and calculating the difference in mass. The organic matter content was assessed by calcining 100.0 g of the sludge at 800 °C, with the weight of the remaining solid corresponding to ash and the mass loss representing organic matter. The sludge samples were analyzed in triplicate.

Chemical analysis of the sludge was carried out using attenuated total reflection infrared spectroscopy (Shimadzu IRSpirit FTIR, Japan) and atomic absorption spectroscopy (Perkin Elmer PinAAcle 900 Series AA). The crystalline phases in the sludge were analyzed by X-ray diffraction (Bruker D8 Advance, CuKα, Germany). To verify the presence of heavy metals in the sludge, inductively coupled plasma mass spectrometry (ICP-MS) was used (Perkin Elmer Elan 9000 ICP/MS, USA).

2.3. Preparation of Biomaterial

2.3.1. Fungal Inoculum

The spawn of *Trametes versicolor* was selected for its performance in biomass generation and hyphal extension, as described by Jones et al. [32]. This fungus is known for degrading lignin and cellulose at higher rates than other fungi, making it particularly suitable for the present study. Its proven ability to grow on Eucalyptus and pine trees, used in manufacturing cellulose pulp at the plant that provided the PPMS, further justified its use in this study. For this study, the spawn of *Trametes versicolor* AC9911 was sourced from Agrocontinental Chile S.A. and stored at 4 °C in a breathing bag.

2.3.2. Substrate Preparation

The pulp and paper mill sludge was dehydrated at 105 °C (ECOCELL 111, MMM group, Germany) for 24 h and stored dry in airtight vacuum bags. To be used as a substrate, the sludge was rehydrated with excess freshwater, blended for 2 min, and soaked overnight. The excess water was removed, and the sludge was sterilized in an autoclave (All America,

25X-2, USA) at 121 °C for 20 min using a polypropylene container. Figure 1 summarizes the substrate preparation carried out in this work.

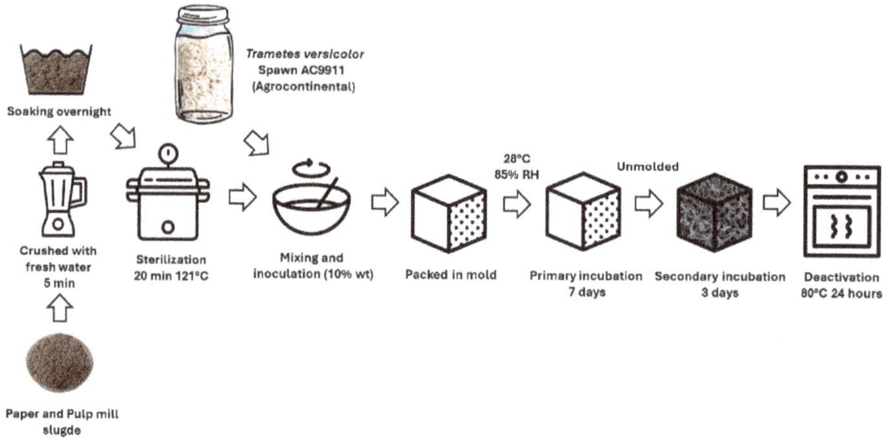

Figure 1. Schematic representation of mycelium-based material fabrication.

2.3.3. Mycelium Cultivation and Deactivation

The biomaterials were manufactured by mixing 10 wt.% of *Trametes versicolor* with 90% of the substrate, which had been sterilized in an autoclave using molds. The molds were filled layer by layer, compressing and filling to achieve a 1.015 g/cm³ density. The prismatic molds used were 160 × 40 × 40 mm, designed in 3D using Thinkercad (http://www.tinkercad.com, accessed on 27 August 2024), and printed in PLA (Creality Ender-3 V2, eSUN 1.75 mm). They featured easy-release mechanisms and holes to facilitate gas exchange between the substrate and the atmosphere.

The samples were then incubated at 28 °C and 85% relative humidity (BIOBASE BJPX-M150) for 7 days inside the molds, followed by 3 days outside to allow the growth of a white mycelium covering. All samples were dried in a convection oven (ECOCELL 111) at 80 °C for 12 to 24 h until their weights stabilized, ensuring a low humidity unsuitable for further fungal development. The mycelium's weight, diameter, and height were measured for each sample before and after drying. The percent shrinkage was calculated by subtracting the dry volume from the wet volume and then dividing this shrinkage by the wet volume. Figure 2 shows photographs of the mycelium-based material fabricated using *Trametes versicolor* on PMMS.

Figure 2. Photograph of biomaterials obtained using *Trametes versicolor* on PPMS.

2.4. Hyphae Growth Rate

To measure the hyphal growth rate, malt extract agar (MEA) (48 g/L, Merck) and a solid medium composed of sludge from the pulp and paper industry and agar-agar (Winkler) (53 g of dry PMMS and 20 g of agar-agar in 1000 mL of distilled water) were used. The media were sterilized in an autoclave at 121 °C for 20 min. Molten agar was aseptically poured into 90 mm Petri dishes under a laminar flow hood and allowed to solidify. A single inoculum was placed at the edge of each Petri dish, contacting the wall of the dish. Each experiment was performed in triplicate. The plates were individually sealed with Parafilm and incubated at 28 °C and 85% relative humidity in the dark. The radial growth (mm) was measured daily from the center of the inoculum to the tip of the longest hypha at the same time each day for 11 days. The plates were photographed and processed with the ImageJ software (1.54i version) [33].

2.5. Characterization of MBM

The density of the biomaterials was evaluated following the ISO standard 9427:2003 [34], considering the biomaterial's dry mass ratio to its volume. The moisture content was also determined based on ISO 16979:2003 [35]. The water absorption rate was also evalua-Japanted following the ASTM standard C 1585 [36].

The biomaterials were chemically analyzed by attenuated total reflectance spectroscopy (Shimadzu IR Spirit FTIR, Japan) at 24 °C with 40% relative humidity, averaging at 10 scans with a resolution of 1 cm^{-1}. Additionally, surface analysis of the biomaterials was performed using field-emission scanning electron microscopy (FE-SEM, Zeiss GeminiSEM 360, Germany) operated at an EHT of 5 keV, which also allowed for a relative chemical analysis using EDX (Oxford max ultra 40).

2.6. Thermal and Acoustic Analysis

The thermal conductivity of the biomaterials was analyzed using a KD-2 PRO thermal analyzer with the KS1 probe. Using the transient linear heat source or hot needle method, the KD2-Pro device was used for thermal analysis and complied with references [37,38]. In the traceability process of the thermal analyzer KD2-Pro with the KS-1 sensor, several thermal conductivity measurements were performed on glycerin (CAS 56-81-5), whose thermal conductivity was 0.282 W/mK at 20 °C.

The acoustic conductivity of the biomaterials was studied through an acoustic We added this reference to the reference section and modified the reference numbers impedance, according to references [39,40], using a sonometer located 5 cm from the tube entrance, and then a white noise signal was applied and adjusted to 90 Dba. The sound spectrum was also adjusted using an equalizer for 1/8 frequency bands between 63 Hz and 5000 Hz. Before the analysis, an initial test was conducted using an acoustic isolation foam sample with the exact dimensions of the studied samples. When the sound absorption measures were similar to those reported by the material supplier or in bibliographic references, each sample was analyzed. To verify the data consistency, every sample was registered up to 6 times for 5 s at 7 different microphone positions.

2.7. Data Analysis

In this work, the OriginPRO 2016 software, Version 93E (Origin Lab Corporation, Northampton, MA, USA) was used to process the raw data using different techniques.

3. Results and Discussion

3.1. Physicochemical Characterization of Sludge

Table 2 shows the main physicochemical properties of the sludge used for manufacturing the mycelium-based materials in this work.

Table 2. Physicochemical parameters of sludge obtained from a local pulp and paper mill.

Humidity (%)	pH	Organic Matter Content (%)	Ash Content (%)	Color
85 ± 2	7.49 ± 0.07	11.45 ± 0.02	1.44 ± 0.1	Grayish

Figure 3 shows the ATR FT-IR spectrum of the sludge and the main infrared transmittance bands of the chemical components of the sludge, where the presence of characteristic cellulose signals can be observed. In particular, the stretches characteristic of the O-H bond specific to cellulose are present between 3595 cm^{-1} and 3865 cm^{-1} (peaks 8, 9, 10, and 11). Also, the peaks between 2800 cm^{-1} and 2937 cm^{-1} (peaks 6 and 7) correspond to the asymmetric stretches of the C-H bonds of the aromatic and aliphatic groups found in cellulose, hemicellulose, and lignin. The presence of cellulose is further supported by the peaks at 1054 cm^{-1} and 1030 cm^{-1} (peaks 1 and 2), corresponding to the asymmetric stretches of the C-O-C bonds present in cellulose [41,42]. The transmittance band at 1684 cm^{-1} (peaks 4 and 5) corresponds to the OH group likely associated with water absorbed in the sludge. Finally, peak 3 at 1512 cm^{-1} is characteristic of calcium carbonate, which is used in the Kraft process of cellulose production [43,44].

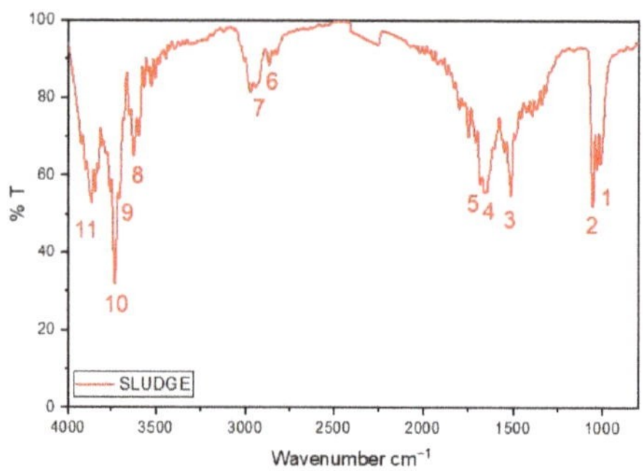

Signal N°	Wavenumber (cm^{-1})	Functional Group
1	1030	C-O-C bond and glucose C-O and OH stretch
2	1054	C-O-C bond and glucose C-O and OH stretch
3	1512	$CaCO_3$
4	1651	C-O aldehyde group
5	1684	OH group
6	2863	C-H stretch of aromatic and aliphatic groups
7	2937	C-H stretch of aromatic and aliphatic groups
8	3595	-OH of cellulose
9	3626	-OH of cellulose
10	3730	-OH of cellulose
11	3865	-OH of cellulose

Figure 3. ATR FT-IR spectrum of sludge.

Table 3 summarizes the contents of elements detected by ICP-MS. It reveals several heavy metal ions, such as Cd, V, and Hg, but their concentrations are insignificant. Therefore, manipulating the sludge in manufacturing mycelium-based materials would be safe for humans. Additionally, the high contents of calcium, sodium, and potassium ions and the absence or low concentration of toxic metals favor mycelial growth, indicating that the substrate is adequate.

Table 3. Chemical composition of sludge (in mm/g) by atomic absorption spectroscopy and inductively coupled plasma mass spectrometry (ICP-MS).

Ca	Na	K	As	Ba	B	Cd	V	Hg
1100.2 ± 5.7	3730 ± 1.7	1378 ± 4.0	4.3 ± 0.1	781.4 ± 2.2	135.5 ± 1.2	0.5 ± 0.01	38.7 ± 0.6	<0.25
Co	Cr	Cu	Mo	Ni	Pb	Se	Zn	Au
1.3 ± 0.01	5.6 ± 0.02	12.3 ± 0.1	3.2 ± 0.04	9.3 ± 0.2	1.5 ± 0.02	<0.22	94.7 ± 0.2	<0.23

Figure 4 shows the sludge's X-ray diffraction patterns, revealing the presence of a crystalline cellulose phase [41,45], with diffraction peaks located at 2θ = 15.04°, 16.88°, and 22.74°, associated with the (1–10), (110), and (200) planes [46], highlighted in red. It has been reported that cellulose favors the development of mycelia, improving the mechanical resistance that allows them to spread and explore their environment [21,47].

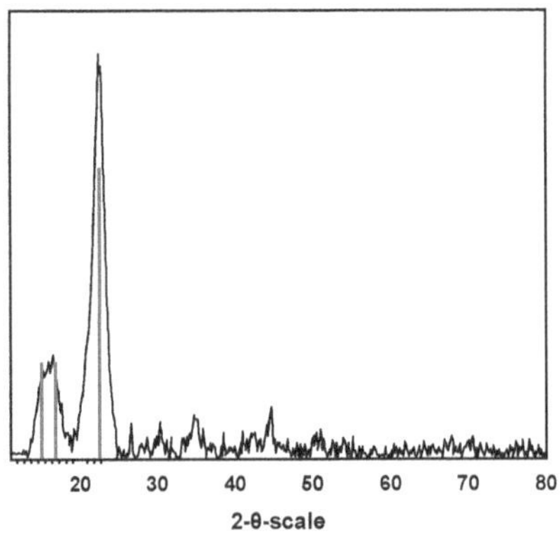

Figure 4. X-ray diffraction pattern of sludge sample. The red lines highlight the diffraction peaks related to the crystalline cellulose phase, as mentioned in the previous paragraph.

3.2. Physicochemical Analysis of Biomaterials

The mycelium-based materials were characterized by scanning electron microscopy to determine the presence of hyphal fibers, which act as a binding agent in the biomaterial. Figure 5 shows scanning electron microscopy images of the biomaterials formed from Trametes versicolor and the substrate formed with the sludge. These images reveal the formation of filaments of different sizes corresponding to the hyphae of *T. versicolor.*

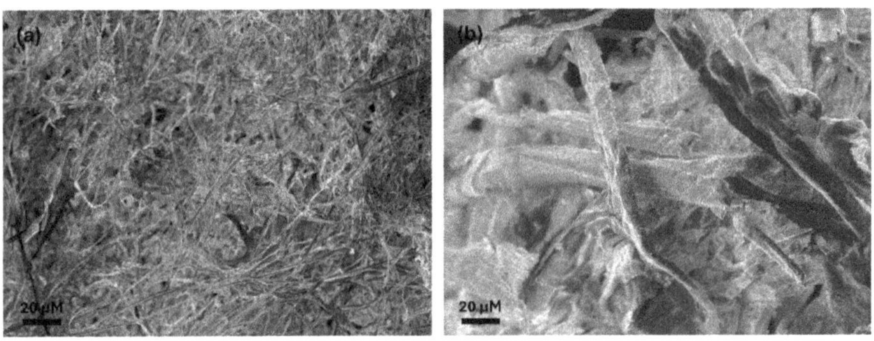

Figure 5. SEM images of mycelium-based biomaterials using *Trametes versicolor*. (**a**) Exterior face; (**b**) interior face.

The hyphal extension rate of *Trametes versicolor* was evaluated to assess its potential as a biomaterial for construction applications. Figure 6 shows the radial extension in both media, MEA and PMMS, revealing that the radial growth rate in MEA increased daily while the growth rate in the PPMS medium remained constant. Additionally, an analysis of the cumulative growth in both media showed that the MEA slope was steeper than that of PPMS. This indicates a difference in nutrient availability between the two media. MEA provides simple molecules like sugars, readily available for fungal assimilation. In contrast, PPMS contains more complex molecules that require biochemical processes to be broken down and utilized by the fungi. A 6.02 mm/day growth rate was observed, slightly higher than the 5.27 mm/day reported in a previous study [48]. This suggests that *Trametes versicolor* could be a promising candidate for biomaterial production using PPMS, as its rapid hyphal growth would allow for the quick colonization of the support material and increased biomaterial production.

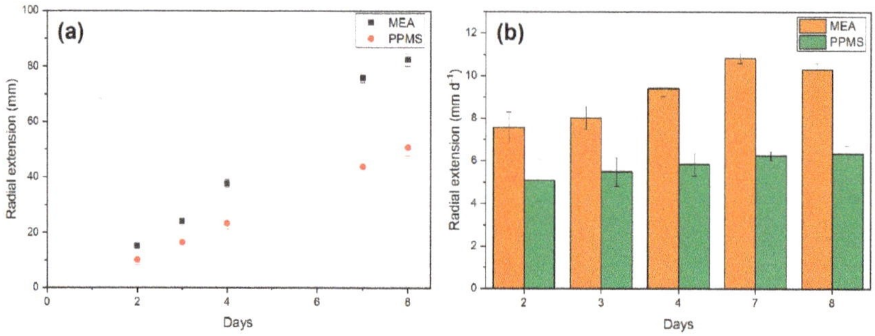

Figure 6. Radial extension of *Trametes versicolor* hyphae in MEA and PPMS. (**a**) Cumulative growth; (**b**) daily growth rate.

The hyphal diameter was also measured using SEM micrography and the ImageJ software. The hyphal mean diameter was close to 1.7 μm, as shown in Figure 7, higher than the hyphae diameter reported for *Trametes versicolor* in other substrates [49]. Thinner hyphae could affect the mechanical properties of the biomaterials, which will be evaluated in future studies.

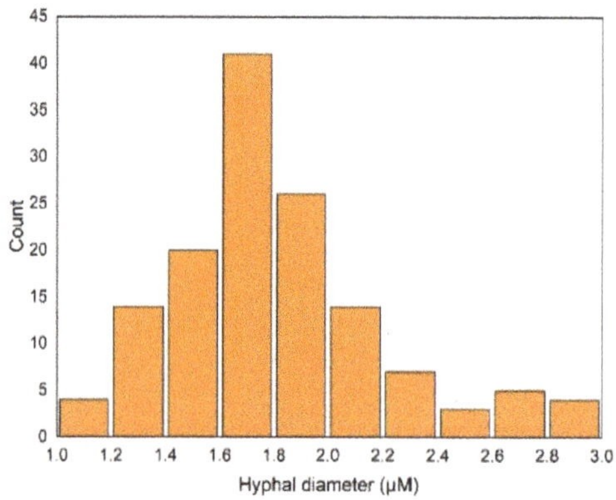

Figure 7. Hyphal diameter of biomaterials obtained using *Trametes versicolor* on PMMS.

Chemical characterization of biomaterials was carried out using FT IR-ATR spectroscopy to confirm the formation of a new biomaterial. Figure 8 shows the infrared spectrum of the biomaterials using the sludge and *Trametes versicolor*, revealing transmittance bands that disappear and appear compared with the substrate's infrared spectrum, as seen in Figure 8. The characteristic bands in the range between 3595 cm^{-1} and 3865 cm^{-1} can be related to the O-H bond stretching due to the presence of cellulose. The peaks between 2800 cm^{-1} and 2937 cm^{-1} correspond to the asymmetric stretching of the C-H bond of the aromatic and aliphatic groups found in cellulose, hemicellulose, and lignin. The transmittance band at 1684 cm^{-1} is related to the OH group of water absorbed in the sludge. The peaks at 1054 cm^{-1} and 1030 cm^{-1} correspond to cellulose's asymmetric stretching of the C-O-C bonds [41,42].

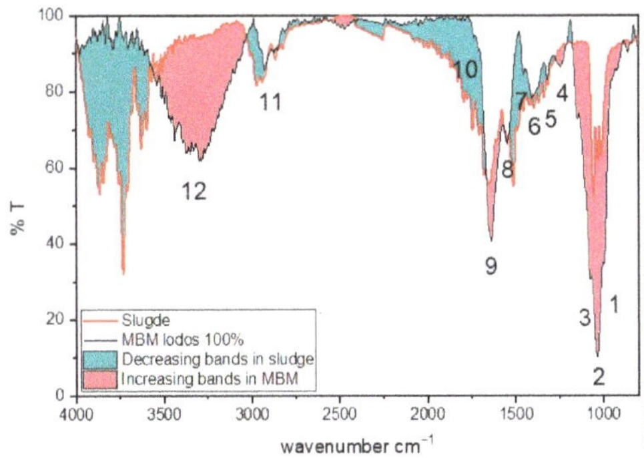

Signal N°	Wavenumber (cm^{-1})	Functional Group
1	1030	Beta-glucans
2	1073	(1→3) and (1→6)-beta-glucans
3	1147	(1→3) and (1→6)-beta-glucans
4	1244	Lignin and xylan degradation
5	1319	Chitin
6	1370	CH stretching in cellulose and hemicellulose, chitin
7	1451	C-O stretching, C=C stretching vibration in the aromatic ring
8	1545	C=C stretching of the aromatic ring in lignin
9	1641	Peptides and secondary metabolites (aromatic rings and conjugated alkenes)
10	1733	C=O stretching in xylans (hemicellulose)
11	2924	C-H stretching (Chitin and ergosterol)
12	3285	Wastewater and trapped CO_2 (O-H and N-H stretching)

Figure 8. ATR FT-IR spectrum of *Trametes versicolor* on PMMS.

3.3. Thermoacoustic Analysis of Biomaterials

The biomaterials obtained using *T. versicolor* on PMMS were analyzed using acoustic impedance, according to reference [40]. Table 4 shows the values of the sound absorption coefficients expressed as noise reduction indices (NRCs) and sound absorption (SAA), which were obtained for the biomaterial using the sludge as a substrate and *Trametes versicolor*. The biomaterial manufactured with *T. versicolor* on PMMS presented similar values to those reported for commercially available thermoacoustic insulators, such as expanded polyurethane and extruded polyurethane. However, its density was higher than that of traditional isolator materials [50], suggesting that it has potential as an alternative to commonly used acoustic insulation materials.

Table 4. Sound absorption coefficients of materials.

Material	Density (g/cm^3)	Noise Reduction Coefficient (NRC)	Sound Absorption (SAA)
Sludge biomaterial with *T. versicolor*	0.689	0.25 ± 0.02	0.25 ± 0.02
Expanded polyurethane	0.028	0.50	-
Extruded polyurethane	0.025	0.33	-

Table 5 shows the thermal conductivity of the biomaterials obtained using the KD-2 PRO thermal analyzer. These results reveal that the biomaterial had similar thermal values and conductivity coefficients lower than 0.1 W/m K, regardless of the type of substrate used. In addition, all biomaterials recorded values of thermal conductivity slightly higher than those of commonly used thermal insulation materials, such as fiberglass and extruded polystyrene [51,52]. These results demonstrate the potential of this type of biomaterial as a thermal insulator, which may represent a significant challenge that must be overcome to gain traction as a viable alternative in the construction industry.

Table 5. Thermal absorption coefficients of biomaterials based on *Trametes versicolor*.

Material	Substrate	Thermal Conductivity Coefficient (W/m K)
Mycelium-based material	Sludge	0.052 ± 0.003
Fiberglass	Not applicable	0.033–0.045
Extruded polystyrene	Not applicable	0.025–0.035

3.4. Prospect for Large-Scale Industrial Use

This study demonstrated that mycelium-based materials produced from forestry waste, like PPMS, possess significant potential for large-scale industrial applications. These biomaterials exhibit high porosity, low density, excellent thermal insulation, and satisfactory sound absorption properties. Furthermore, they are fully compostable at the end of their lifecycle, promoting environmental sustainability [53]. However, for practical use at a large scale, there are still some challenges to be addressed:

1. The selection of fungal species: Using native fungi adapted to local conditions could make the production process more efficient and sustainable [32,54]. These fungi thrive in their local environment, leading to better and more consistent growth on forest debris. Alternatively, using economically damaging fungi such as *Trametes versicolor* [55], known for their efficiency in breaking down lignocellulosic materials, could reduce the energy required for incubation.
2. The optimization of growth parameters: The optimization of parameters such as temperature, humidity, aeration, and modularity to allow the exchange of energy and matter during the growth process could improve the properties and consistency of the final product [56].
3. Substrate quality: The contaminant and metal concentration, as well as the additional processing of the forest waste substrate, such as size reduction and pretreatment, could improve the homogeneity and performance of biomaterials [31,57,58].
4. Sterilization: For large-scale production, it is essential to investigate methods that enable the rapid sterilization of substrates, such as nuclear radiation, to ensure the development of microbiologically contamination-free biomaterials [59].

Further research and development are essential to overcome these challenges and enable large-scale industrial use. Optimizing growth parameters, improving substrate quality, maintaining homogeneity, and developing scalable sterilization techniques will be crucial for successfully implementing mycelium-based materials from forestry waste in various applications, such as thermal and acoustic insulation, in building construction.

4. Conclusions

Sludge generated by the pulp and paper industry is now being used as a raw material for sustainable construction biomaterials based on the mycelium of the fungus *Trametes versicolor*. Chemical, physical, microstructural, and surface analyses confirmed the formation of these new biomaterials. Acoustic impedance analyses revealed that these materials have a noise reduction coefficient and sound absorption average comparable to extruded polystyrene and polyurethane. In addition, the material's thermal conductivity is near that of sheep wool. These mycelium-based materials produced from forestry waste possess significant potential for large-scale industrial applications due to their high porosity, low density, excellent thermal insulation, and satisfactory sound absorption properties. Therefore, they are an ecological and promising alternative to conventional construction materials.

Author Contributions: Conceptualization, H.M., P.M., F.C. and M.S.; methodology, H.M., P.M., I.A.U.-P., F.C., G.R.-G. and M.S.; validation, P.M., D.A.V. and M.S.; formal analysis, H.M., P.M., D.A.V. and M.S.; investigation, H.M., P.M. and M.S.; resources and data curation, H.M., I.A.U.-P. and P.M.; writing—original draft preparation, H.M., P.M., M.W. and M.S.; writing—review and editing, H.M., P.M., M.W. and M.S.; visualization, H.M., P.M. and M.S.; supervision, P.M., M.W. and M.S.; project administration, P.M. and M.S.; funding acquisition, M.S. All authors have read and agreed to the published version of the manuscript.

Funding: This research was funded by ANID BASAL FB210015 CENAMAD.

Institutional Review Board Statement: Not applicable.

Informed Consent Statement: Not applicable.

Data Availability Statement: Data is unavailable due to privacy related to the Funder.

Acknowledgments: The authors thank ANID BASAL FB210015 CENAMAD and ANID Becas Doctorado Nacional 21231394.

Conflicts of Interest: The authors declare no conflicts of interest.

References

1. Ardente, F.; Beccali, M.; Cellura, M.; Mistretta, M. Building Energy Performance: A LCA Case Study of Kenaf-Fibres Insulation Board. *Energy Build.* **2008**, *40*, 1–10. [CrossRef]
2. Kulakovskaya, A.; Wiprächtiger, M.; Knoeri, C.; Bening, C.R. Integrated Environmental-Economic Circular Economy Assessment: Application to the Case of Expanded Polystyrene. *Resour. Conserv. Recycl.* **2023**, *197*, 107069. [CrossRef]
3. Vila-Cortavitarte, M.; Lastra-González, P.; Calzada-Pérez, M.Á.; Indacoechea-Vega, I. Analysis of the Influence of Using Recycled Polystyrene as a Substitute for Bitumen in the Behaviour of Asphalt Concrete Mixtures. *J. Clean. Prod.* **2018**, *170*, 1279–1287. [CrossRef]
4. Oswald, F. Recycling and Reutilisation Opportunities and Techniques of the Insulation Material EPS (Expanded Polystyrene) after the Lifetime of EIFS (Exterior Insulation Finishing System). In *Sustainability and Toxicity of Building Materials—Manufacture, Use and Disposal Stages*; Elsevier: Amsterdam, The Netherlands, 2024; ISBN 9780323983365.
5. Cerimi, K.; Akkaya, K.C.; Pohl, C.; Schmidt, B.; Neubauer, P. Fungi as Source for New Bio-Based Materials: A Patent Review. *Fungal Biol. Biotechnol.* **2019**, *6*, 1–10. [CrossRef]
6. Al-Qahtani, S.; Koç, M.; Isaifan, R.J. Mycelium-Based Thermal Insulation for Domestic Cooling Footprint Reduction: A Review. *Sustainability* **2023**, *15*, 13217. [CrossRef]
7. Volk, R.; Schröter, M.; Saeidi, N.; Steffl, S.; Javadian, A.; Hebel, D.E.; Schultmann, F. Life Cycle Assessment of Mycelium-Based Composite Materials. *Resour. Conserv. Recycl.* **2024**, *205*, 107579. [CrossRef]
8. Charpentier-Alfaro, C.; Benavides-Hernández, J.; Poggerini, M.; Crisci, A.; Mele, G.; Della Rocca, G.; Emiliani, G.; Frascella, A.; Torrigiani, T.; Palanti, S. Wood-Decaying Fungi: From Timber Degradation to Sustainable Insulating Biomaterials Production. *Materials* **2023**, *16*, 3547. [CrossRef]
9. Ghazvinian, A.; Farrokhsiar, P.; Vieira, F.; Pecchia, J.; Gursoy, B. Mycelium-Based Bio-Composites for Architecture. *Mater. Res. Innov.* **2019**, *2*, 505–514.
10. Attias, N.; Reid, M.; Mijowska, S.C.; Dobryden, I.; Isaksson, M.; Pokroy, B.; Grobman, Y.J.; Abitbol, T. Biofabrication of Nanocellulose–Mycelium Hybrid Materials. *Adv. Sustain. Syst.* **2021**, *5*, 2000196. [CrossRef]
11. Teixeira, J.L.; Matos, M.P.; Nascimento, B.L.; Griza, S.; Holanda, F.S.R.; Marino, R.H. Production and Mechanical Evaluation of Biodegradable Composites by White Rot Fungi. *Ciênc. Agrotecnol.* **2018**, *42*, 676–684. [CrossRef]
12. Huang, D.; Guo, X.; Peng, Z.; Zeng, G.; Xu, P.; Gong, X.; Deng, R.; Xue, W.; Wang, R.; Yi, H.; et al. White Rot Fungi and Advanced Combined Biotechnology with Nanomaterials: Promising Tools for Endocrine-Disrupting Compounds Biotransformation. *Crit. Rev. Biotechnol.* **2018**, *38*, 671–689. [CrossRef]
13. Harshvardhan, A.; Saikia, P. Microbial Bioremediation: A Sustainable Approach for Restoration of Contaminated Sites. In *Microbes in Agri-Forestry Biotechnology*; CRC Press: Boca Raton, FL, USA, 2022; ISBN 9789811518119.
14. Nussbaumer, M.; Van Opdenbosch, D.; Engelhardt, M.; Briesen, H.; Benz, J.P.; Karl, T. Material Characterization of Pressed and Unpressed Wood–Mycelium Composites Derived from Two Trametes Species. *Environ. Technol. Innov.* **2023**, *30*, 103063. [CrossRef]
15. Xing, Y.; Brewer, M.; El-Gharabawy, H.; Griffith, G.; Jones, P. Growing and Testing Mycelium Bricks as Building Insulation Materials. *IOP Conf. Ser. Earth Environ. Sci.* **2018**, *121*, 022032. [CrossRef]
16. Elsacker, E.; Vandelook, S.; Damsin, B.; Van Wylick, A.; Peeters, E.; De Laet, L. Mechanical Characteristics of Bacterial Cellulose-Reinforced Mycelium Composite Materials. *Fungal Biol. Biotechnol.* **2021**, *8*, 1–14. [CrossRef] [PubMed]
17. Zhang, X.; Hu, J.; Fan, X.; Yu, X. Naturally Grown Mycelium-Composite as Sustainable Building Insulation Materials. *J. Clean. Prod.* **2022**, *342*, 130784. [CrossRef]
18. Bhagavathi, L.R.; Deshpande, A.P.; Ram, G.D.J.; Panigrahi, S.K. Effect of Cellulosic Fillers Addition on Curing and Adhesion Strength of Moisture-Curing One-Component Polyurethane Adhesives. *Int. J. Adhes. Adhes.* **2021**, *108*, 102871. [CrossRef]
19. Marcuello, C.; Foulon, L.; Chabbert, B.; Molinari, M.; Aguié-Béghin, V. Langmuir-Blodgett Procedure to Precisely Control the Coverage of Functionalized AFM Cantilevers for SMFS Measurements: Application with Cellulose Nanocrystals. *Langmuir* **2018**, *34*, 9376–9386. [CrossRef]
20. Blauwhoff, D.; La Bianca, I. *From Biomass to Mycelium Composite: An Exploration on Cellulose and Weed Residues*; Stichting Toegepast Onderzoek Waterbeheer (STOWA): Amersfoort, The Netherlands, 2019.
21. Sayfutdinova, A.; Samofalova, I.; Barkov, A.; Cherednichenko, K.; Rimashevskiy, D.; Vinokurov, V. Structure and Properties of Cellulose/Mycelium Biocomposites. *Polymers* **2022**, *14*, 1519. [CrossRef]
22. Ghazvinian, A.; Gürsoy, B. Mycelium-Based Composite Graded Materials: Assessing the Effects of Time and Substrate Mixture on Mechanical Properties. *Biomimetics* **2022**, *7*, 48. [CrossRef]
23. Sun, W.; Tajvidi, M.; Hunt, C.G.; McIntyre, G.; Gardner, D.J. Fully Bio-Based Hybrid Composites Made of Wood, Fungal Mycelium and Cellulose Nanofibrils. *Sci. Rep.* **2019**, *9*, 3766. [CrossRef]

24. Quintana, E.; Valls, C.; Roncero, M.B. Valorization of Waste Paper Sludge as a Sustainable Source for Packaging Applications. *Polym. Bull.* **2024**, *81*, 9321–9345. [CrossRef]
25. Faubert, P.; Barnabé, S.; Bouchard, S.; Côté, R.; Villeneuve, C. Pulp and Paper Mill Sludge Management Practices: What Are the Challenges to Assess the Impacts on Greenhouse Gas Emissions? *Resour. Conserv. Recycl.* **2016**, *108*, 107–133. [CrossRef]
26. Turner, T.; Wheeler, R.; Oliver, I.W. Evaluating Land Application of Pulp and Paper Mill Sludge: A Review. *J. Environ. Manag.* **2022**, *317*, 115439. [CrossRef]
27. Hazarika, J.; Khwairakpam, M. Evaluation of Biodegradation Feasibility through Rotary Drum Composting Recalcitrant Primary Paper Mill Sludge. *Waste Manag.* **2018**, *76*, 275–283. [CrossRef]
28. Jackson, M.J.; Line, M.A. Organic Composition of a Pulp and Paper Mill Sludge Determined by FTIR, 13 C CP MAS NMR, and Chemical Extraction Techniques. *J. Agric. Food Chem.* **1997**, *45*, 2354–2358. [CrossRef]
29. Kim, J.S.; Lee, Y.Y.; Park, S.C. Pretreatment of Wastepaper and Pulp Mill Sludge by Aqueous Ammonia and Hydrogen Peroxide. *Appl. Biochem. Biotechnol.* **2000**, *84*, 129–139. [CrossRef]
30. Aiduang, W.; Chanthaluck, A.; Kumla, J.; Jatuwong, K.; Srinuanpan, S.; Waroonkun, T.; Oranratmanee, R.; Lumyong, S.; Suwannarach, N. Amazing Fungi for Eco-Friendly Composite Materials: A Comprehensive Review. *J. Fungi* **2022**, *8*, 1–28. [CrossRef]
31. Elsacker, E.; Vandelook, S.; Brancart, J.; Peeters, E.; De Laet, L. Mechanical, Physical and Chemical Characterisation of Mycelium-Based Composites with Different Types of Lignocellulosic Substrates. *PLoS ONE* **2019**, *14*, e0213954. [CrossRef] [PubMed]
32. Jones, M.; Huynh, T.; John, S. Inherent Species Characteristic Influence and Growth Performance Assessment for Mycelium Composite Applications. *Adv. Mater. Lett.* **2018**, *9*, 71–80. [CrossRef]
33. Collins, T.J. ImageJ for Microscopy. *Biotechniques* **2007**, *43*, S25–S30. [CrossRef]
34. *ISO 9427:2003*; Wood-Based Panels—Determination of Density. The International Organization for Standardization: Geneva, Switzerland, 2003.
35. *ISO 16979:2003*; Wood-Based Panels—Determination of Moisture Content. The International Organization for Standardization: Geneva, Switzerland, 2003.
36. *ASTM C1585-04*; Standard Test Method for Measurement of Rate of Absorption of Water by Hydraulic-Cement Concretes. ASTM International: West Conshohocken, PA, USA, 2010.
37. *ASTM D5334-08*; Standard Test Method for Determination of Thermal Conductivity of Soil and Soft Rock by Thermal Needle Probe Procedure. ASTM International: Conshohocken, PA, USA, 2005.
38. Abdullah, A.; Mohamad, I.S.; Bani Hashim, A.Y.; Abdullah, N.; Wei, P.B.; Isa, M.H.M.; Abidin, S.Z. Thermal conductivity and viscosity of deionised water and ethylene glycol-based nanofluids. *J. Mech. Eng. Sci.* **2016**, *10*, 2249–2261. [CrossRef]
39. *ASTM E1050-24*; Standard Test Method for Impedance and Absorption of Acoustical Materials Using a Tube, Two Microphones and a Digital Frequency Analysis System. ASTM International: Conshohocken, PA, USA, 2024.
40. Tao, J.; Wang, P.; Qiu, X.; Pan, J. Static flow resistivity measurements based on the ISO 10534.2 standard impedance tube. *Build. Environ.* **2015**, *94*, 853–858. [CrossRef]
41. Hernández -Hernández, R.N.; Vázquez-García, R.Á.; Hernández-Ávila, J.; Veloz-Rodríguez, M.A.; Villagomez-Ibarra, J.R. Obtención de Celulosa Cristalina de Hoja de Maíz de Metztitlán Hidalgo. *Pädi Bol. Cient. Cienc. Básicas Ing. ICBI* **2021**, *9*, 7–12. [CrossRef]
42. Yaras, A.; Arslanoğlu, H. Valorization of Paper Mill Sludge as Adsorbent in Adsorption Process of Copper (II) Ion from Synthetic Solution: Kinetic, Isotherm and Thermodynamic Studies. *Arab. J. Sci. Eng.* **2018**, *43*, 2393–2402. [CrossRef]
43. Han, J.S.; Kang, D.S.; Seo, Y.B. Application of In Situ Calcium Carbonate Process for Producing Papermaking Fillers from Lime Mud. *ACS Omega* **2021**, *6*, 3884–3890. [CrossRef] [PubMed]
44. Clausi, M.; Pinto, D. Valorisation of Water Potabilization Sludges as Precursors for Alkali-Activated Binders: Characterization and Feasibility Study. *Materials* **2023**, *16*, 1998. [CrossRef]
45. Padmanabhan, S.K.; Lionetto, F.; Nisi, R.; Stoppa, M.; Licciulli, A. Sustainable Production of Stiff and Crystalline Bacterial Cellulose from Orange Peel Extract. *Sustainability* **2022**, *14*, 2247. [CrossRef]
46. Ren, W.; Guo, F.; Zhu, J.; Cao, M.; Wang, H.; Yu, Y. A Comparative Study on the Crystalline Structure of Cellulose Isolated from Bamboo Fibers and Parenchyma Cells. *Cellulose* **2021**, *28*, 5993–6005. [CrossRef]
47. Sydor, M.; Cofta, G.; Doczekalska, B.; Bonenberg, A. Fungi in Mycelium-Based Composites: Usage and Recommendations. *Materials* **2022**, *15*, 6283. [CrossRef]
48. Schoder, K.A.; Krümpel, J.; Müller, J.; Lemmer, A. Effects of Environmental and Nutritional Conditions on Mycelium Growth of Three Basidiomycota. *Mycobiology* **2024**, *52*, 124–134. [CrossRef]
49. Alves, R.M.; Alves, M.L.; Campos, M.J. Morphology and Thermal Behaviour of New Mycelium-Based Composites with Different Types of Substrates. In *Progress in Digital and Physical Manufacturing, Proceedings of the ProDPM'19, Leiria, Portugal, 2–4 October 2019*; Springer International Publishing: Berlin/Heidelberg, Germany, 2020; pp. 189–197, ISBN 9783030290405.
50. Pedroso, M.; de Brito, J.; Silvestre, J.D. Characterization of Eco-Efficient Acoustic Insulation Materials (Traditional and Innovative). *Constr. Build. Mater.* **2017**, *140*, 221–228. [CrossRef]
51. Yucel, K.T.; Basyigit, C.; Ozel, C. Thermal Insulation Properties of Expanded Polystyrene as Construction and Insulating Materials. In Proceedings of the 15th Symposium on Thermophysical Properties, Boulder, CO, USA, 22–27 June 2003; pp. 54–66.
52. Lafond, C.; Blanchet, P. Technical Performance Overview of Bio-Based Expanded Polystyrene. *Buildings* **2020**, *10*, 81. [CrossRef]

53. Shakir, M.A.; Ahmad, M.I. Bioproduct Advances: Insight into Failure Factors in Mycelium Composite Fabrication. *Biofuels Bioprod. Biorefining* **2024**, 1–16. [CrossRef]
54. Livne, A.; Pearlmutter, D.; Gal, E.; Wösten, H.A.B. Increased CO_2 Fixation and Reduced Embodied Energy of Mycelium Bio-Composite Materials Grown on a Mixed Substrate over Diurnal Temperature Cycles. *Constr. Build. Mater.* **2024**, *421*, 135566. [CrossRef]
55. Tišma, M.; Žnidaršič-Plazl, P.; Šelo, G.; Tolj, I.; Šperanda, M.; Bucić-Kojić, A.; Planinić, M. Trametes Versicolor in Lignocellulose-Based Bioeconomy: State of the Art, Challenges and Opportunities. *Bioresour. Technol.* **2021**, *330*, 124997. [CrossRef]
56. Bonga, K.B.; Bertolacci, L.; Contardi, M.; Paul, U.C.; Zafar, M.S.; Mancini, G.; Marini, L.; Ceseracciu, L.; Fragouli, D.; Athanassiou, A. Mycelium Agrowaste-Bound Biocomposites as Thermal and Acoustic Insulation Materials in Building Construction. *Macromol. Mater. Eng.* **2024**, *309*, 2300449. [CrossRef]
57. Manan, S.; Ullah, M.W.; Ul-Islam, M.; Atta, O.M.; Yang, G. Synthesis and Applications of Fungal Mycelium-Based Advanced Functional Materials. *J. Bioresour. Bioprod.* **2021**, *6*, 1–10. [CrossRef]
58. Van Wylick, A.; Elsacker, E.; Yap, L.L.; Peeters, E.; de Laet, L. Mycelium Composites and Their Biodegradability: An Exploration on the Disintegration of Mycelium-Based Materials in Soil. *Constr. Technol. Archit.* **2022**, *1*, 652–659.
59. Williams, E.; Cenian, K.; Golsteijn, L.; Morris, B.; Scullin, M.L. Life Cycle Assessment of MycoWorks' Reishi^{TM}: The First Low-Carbon and Biodegradable Alternative Leather. *Environ. Sci. Eur.* **2022**, *34*, 120. [CrossRef]

sustainability

MDPI

Article

Sustainable Cement Paste Development Using Wheat Straw Ash and Silica Fume Replacement Model

Bryan Bastías [1,2], Marcelo González [1], Juan Rey-Rey [3], Guillermo Valerio [1] and Pablo Guindos [2,3,4,*]

1 Department of Engineering Construction and Management and UC Concrete Innovation Center (CIHUC), Pontificia Universidad Católica de Chile, Santiago 8320165, Chile; brbastias@uc.cl (B.B.); magonza7@ing.puc.cl (M.G.); gavalerio@uc.cl (G.V.)
2 Centro Nacional de Excelencia para la Industria de la Madera (CENAMAD), Pontificia Universidad Católica de Chile, Santiago 8320165, Chile
3 Faculty of Architecture, Universidade da Coruña, 15001 A Coruña, Spain; j.rey.rey@udc.es
4 Centro de Innovación Tecnolóxica en Edificación e Enxeñería Civil (CITEEC), Universidade da Coruña, 15001 A Coruña, Spain
* Correspondence: pablo.guindos@udc.es

Abstract: Conventional cement production is a major source of carbon dioxide emissions, which creates a significant environmental challenge. This research addresses the problem of how to reduce the carbon footprint of cement paste production using agricultural and industrial waste by-products, namely wheat straw ash (WSA) and silica fume (SF). Currently, accurate models that can predict the mechanical properties of cement pastes incorporating these waste materials are lacking. To fill this gap, our study proposes a model based on response surface methodology and Box-Behnken design, designed to predict the strength of cement pastes with partial substitutions of WSA and SF. Through mechanical and characterization tests, the model demonstrated high accuracy in predicting the strength of the pastes, validated with three mixes, which showed maximum errors of less than 6% at different ages (7, 28, and 56 days). Response surface analysis revealed that replacing cement with 0–20% WSA and more than 5% SF can effectively reduce the carbon footprint by maximizing waste incorporation. This model allows for the calculation of optimal cement substitution levels based on the required strength, thus promoting sustainability in the construction industry through the use of local waste/resources.

Keywords: sustainable concrete; cement paste; design of experiment; silica fume; wheat straw ash; low carbon; supplementary cementitious material; Box-Behnken Design

1. Introduction

According to data provided by the Global Cement and Concrete Association, 14 billion m^3 of Portland cement concrete (PCC) were produced in 2020 [1], and it is estimated that this production was responsible for 5–8% of global anthropogenic carbon dioxide emissions [2]. Because conventional PCC is a mixture of cement, water, admixtures, and mineral aggregates, its most polluting aspect is the Portland cement manufacturing process, with emissions of approximately 866 kg CO_{2eq} per ton [3], and currently about 4.5 billion tons of cement are yearly produced worldwide [4]. On the other hand, due to an increase in housing and infrastructure demands, especially in developing countries, it is estimated that cement production will continue to increase to 5–6 billion tons per year by 2050 [5]. For example, in Chile, the housing shortage in 2020 was 3.3%, equivalent to 641,421 houses [6], and in other South American countries, similar amounts are demanded. Among them, the Brazilian housing deficit stands out with a shortage of more than 6 million dwellings [7]. In Colombia, during 2021, the housing deficit was estimated for the first time, and the results showed a deficit of 5.24 million households, equivalent to 31% of the country's households [8]. Likewise, in 2015, the Argentine Chamber of Construction estimated the housing deficit based on data from the 2010 census at 1.8 million homes, equivalent to 16.1% of the total housing stock [9]. This growing demand for housing will result in an increased demand for concrete production to meet the housing deficit.

For the previous reasons, the concrete industry must move towards sustainability, and one of the significant measures of such would be reducing CO_2 emissions related to the cement content per cubic meter of PCC acting in the reduction of the clinker factor, which may be accomplished by using supplementary cementitious materials (SCMs) because such materials have been shown to be effective in significantly reducing emissions [10–12]. These replacements have been used either on site or in cement production and have been extensively studied. Among the best known are industrial by-products such as fly ash, silica fume (SF), ground granulated blast furnace slag, limestone dust, and metakaolin, among others [13–17]. Micro-silica, for example, is an industrial by-product from ferrosilicon alloys or silicon industries [18], which has been widely studied and used to improve the physical and chemical properties of concrete because it accelerates the hydration process of cement at early ages given its fine particle size [19]. On the other hand, a major source of waste in the world is agriculture since 9.4 billion tons of primary crops such as sugarcane, corn, wheat, and rice are produced annually to feed the population, and all these crops have by-products [20]. Many of these by-products can be converted into SCMs, given the large amounts of biomass residues derived from the harvest, such as leaves, stalks, husks, or straw [21]. Among the best known are rice husk ash [22–31].

According to cereal production data, wheat is one of the three most-produced cereals, with an estimated production from 2022 to 2023 of 776.7 million tons, and its consumption is expected to continue increasing as a result of population growth [32]. On the other hand, to obtain 1 kg of wheat grain, 1.3–1.4 kg of wheat straw is generated [33]. Although wheat straw is sometimes used to feed livestock, a large amount is still considered waste, so, in many countries, it is typically burned in the fields because it is a fast and economical method to eliminate it. However, this method brings negative consequences such as oxidation of soil organic matter; generation of greenhouse gases; increased water erosion; air pollution with particulate matter, which leads to serious health problems; increased risk of fires; and decreased water permeability, among others [34]. Even though there are alternatives to burning, none have completely solved the problem [35]. In Chile, this problem persists given that there is a planted area of more than 623,000 ha, which translates into more than 5 million tons of total stubble throughout the country [36], including crops such as wheat, oats, malting barley, corn, rice, beans, and lentils. One of the most relevant research projects carried out in Chile on the addition of agricultural residues in construction used grapevine shoots with the aim of improving the thermal conductivity of fired clay bricks [37].

Taking into consideration the previous context, this research considered the use of WSA, started by [38], who demonstrated that this material possessed pozzolanic activity and triggered the study of this new potential SCM. It is known that the chemical composition of WSA changes depending on different factors, some related to its place of origin, such as variations in soil conditions and climate, thus directly affecting its pozzolanic efficiency [39]. Furthermore, calcination temperature and particle size are also important when using WSA as an SCM [40]. Early research investigated the controlled combustion of WSA for 5 h at temperatures of 570 °C and 670 °C [38] and showed that WSA obtained at 670 °C had better pozzolanic behavior. It was then reported that the mechanical properties "of autoclaved mortar specimens" were improved with WSA obtained by burning straw at 650 °C for 20 h [41]. Since those studies, several other similar studies have concluded that the optimum WSA calcination temperature is roughly 600 °C, evidencing that upper temperatures tend to transform the amorphous silica into crystalline silica, thus reducing the pozzolanic efficiency [42,43].

Although the WSA is a promising material, its incorporation into cement pastes presents some negative impacts, such as the loss of strength, which can be improved using another type of SCM such as SF. To see the relationship between the incorporation of SCM and changes in mechanical characteristics, response surface method studies have been carried out on concretes incorporating industrial and agricultural wastes to evaluate the mechanical characteristics, but such studies have not been carried out with WSA and SF. Therefore, the objectives of this research were to predict the compressive strength as a function of the percentage of cement replaced by SCM, to analyze the physical-chemical behavior of the materials, to propose ranges that maximize the substitution of cement by SCM and minimize the loss of strength, and to quantify the reduction in the carbon footprint of the mixes defined in the research. According to the findings of this research, up to 30% of the cement can be replaced by the combination of WSA and SF at the cost of losing up to 24% of compressive strength but achieving an 11% decrease in CO_2 emissions from the mix.

In relation to the limitations of the research, it is worth mentioning, on the one hand, the inherent variability of the WSA nature depending on its origin. On the other hand, it should be noted that the WSA and SF replacement rates have been set at fixed values: (0, 10, 20) for the WSA and (0, 5, 10) for the SF.

2. Materials and Methods

2.1. Materials

2.1.1. Research Workflow

Figure 1 below presents the research workflow in a simplified form with the main objectives of the research.

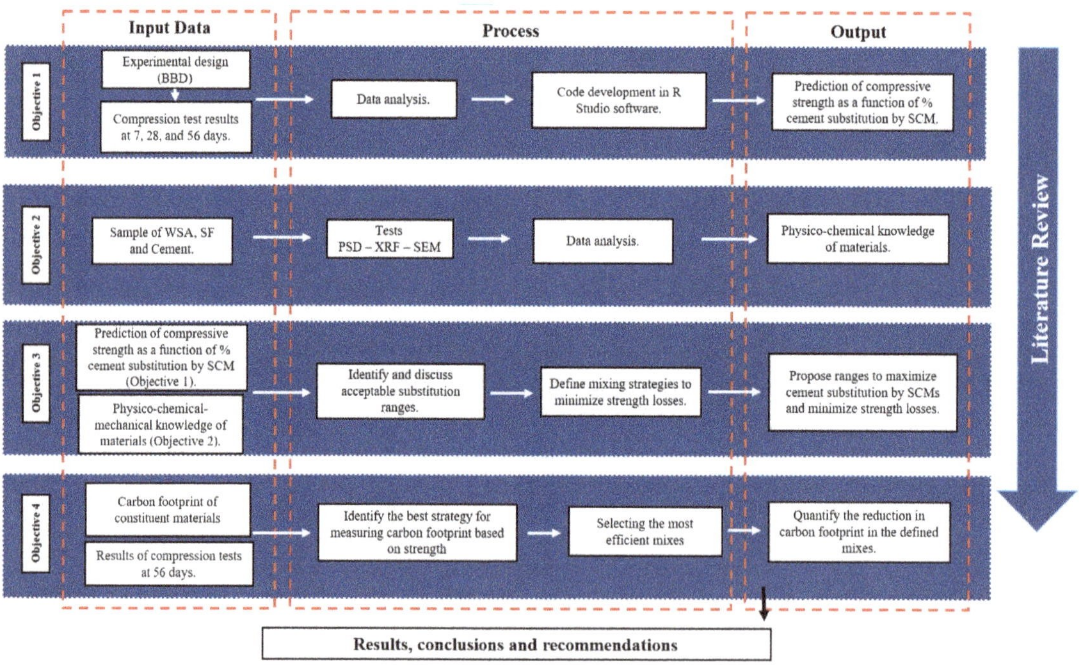

Figure 1. Research workflow.

2.1.2. Raw Materials

For this study, high-strength pozzolanic Portland cement supplied by Melon S.A., Santiago, Chile, in accordance with ASTM C595 [44] was used with a density of 3000 kg/m^3. WSA was prepared in the lab according to the procedure explained next. SF use in the research was obtained from a supplier (Sika, Baar, Switzerland) that corresponded to the product SikaFume®, which has a density of 2200 kg/m^3.

The CO_{2eq} values of the raw materials were obtained from the literature and are shown in Table 1; specifically, values for the WSA were not found, so an approximate value for biomass ash is shown.

Table 1. Emissions of CO_2 in kg per kg of raw material.

Material	CO_2eq (kg/kg)	Reference
Cement	0.813	(Maddalena et al., 2018) [40]
Biomass ash	0.05	(González-Kunz et al., 2017) [45]
SF	0.00001	(Maddalena et al., 2018) [40]

2.1.3. Preparation of WSA

The straw waste used in this investigation was obtained from durum wheat (*Triticum turgidum* var. *Durum* L.) harvested in Linares, a central region of Chile with a mild Mediterranean climate, Csb according to the Köppen climate classification. The wheat straw was burned in an electric furnace with an initial temperature of 50 °C and subsequent heating at a rate of 10 °C/min up to 600 °C for 120 min, following previous pozzolanic efficiency studies [43,46,47]. Once burned, the wheat straw was rapidly cooled by setting the ashes at a controlled room temperature of 20 °C. For each calcination, a consistent ash yield of 7% of the original straw weight content was obtained. The resulting ash was then filtered through a #50 sieve to reduce the particle size and thus increase surface area since fineness has been reported to yield higher pozzolanic performance [39]. The density of this material was 2210 kg/m^3 [46].

2.1.4. Definitions of the Design of Mixtures

The definitions of the input levels of the different variables necessary for setting up the theoretical model were determined from the literature so that calculations could be made to properly infer intermediate values not studied experimentally [48]. The proportion of cement to be replaced in mass was based on a literature review so that the maximum WSA replacement ranged up to 20%, beyond which several authors reported significant decreases in the pozzolanic reaction [39,49]. Likewise, the maximum SF replacement percentage was set up to 10% because multiple studies demonstrated a relevant worsening of the mechanical properties beyond this value [50–52]. Additionally, the water content represented by the water-binder ratio (W/B) ranged from 0.4 to 0.6 to obtain complete hydration of the cement particles. The designed mixtures were classified in three categories or combinations: 100% cement (mono), binary, and ternary mixes using the two SCMs. In conclusion, the final selected variants entailed W/B ratios of 0.4, 0.5, and 0.6, WSA 0%, 10%, and 20%, and SF replacements of 0%, 5%, and 10%.

2.2. Methods

2.2.1. Experimental Design and Predicting Modeling/Carbon Footprint

The study by Ferreira et al. validates the use of the Box-Behnken design as an efficient methodology for the optimization of analytical methods. Through various practical examples, the authors demonstrate that the Box-Behnken design can provide precise results at intermediate levels of selected ranges, minimizing the number of necessary experiments and improving the efficiency of the experimental process. This ability to infer results at intermediate levels supports the application of the Box-Behnken design in a wide range of scientific and technological research, confirming its validity and utility in the optimization of complex analytical methods. Figure 2 presents an illustration of the BBD model experimental design space. The BBD allows for fewer experimental runs than does a central composite design (CCD) because the BBD needs only three levels defined ($-1, 0, 1$), whereas the CCD needs five ($-\alpha, -1, 0, 1, \alpha$). This methodology was used to predict the response variable with three input variables, in which the coefficients were estimated and adjusted to a mathematical model capable of being adjusted to check the effectiveness of the model [48]. The response surface methodology (RSM) allowed for the correct multivariable analysis, in which the variables of interest were affected by others. The objective of setting the BBD was that the experiments and thus the data collection would permit a model that fit in such a way that the response variable obtained accurate values, thus predicting the behavior of the variable in question. An example would be the determination of local tendencies, maxima, or minima [53], with the objective of improving the processes involved. The mathematical formulation of the BDD can be expressed as a quadratic or higher-order polynomial, as shown in Equation (1):

$$Y = \beta_0 + \sum_{i=1}^{n} \beta_i x_i + \sum_{i=1}^{n} \beta_i x_i^2 + \sum_{j=1}^{n} \sum_{i=1}^{n} \beta_{ij} x_i x_j + \varepsilon \quad (1)$$

where Y represents the variable to be predicted, x the values of each decision variable, β the regression coefficients, β_0 the expected value of Y if all the variables associated with the response are equal to 0, n the number of variables and i, and j and ε a random error of the system [52].

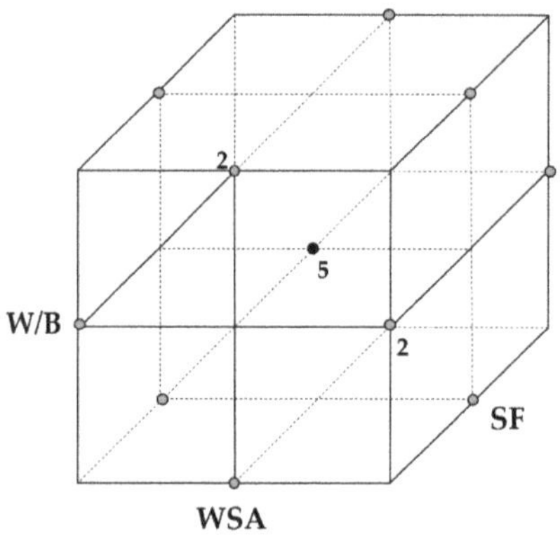

Figure 2. BBD model.

To calculate the impact of using SCMs on the reduction of the carbon footprint, CO_2eq/m^3 was calculated using the CO_2eq values (kg/kg) obtained from the literature with the mix ratios used on the design considered above.

It can be seen how this value was obtained from Equation (2):

$$CO_{2,eq} = \rho_c \cdot CO_{2,c} + \rho_{wsa} \cdot CO_{2,wsa} + \rho_{sf} \cdot CO_{2,sf} \quad (2)$$

where $CO_{2,eq}$ refers to the kg/m^3 of CO_2 equivalent of the ternary mixture, ρ_c refers to the cement density of the mixture in kg per m^3, $CO_{2,c}$ are the kg of CO_2 emitted by kg of cement, ρ_{wsa} refers to the wheat straw ash density of the mixture in kg per m^3, $CO_{2,wsa}$ are the kg of CO_2 emitted by kg of wheat straw ash, and ρ_{sf} and $CO_{2,sf}$ are, respectively, the density and CO_2 emission related to silica fume. As Equation (2) shows, increasing the replacement with both WSA and SF will decrease the CO_2eq/m^3. Therefore, in order to compare how the resistance obtained was related to the carbon footprint emitted, the intensity indicator was analyzed, which allowed the estimation of the global warming potential in the formulations by calculating the amount of binder required (kg/m^3) to provide 1 MPa of resistance and was expressed as $CO_2eq/m^3/MPa$.

In order to determine the compressive strength of the 15 mixes (combinations), Table 2 presents how the different levels (−1, 0, 1) were distributed for the three studied variables, W/B, WSA, and SF, where the last two show the percentage of cement replacement by weight; for example, for the first mix, the levels were −1, −1, and 0, which means a water-cement ratio of 0.4, with 0% WSA cement replacement and 5% SF cement replacement.

Table 2. Summary of the different paste mixes with cement replacement by WSA and SF.

	Levels in the BBD Model			Experimental Composition		
Mixes Nomenclature	**W/B**	**WSA**	**SF**	**W/B**	**WSA (%)**	**SF (%)**
B0.4W0S5	−1	−1	0	0.4	0	5
B0.6W0S5	1	−1	0	0.6	0	5
B0.4W20S5	−1	1	0	0.4	20	5
B0.6W20S5	1	1	0	0.6	20	5
B0.4W10S0	−1	0	−1	0.4	10	0
B0.6W10S0	1	0	−1	0.6	10	0
B0.4W10S10	−1	0	1	0.4	10	10
B0.6W10S10	1	0	1	0.6	10	10
B0.5W0S0	0	−1	−1	0.5	0	0
B0.5W20S0	0	1	−1	0.5	20	0
B0.5W0S10	0	−1	1	0.5	0	10
B0.5W20S10	0	1	1	0.5	20	10
B0.5W10S5-1	0	0	0	0.5	10	5
B0.5W10S5-2	0	0	0	0.5	10	5
B0.5W10S5-3	0	0	0	0.5	10	5

In the mixture's nomenclature of BBD experimental design, B corresponds to the W/B ratio, W to WSA, and S to SF.

2.2.2. Mixing and Preparation of Specimens

The materials were mixed in accordance with the ASTM C305-20 standard [54] using a mechanical overhead stirrer to homogenize the dry components prior to the addition of water. Cubic specimens were manufactured following ASTM C109-21 [55], with dimensions of 20 mm to optimize resource utilization. A total of nine specimens were cast for each cement paste mixture with the aim of assessing the evolution of compressive strength after 7, 28, and 56 days of curing (three identical specimens for each time interval). The specimens were stored in a curing chamber, maintaining temperature and relative humidity conditions as prescribed by ASTM C511-13 [56] for a duration of 24 h. Subsequently, the specimens were demolded and immersed in water until testing, as can be seen in Figure 3.

Figure 3. Curing of specimens in water.

2.2.3. Particle Size Distribution

The particle size distribution was performed in the raw materials by wet method laser diffraction using a Mastersizer 2000 (Malvern Instruments, Malvern, UK) Hydro 2000S wet sample dispersion unit with a recirculating pump. The dispersant used was isopropyl alcohol. The particle size distribution (PSD) is summarized in Figure 4. The cement particles had an average particle size (D_{50}) of 14.22 μm. WSA had a D_{50} of 32.2 μm and presented a wider distribution, which therefore could have improved the packing density of the blends [57]. SF was the finest material, with a D_{50} of 0.674 μm; however, in Figure, it appears as the coarsest material, with a D_{50} of 113 μm, because particles were agglomerated, so the properties of blended cement pastes may have been controlled by the agglomerated size of this particle, which led to the pozzolanic reactivity potential of this SCM decreasing considerably, resulting in an increase in the late-age strength of the mixtures [58].

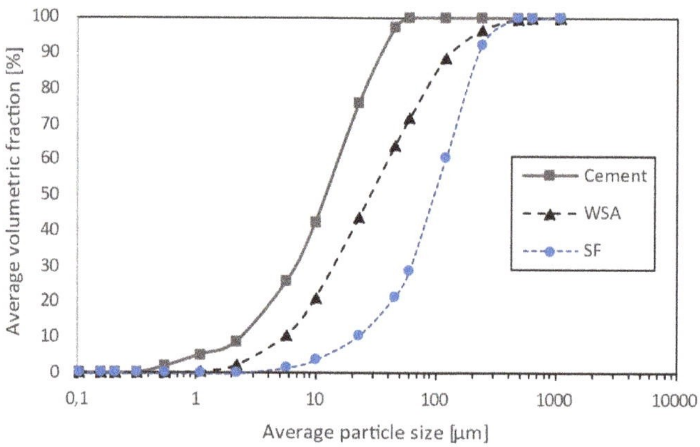

Figure 4. Particle size distribution of raw materials.

2.2.4. Compressive Strength Tests

As per the ASTM C109-21 [55] standard, compression tests were performed on cubic specimens using a 300 kN-capacity universal testing machine, Laryee UE34300 (Laryee, Beijing, China), at a loading rate of 2.4 kN/s at 7, 28, and 56 days old, with three specimens tested at each age. Figure 5 shows the different samples that were tested in compression.

Figure 5. Compression test of samples.

2.2.5. Field Emission Scanning Electron Microscopy/Energy-Dispersive X-Ray Spectroscopy

The field emission scanning electron microscopy (FE-SEM) analysis of all paste mixes was performed on crushed specimens at 56 days old. Samples of raw materials (cement, WSA, SF) were also analyzed using a Quanta FEG250 microscope (Federation Equestre International, Lausanne, Switzerland). All specimens were examined for changes in morphology and composition, and magnification scales were chosen according to representative particle sizes. All the samples were spread on carbon tape and coated with a gold layer approximately 5 nm thick to improve sample conductance.

2.2.6. Isothermal Calorimetry

The isothermal calorimetry was performed in the paste mixtures according to ASTM C1679–14 [59] using a TAM Air isothermal calorimeter at 23 °C for 7 days with duplicated samples for each mixture. The fresh mixtures were weighted with an ampoule and directly placed in the calorimeter. The heat-release rate was normalized according to the weight of cementitious materials.

2.2.7. X-Ray Fluorescence

The analysis of the samples was conducted using a wavelength dispersive sequential spectrometer (WDX, S4 equipment). The measurement was performed using the Fast-Vac34 method.

3. Results and Discussion

3.1. Discussion of the Experimental Results

3.1.1. Field Emission Scanning Electron Microscopy

The morphological characteristics of the materials were visualized by Field Scanning Electron Microscopy (FE-SEM). It was observed that the SF was in an agglomerated state with a large spherical shape (see Figure 6a,b) by a large number of fine SF particles, approximately <1 μm (see Figure 6c), which was in concordance with the PSD test shown in figure.

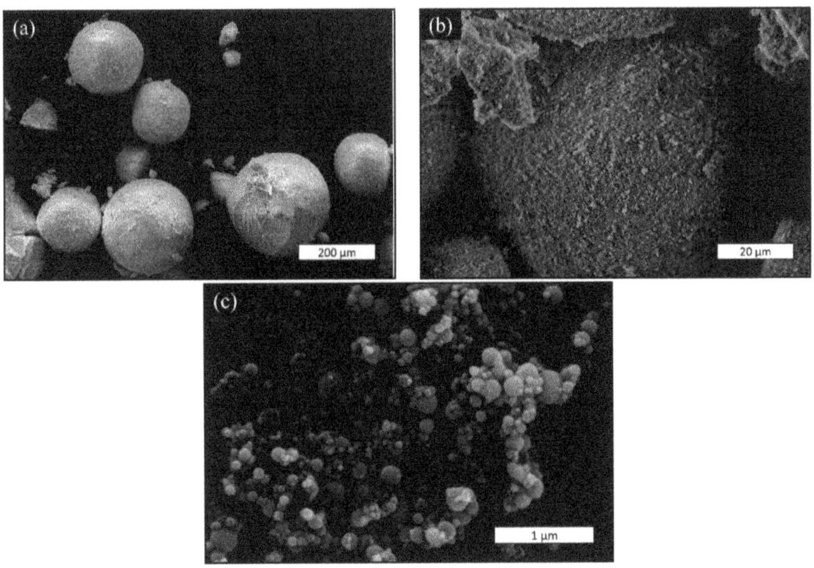

Figure 6. (**a**) Morphology of silica fume by electron microscope, (**b**) agglomeration, and (**c**) silica fume.

This size of the agglomerated silica particles directly affected the strength of the cementitious mixes. Since the pozzolanic reaction would only take place at the agglomeration surface and at early ages, the strength of cement pastes would be most affected because, for conventional concrete mixes, the aggregates are able to separate these particles [60,61].

The FE-SEM of the rest of the constituents at a scale of 20 μm and 200 μm is shown in Figure 7. As can be seen in this figure, cement particles were the smallest, followed by WSA and SF, respectively. Figure 7a shows the cement with particles of different sizes, where an angular morphology can be appreciated. Figure 7b,c show two distinct structures: on the one hand, elongated and flat structures in the longitudinal direction and, on the other hand, a honeycomb structure corresponding to a cross-section of the WSA particle. Moreover, as can be observed, these ashes are very porous, thus explaining the decrease in workability in the mixtures because the pores absorbed the mixing water [62].

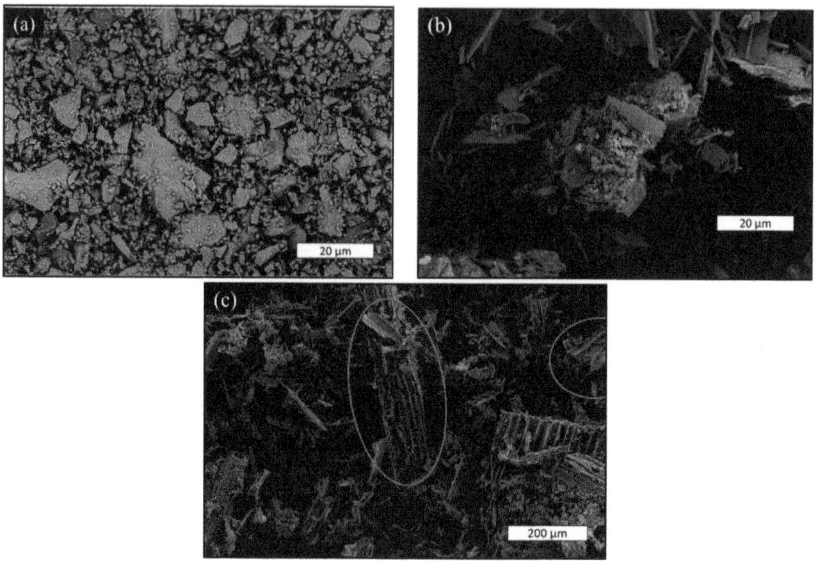

Figure 7. Morphology of cement and WSA. (**a**) cement particles, (**b**) and (**c**) WSA.

3.1.2. X-Ray Fluorescence of Raw Materials

As shown in Table 3, the cement was mainly composed of CaO, SiO_2, Al_2O_3, and Fe_2O_3, which were present in silicate and aluminate forms. SF was composed almost entirely of SiO_2 in 95.5% of its weight. Finally, the WSA obtained was mainly composed of SiO_2, K_2O, and CaO. However, the oxides did not total 100% because there was organic matter that was calcined in the form of carbon, and this was not detected by the equipment.

Table 3. Chemical composition of raw materials.

Compounds	Cement (%)	WSA (%)	SF (%)
CaO	59.0	5.33	0.18
SiO_2	24.8	27.12	94.50
Al_2O_3	5.80	0.57	-
Fe_2O_3	3.29	0.15	0.14
SO_3	2.88	0.79	-
MgO	1.07	0.52	-
K_2O	0.79	6.81	0.45
TiO_2	0.23	-	-
SrO	0.09	-	-
Cl	-	1.18	-
P_2O_5	-	0.69	-
Na_2O	-	0.42	-
MnO	-	0.17	-
LOI	2.01	-	4.76

3.1.3. Isothermal Calorimetry of the Mixtures

The normalized hydration heat and the normalized heat flow from selected mixtures at 48 h are presented in Figures 8 and 9, respectively. In the mixes nomenclature of BBD experimental design, B corresponds to the W/B ratio, W to WSA, and S to SF. A water-cement ratio of W/B = 0.5 was used for all the mixtures in the calorimetry test. The mixtures used for the analysis corresponded to a cement-only mixture (B0.5W0S0), a binary mixture with WSA (B0.5W20S0), and another with SF (B0.5W0S10). The B0.5W0S10 mix showed the highest value of hydration heat when the calorimetry data were normalized by cement. This result indicated that the B0.5W0S10 mix produced the highest number of hydration products during the first 24 h.

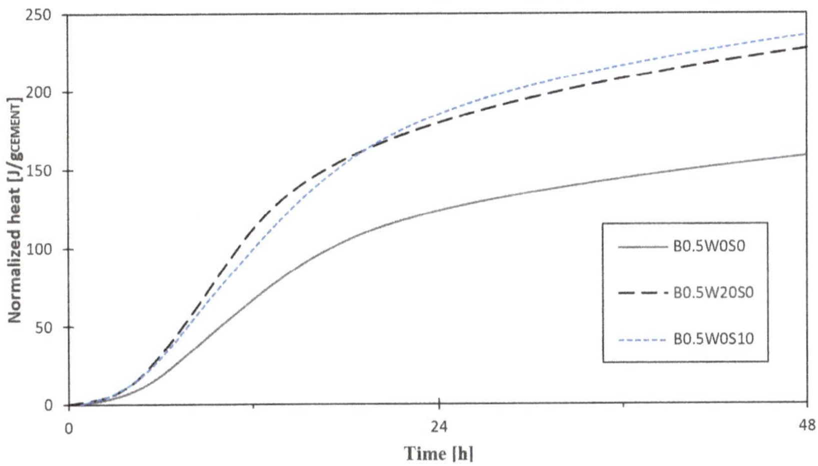

Figure 8. Normalized hydration heat.

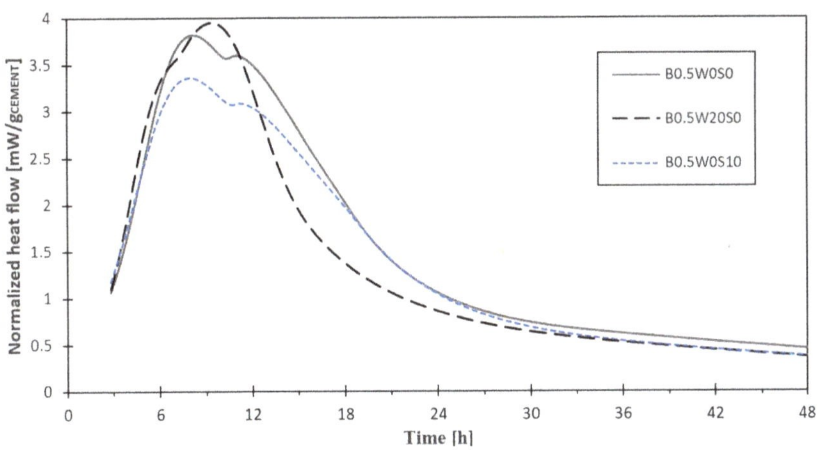

Figure 9. Normalized heat flow.

On the other hand, with respect to the normalized heat flow of the mixes, the second peak heat flow is generally related to cement properties such as initial and final setting times and late compressive strength, among others. In Figure 9, it can be observed that the mix with the highest heat flow corresponds to the sample with WSA, which has the lowest compressive strength at 56 d, while the mix with SF has the lowest heat flow and the highest compressive strength at 56 d.

3.1.4. Compressive Strength of Hardened Samples

Figure 10 presents the results of compressive strengths for each mix at 7, 28, and 56 days of curing and representative photographs of the failure mechanism. As shown in Figure 10, cone failure was predominant in the tested specimens, which indicates a homogeneous load application and good mechanical behavior. In general, compressive strength increases with age. These results are used to build the model and the regression analysis is presented in the next section.

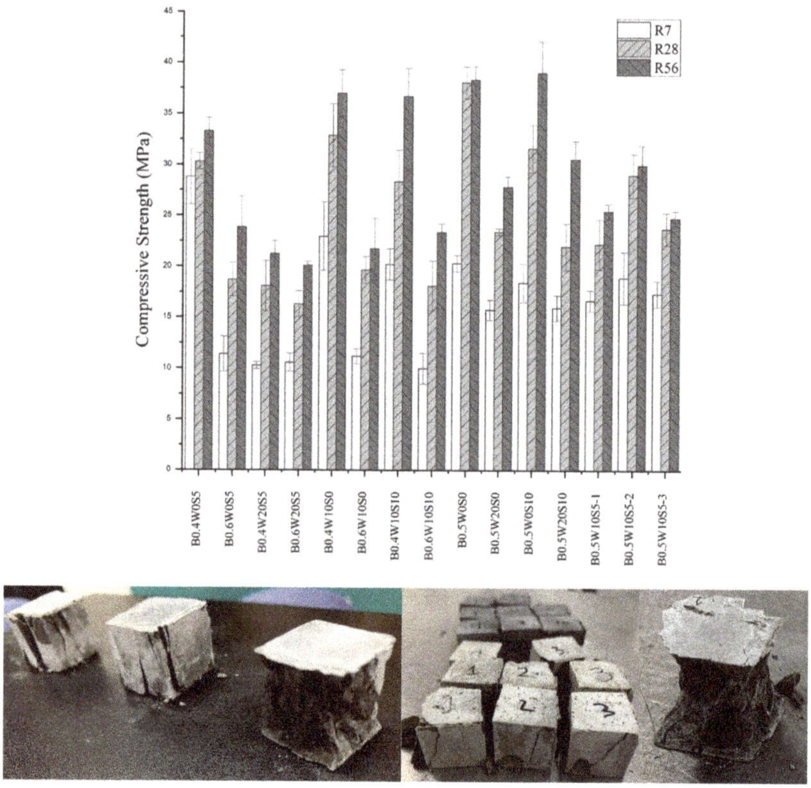

Figure 10. Results of compressive strength (**above**) and illustration of the failure mechanism (**below**).

As evidenced in Figure 10, the compressive strength of the different combinations presented some variability as expected, and some combinations did not have high strength. The agglomerated silica particles may have directly affected the strength of the cementitious mixes. Since the pozzolanic reaction only took place at the agglomeration surface and at early ages, the strength of cement pastes would have been most affected because, for conventional concrete mixes, the aggregates were able to separate these particles [55,56]. Therefore, based on the literature and results obtained by SEM and PSD, in the mixtures in this study, the particle size of the agglomerated micro-silica was the one that controlled the mechanical properties of the cement pastes.

The microstructure of samples at 56 days obtained by FE-SEM is shown in Figure 11. By comparing Figure 11a,b, it can be noticed that the latter shows more amorphous structures due to the pozzolanic reaction of SF with unreacted compounds of the cement to form secondary calcium silicates. In addition, the pore size decreased due to the nucleation of the cement, causing a more compact microstructure and therefore increasing the strength of the cement pastes. Figure 11d shows how WSA was incorporated into the mixture, whereas

Figure 11c, being on the same scale as the control paste, showed a decrease in pore size, an increase in hydration products, and a densification of the matrix.

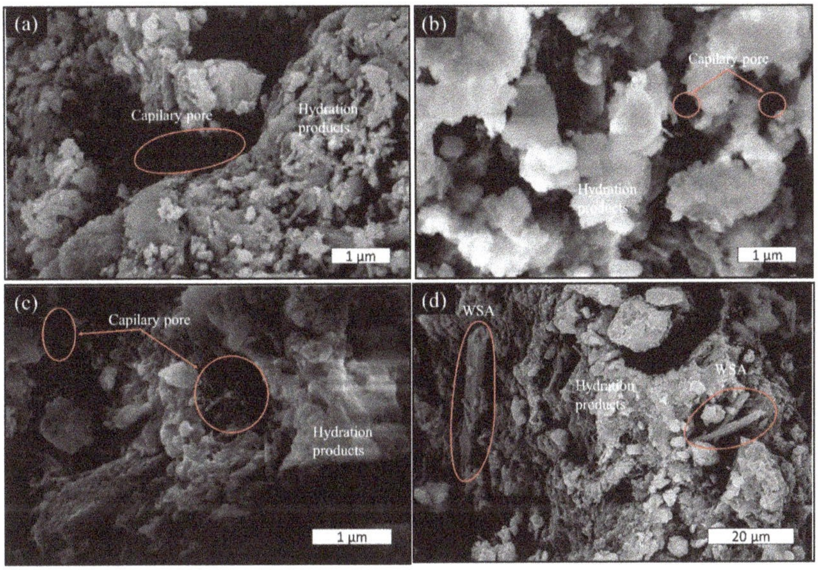

Figure 11. Exemplary FE-SEM images of the B0.6W10S10 paste (28-day average compressive strength of about 23 MPa) at 56 days. (**a**) cement, (**b**) SF, and (**c**) WSA at 1 µm, (**d**) WSA at 20 µm.

In summary, the results obtained in the simple compression tests of the cement pastes with partial substitution of WSA and SF have been satisfactory, obtaining values at 28 days ranging between 20 and 39 MPa, compared to the 38 MPa obtained with the cement paste without substitutions.

A more detailed analysis of the results obtained allows us to draw some preliminary conclusions and highlight some significant data. On the one hand, the reference results, those obtained for a cement paste without substitutions, are situated in compressive strengths of around 38 MPa (very similar values for 28 and 56 days), with strengths at 7 days slightly above 20 MPa. This is the specimen with the smallest increase in strength between 28 and 56 days. On the contrary, the greatest differences in strength between these ages are found in those specimens with a higher proportion of SF in the mix.

The specimens with the highest results in the simple compression test were those without WSA substitution and with the maximum proportion of SF substitution, slightly improving the results at 56 days with respect to the reference specimen, although the results at 28 days are significantly lower.

The specimens with the lowest results are in general those with maximum WSA substitution, as expected, although with appropriate adjustments in the SF substitution percentages and in the W/B mix, satisfactory results are obtained.

3.2. Multiple Regression Analysis and Design Model

3.2.1. Data Analysis and Model Adjustment

The results of the Shapiro–Wilk test, which is a statistical test used to determine whether a set of data comes from a normal distribution, are presented in Table 4. As shown in the table, a p value > 0.05 was obtained at all ages; therefore, the null hypothesis could not be rejected, and the population of data were normally distributed. In addition, the quantile-quantile graphs shown in Figure 12 for all ages and mixtures evidenced a tendency towards a straight line, further confirming their normal distribution.

Table 4. Shapiro–Wilk normality test.

Age	W	*p* Value
7 days	0.97	0.86
28 days	0.97	0.89
56 days	0.97	0.85

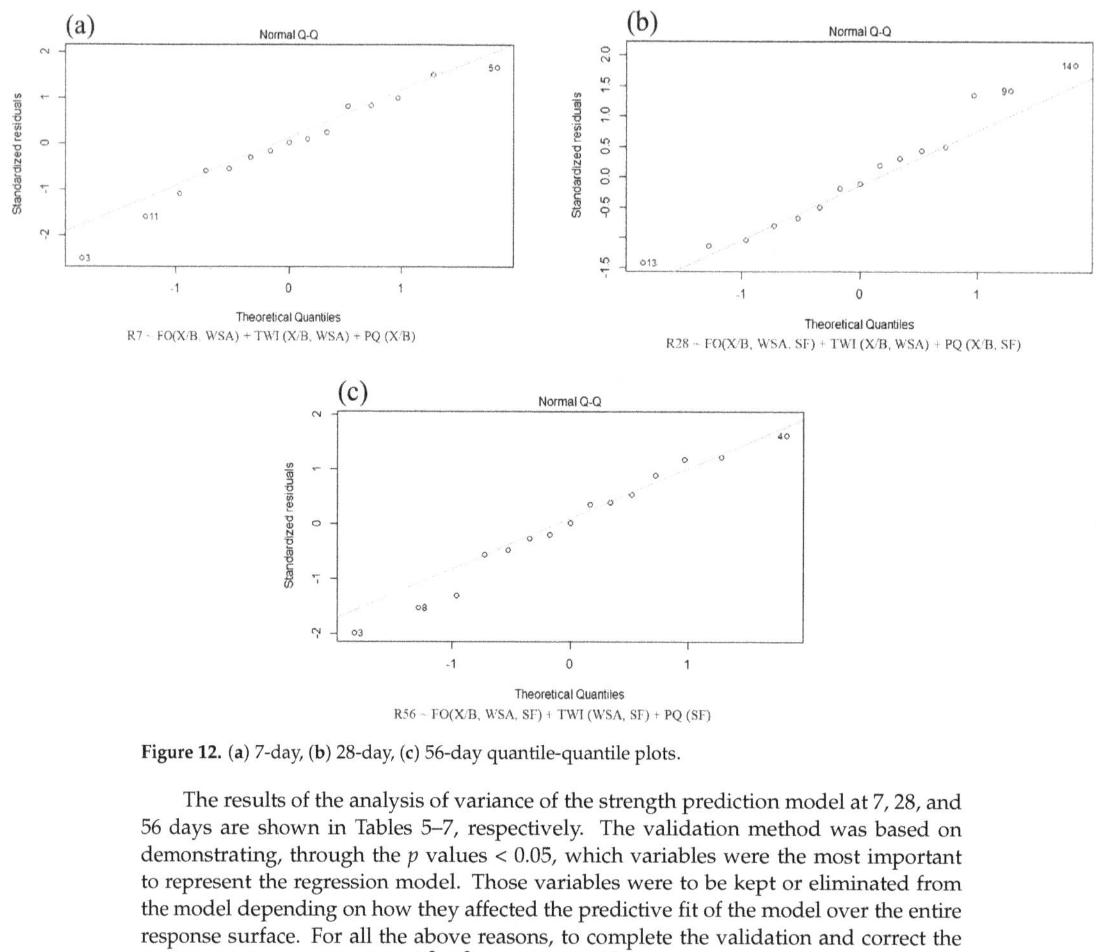

Figure 12. (**a**) 7-day, (**b**) 28-day, (**c**) 56-day quantile-quantile plots.

The results of the analysis of variance of the strength prediction model at 7, 28, and 56 days are shown in Tables 5–7, respectively. The validation method was based on demonstrating, through the p values < 0.05, which variables were the most important to represent the regression model. Those variables were to be kept or eliminated from the model depending on how they affected the predictive fit of the model over the entire response surface. For all the above reasons, to complete the validation and correct the model, we also computed the R^2, R^2_{Adj}, and lack of fit.

The R^2 obtained from the model already adjusted for the three ages (7, 28, and 56) corresponded to $R^2_{7days} = 92\%$, $R^2_{28days} = 93\%$, and $R^2_{56days} = 83\%$, respectively. This indicated a good fit of the regression model and explained more than 80% of the variability of the data, as well as a significant relationship between predicted and calculated values for all ages. By taking into account the predictor variables of the three models, we obtained the R^2_{Adj} for the three ages, with values of $R^2_{Adj-7days} = 89\%$, $R^2_{Adj-28days} = 88\%$, and $R^2_{Adj-56days} = 77\%$, meaning that the parameters investigated explained a large percentage of the variation in the sample while the difference was the overall variance not explained by the model. The predictor variables were chosen through a process of iteration that allowed the improvement of the model at a global level, with the criteria described above. Finally, to complete the validation, we analyzed the lack-of-fit, yielding a p $\text{value}_{7\text{days}} = 0.065$,

p value$_{28days}$ = 0.92, and p value$_{56days}$ = 0.47, which is non-significant, implying that the predictions made by the model were as good as if additional experiments were performed (maintaining the experimental conditions).

Table 5. Analysis of variance and estimated coefficients of compressive strength for the 7-day model.

Factor	Coefficient	Std. Error	*t* Value	Pr (>|t|)
Constant	17.62	0.65	27.07	0.0000000001093 ***
W/B	−4.69	0.61	−7.7	0.01649 ***
WSA	−3.3	0.61	−5.14	0.0004356 ***
W/B:WSA	4.13	0.86	4.8	0.0007220 ***
$(W/B)^2$	−2.14	0.89	−2.4	0.0373727 *

Note: significance codes: 0 '***' 0.001 '*' 0.05.

Table 6. Analysis of variance and estimated coefficients of compressive strength for the 28-day model.

Factor	Coefficient	Std. Error	*t* Value	Pr (>|t|)
Constant	25.13	1.14	22.04	0.00001896 ***
W/B	−5.31	0.84	−6.32	0.0002274 ***
WSA	−5.35	0.84	−6.38	0.0002136 ***
SF	−1.53	0.84	−1.82	0.1057541
W/B:WSA	3.48	1.19	2.93	0.0189174 *
$(W/B)^2$	−3.39	1.23	−2.75	0.0249159 *
$(SF)^2$	3.52	1.23	2.86	0.0211537 *

Note. Significance codes: 0 '***' 0.001 '*' 0.05.

Table 7. Analysis of variance and estimated coefficients of compressive strength for the 56-day model.

Factor	Coefficient	Std. Error	*t* Value	Pr (>|t|)
Constant	26.21	1.25	21.04	0.000001308 ***
W/B	−5.21	1.17	−4.47	0.001203 **
WSA	−5.08	1.17	−4.36	0.001423 **
SF	0.8	1.17	0.68	0.510685
$(SF)^2$	5.52	1.71	3.23	0.008980 **

Note. Significance codes: 0 '***' 0.001 '**' 0.01.

3.2.2. Models and Interactions Effect of Factors

With the adjustments made and the validations obtained, the following equations for the compressive strength regression models could be determined, where the variables were normalized:

$$C_{7days} = 17.62 - 4.69\frac{W}{B} - 3.13WSA + 4.13\frac{W}{B} * WSA - 2.14\left(\frac{W}{B}\right)^2 \quad (3)$$

$$C_{28days} = 25.13 - 5.31\frac{W}{B} - 5.35WSA - 1.53\,SF + 3.48\frac{W}{B} * WSA - 3.39\left(\frac{W}{B}\right)^2 + 3.52\,(SF)^2 \quad (4)$$

$$C_{56days} = 26.21 - 5.21\frac{W}{B} - 5.08WSA + 0.8\,SF + 5.52\,(SF)^2 \quad (5)$$

Once the models for each age were defined, it was possible to identify the behavior of the most important variables and how they affected the compressive strength of the mixes. Figure 13 illustrates the response surface at different ages: (a) shows the interaction among the significant variables of the model at 7 days W/B with WSA, and (b) and (c) show the interaction at 28 and 56 days, respectively, between the variables SF and WSA keeping W/B constant at the mean level of 0.5.

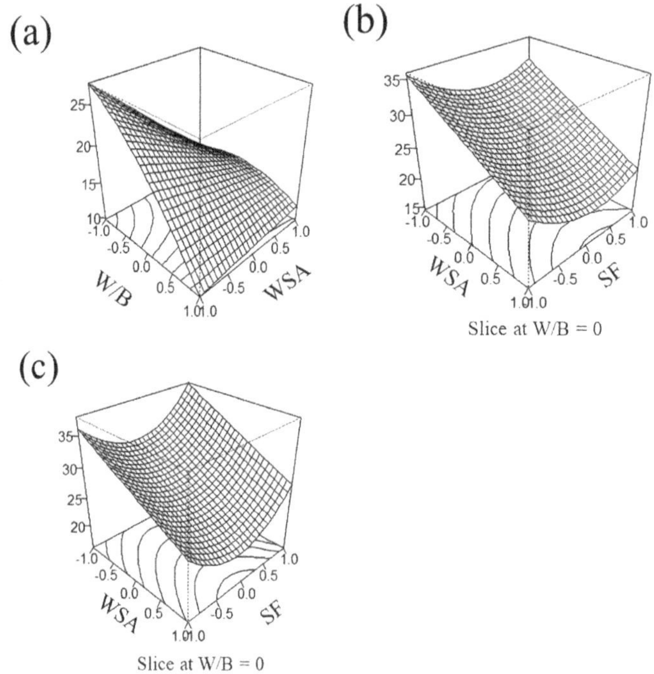

Figure 13. Three-dimensional model of RSM. (**a**) 7, (**b**) 28, and (**c**) 56 days with W/B = 0.5.

In general, the effect of the W/B cementitious water ratio decreased the mechanical compressive strength as its levels increased at all ages, which can be observed in Equations (3)–(5) above, so this ratio was identified as a significant variable in them.

In the 7-day model, we observed that the substitution of cement by WSA was a significant first-order variable that negatively affected the strength throughout the response zone and was observed in the regression equation (see Equation (3)). As shown in Table 5, the interaction between W/B and WSA was significant, mainly because of the variations in strength caused by both variables (see Figure 14a), because by reducing the W/B ratio from 0.55 to 0.4, the replacement of cement by WSA contributed to the loss of strength more directly by replacing more mass.

For the 28-day model, the significance of the variables exposed at 7 days was maintained for both the first-order variables and the interaction between W/B and WSA, but the main difference was that now the second-order terms for the variables W/B and SF were significant. It is important to consider that due to the chemical and morphological composition of SF, it was to be expected that at late ages, it would begin to have relevance in the increase in resistance, which was also confirmed by the regression model in Equation (4). It is also possible to observe the effect of the second-order factors on the response surface in Figure 13b, which shows concavity on the axis associated with SF. In the same line, in the contour plots (see Figure 14) relating WSA to SF (see Figure 13b), it is possible to observe that for the SF axis, approximately from levels −5% to 5%, where the concavity was found, the resistance decreased.

Finally, the 56-day model further confirmed what was described above because, for this model, the second-order coefficient corresponding to SF was the only significant one and largely determined the strength increase at this age (see Equation (5)) because the highest level of substitution (10%) managed to overcome the strength of the cement-only mix (see Figure 11). Likewise, from Figure 14c, we can see that from the intermediate level of replacement, i.e., 5%, up to the highest level of SF, the strength of the mixes increased independently of the amount of WSA replacement, maximizing in this range the total amount of replacement that could be generated with both SCMs.

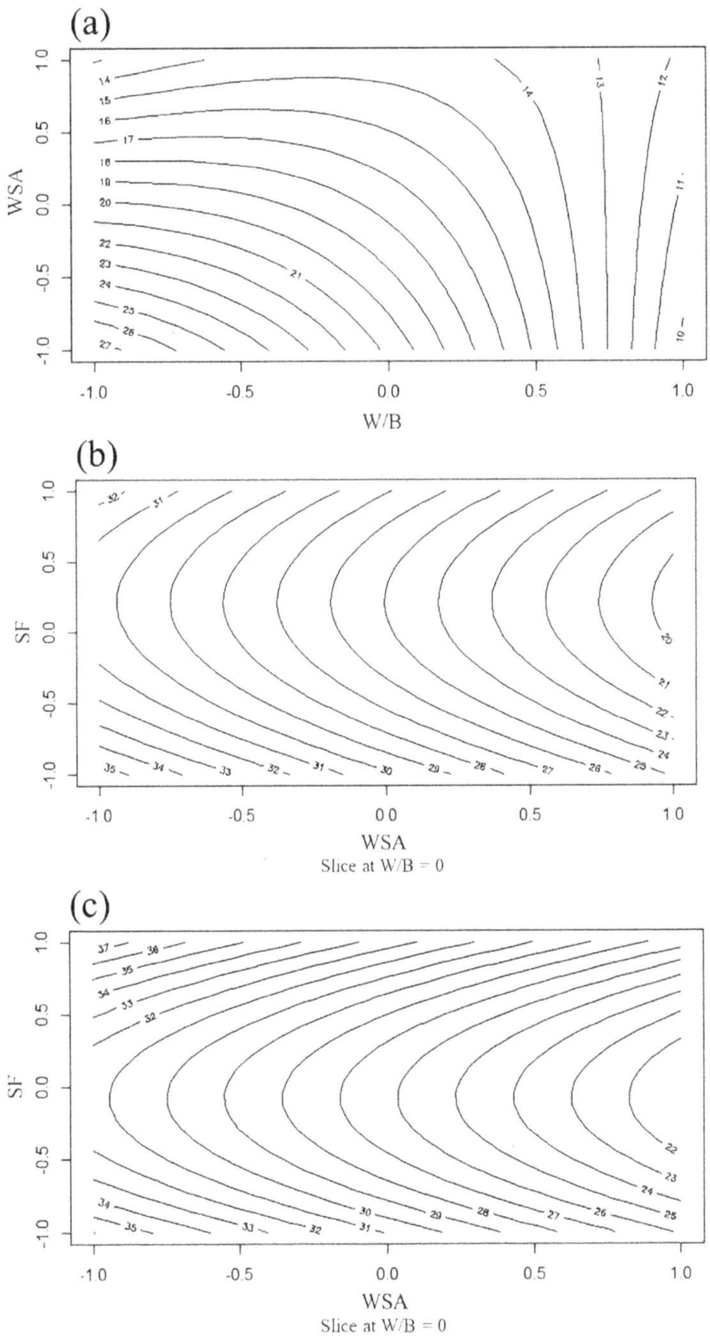

Figure 14. Contour plots for (**a**) 7, (**b**) 28, and (**c**) 56 days with W/B = 0.5.

3.2.3. Model Validation

To validate the RSM, three mixtures were chosen and are shown in Table 8.

Table 8. Validation mixtures to confirm the predictive model.

Mixtures	W/B	WSA	SF	W/B	WSA (%)	SF (%)
B0.5W15S10	0	0.5	1	0.5	15	10
B0.51W12S9	0.1	0.2	0.8	0.51	12	9
B0.5W12S5	0	0.2	0	0.5	12	5

The experimental results can be seen in Figure 15, where the -P mixtures correspond to the values predicted by the model while the -E mixtures correspond to the experimental values obtained at 7, 28, and 56 days. Table 9 shows the model errors with respect to the mean of the experimental results at each age, where a conservative model was observed, because in 90% of the cases, the experimental results showed higher values than the model. Therefore, in addition to accurately approaching the values predicted by the model, the low errors (the maximum was 6.1%) confirmed the use of the BBD and the RSM to optimize the design of mixes with SCM cement substitution.

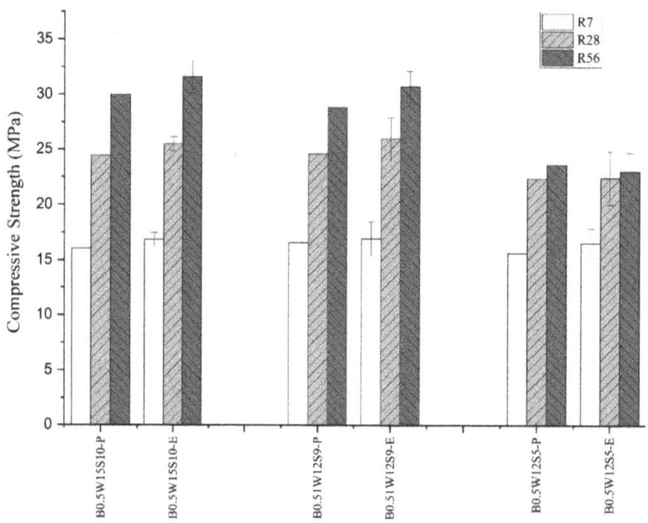

Figure 15. Predicted (-P) and experimental (-E) mixes from RSM.

Table 9. Model prediction errors as per the validation experiments.

Mixtures	Model Error 7 Days (%)	Model Error 28 Days (%)	Model Error 56 Days (%)
B0.5W15S10	4.8	4.1	5.1
B0.51W12S9	2.2	5.4	6.1
B0.5W12S5	5.6	0.3	−2.7

3.2.4. Carbon Footprint

In order to analyze the CO_2eq emissions of pastes, the kg of CO_2 per unit volume (m^3) of pastes with a W/B ratio of 0.5 were calculated since this W/B ratio contained mono, binary, and ternary mixes. First, considering the CO_2eq values (see Table 1), as the replacement of cement by SCM increased, the CO_2/m^3 decreased, which is why we analyzed it on the basis of the intensity index.

Figure 16 shows the compressive strength and the intensity index, which, as described in previous paragraphs, allowed us to measure the amount of binder required for 1 MPa of compressive strength, defined as CO_2eq (kg/m^3)/strength (MPa) at 56 days. In order to have a reduction of CO_2eq emissions with an increase in compressive strength, the intensity index needed to be lower.

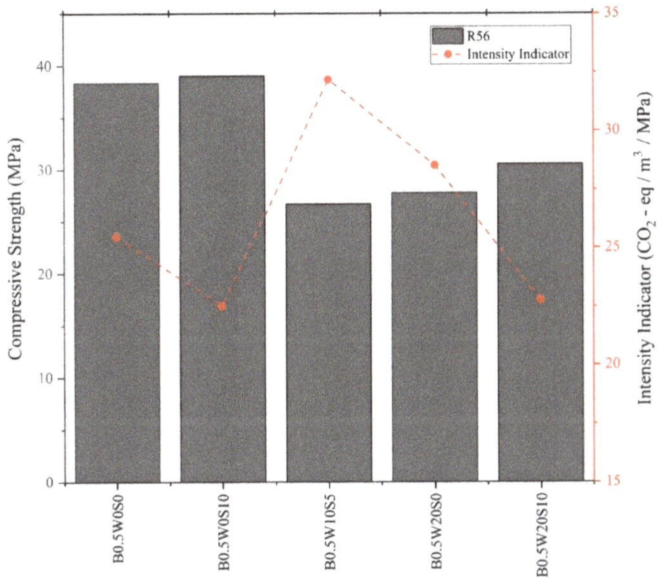

Figure 16. Carbon footprint (intensity index) versus 56-day compressive strength.

According to the results, the ternary mixture B0.5W10S5 had the highest intensity index because, although the total replacement was 15%, the levels used were intermediate, and, as explained in previous paragraphs, it was the most unfavorable region in terms of resistance. However, the opposite scenario was the case with B0.5W20S10 (also ternary), which was in the optimal region of replacement to obtain higher mechanical properties. Comparing both cases, it was possible to determine that the last mix was 29% lower in terms of CO_2eq/strength and had a higher resistance at 56 days, confirming that the proposed region was better in terms of both resistance and reduction of CO_2 emissions. Furthermore, compared to the cement-only mix (B0.5W0S0), B0.5W20S10 was 11% lower in terms of CO_2 emissions.

Finally, it is important to highlight that B0.5W20S10 presented lower CO_2 emissions per m_3, even though its strength was not the highest. At the same time, B0.5W20S0 (only WSA replacement) presented lower compressive strength and the highest CO_2 emissions.

Based on the research carried out, optimal recommended proportions for the mixes are suggested, considering all the variables analyzed and seeking a balance between the resistances obtained and the equivalent CO_2 in each case. In this sense, as has previously been pointed out, the most suitable proportion, based on the results obtained, is the one with 20% WSA and 10% SF substitution.

4. Conclusions

This research demonstrated that the use of agricultural and industrial wastes, specifically wheat straw ash (WSA) and silica fume (SF), as supplementary cementitious materials (SCMs), is technically feasible in certain proportions. Three statistical models were developed using the Box-Behnken design (BBD) that accurately predicted the strength of all possible combinations of W/B variables in the range of 0.4 to 0.6, SF at 0–10%

cement replacement by mass, and WSA at 0–20% with good R^2_{Adj} ($R^2_{Adj-7days}$ = 89%, $R^2_{Adj-28days}$ = 88%, and $R^2_{Adj-56days}$ = 77%). These models explained a large portion of the observed variance, and the lack of fit was non-significant ($p > 0.05$), indicating a good fit over the experimental points (p value$_{7days}$ = 0.065, p value$_{28days}$ = 0.92, and p value$_{56days}$ = 0.47).

The response surface methodology (RSM) allowed maximizing the utilization of WSA and SF, achieving cement paste formulations that meet strength requirements at different ages (7, 28, and 56 days) while allowing lower CO_2 emissions. The physicochemical characterization of the materials facilitated the understanding of the design model, explaining the results. To confirm the model, three mixtures were tested with different proportions of SCM at a W/B ratio = 0.5, resulting in values close to those predicted, with maximum errors of 6% at the three ages.

The results suggest that for any addition of WSA between 0 and 20% and SF between 5 and 10%, the mechanical strength losses due to the incorporation of WSA are reduced, maximizing the incorporation of both SCMs. This model allows predicting the compressive strength at different ages, facilitating the determination of the required strength and the calculation of the maximum possible cement replacements.

It is suggested that any cement paste production center implement the model to calculate optimal cement replacement levels, which will contribute to an effective reduction of the carbon footprint and sustainability of the industry. It is also recommended in agricultural production to promote the collection and use of wheat straw ash as a valuable resource, and in industrial production to use silica fume and other industrial by-products as substitutes in the formulation of cement pastes, thus contributing to sustainability through the use of local waste. Finally, it is recommended to validate this model in larger-scale production plants in order to implement this optimization in a safe and viable manner.

Author Contributions: Conceptualization, M.G. and P.G.; methodology, M.G. and P.G.; software, B.B.; validation, B.B.; formal analysis, B.B. and J.R.-R.; investigation, B.B. and G.V.; resources, P.G. and M.G.; data curation, B.B.; writing—original draft preparation, B.B.; writing—review and editing, M.G., J.R.-R., and P.G.; visualization, B.B.; supervision, M.G. and P.G.; project administration, P.G.; funding acquisition, P.G. All authors have read and agreed to the published version of the manuscript.

Funding: This research was funded by CENAMAD (ANID BASAL FB210015), FONDEF project ID20I20312, and the Concrete Innovation Hub UC (CIHUC) from the Pontificia Universidad Católica de Chile. The APC was funded by Universidade da Coruña.

Institutional Review Board Statement: Not applicable.

Informed Consent Statement: Not applicable.

Data Availability Statement: Data can be available upon request.

Acknowledgments: CENAMAD (ANID BASAL FB210015), Sika and Melon, and Universidade da Coruña for allowing us the Open Access.

Conflicts of Interest: The authors declare no conflicts of interest.

References

1. Global Cement and Concrete Association. *Concrete Future—The GCCA 2050 Cement and Concrete Industry Roadmap for Net Zero Concrete*; Global Cement and Concrete Association: London, UK, 2021; pp. 1–48. Available online: https://gccassociation.org/concretefuture/wp-content/uploads/2021/10/GCCA-Concrete-Future-Roadmap.pdf (accessed on 4 November 2024).
2. Skocek, J.; Zajac, M.; Ben Haha, M. Carbon Capture and Utilization by mineralization of cement pastes derived from recycled concrete. *Sci. Rep.* **2020**, *10*, 1–12. [CrossRef]
3. Summerbell, D.L.; Barlow, C.Y.; Cullen, J.M. Potential reduction of carbon emissions by performance improvement: A cement industry case study. *J. Clean. Prod.* **2016**, *135*, 1327–1339. [CrossRef]
4. Scrivener, K.L.; John, V.M.; Gartner, E.M. Eco-efficient cements: Potential economically viable solutions for a low-CO_2 cement-based materials industry. *Cem. Concr. Res.* **2018**, *114*, 2–26. [CrossRef]
5. Müller, N.; Harnisch, J. *A Blueprint for a Climate Friendly Cement Industry*; A report prepared for the WWF—Lafarge Conservation Partnership; WWF International: Gland, Switzerland, 2008.

6. Centro UC Políticas Públicas, Déficit Cero. Déficit Habitacional: ¿Cuántas Familias Necesitan una Vivienda y en qué Territorios? 2022. Available online: https://deficitcero.cl/uploads/estudios/Minuta_EstimaciondelDeficit.pdf (accessed on 27 October 2024).
7. Habitat for Humanity. *Annual Report 2020*; National Office: Recife, Brazil; Urban Innovation Office: São Paulo, Brazil, 2020; pp. 4–10. Available online: https://habitatbrasil.org.br/wp-content/uploads/2021/06/Annual-Report-2020.pdf (accessed on 27 October 2024).
8. Departamento Administrativo Nacional De Estadística. Boletín Técnico: Déficit Habitacional CNPV 2022. 2018; p. 20. Available online: https://www.dane.gov.co/files/investigaciones/deficit-habitacional/Boletin-tec-deficit-hab-2021.pdf (accessed on 27 October 2024).
9. Lago, F. Estimación de la Evolución del Déficit Habitacional en la Argentina, Buenos Aires. 2016. Available online: https://biblioteca.camarco.org.ar/libro/estimacion-de-la-evolucion-del-deficit-habitacional-en-la-argentina/ (accessed on 27 October 2024).
10. Lothenbach, B.; Scrivener, K.; Hooton, R.D. Supplementary cementitious materials. *Cem. Concr. Res.* **2011**, *41*, 1244–1256. [CrossRef]
11. Su, Y.; Luo, B.; Luo, Z.; Xu, F.; Huang, H.; Long, Z.; Shen, C. Mechanical characteristics and solidification mechanism of slag/fly ash-based geopolymer and cement solidified organic clay: A comparative study. *J. Build. Eng.* **2023**, *71*, 106459. [CrossRef]
12. Jhatial, A.A.; Goh, W.I.; Mohamad, N.; Mo, K.H.; Sohu, S. Incorporation of Palm Oil Fuel Ash and Egg shell Powder as Supplementary Cementitious Materials in Sustainable Foamed Concrete. *Tech. Gaz.* **2020**, *27*, 1394–1402. [CrossRef]
13. Gencel, O.; Nodehi, M.; Yavuz Bayraktar, O.; Kaplan, G.; Benli, A.; Gholampour, A.; Ozbakkaloglu, T. Basalt fiber-reinforced foam concrete containing silica fume: An experimental study. *Constr. Build. Mater.* **2022**, *326*, 126861. [CrossRef]
14. Alyousef, R.; Abbass, W.; Aslam, F.; Gillani, S.A.A. Characterization of high-performance concrete using limestone powder and supplementary fillers in binary and ternary blends under different curing regimes. *Case Stud. Constr. Mater.* **2023**, *18*, e02058. [CrossRef]
15. Alamri, M.; Ali, T.; Ahmed, H.; Qureshi, M.Z.; Elmagarhe, A.; Adil Khan, M.; Ajwad, A.; Sarmad Mahmood, M. Enhancing the engineering characteristics of sustainable recycled aggregate concrete using fly ash, metakaolin and silica fume. *Heliyon* **2024**, *10*, e29014. [CrossRef]
16. Bhagath Singh, G.V.P.; Durga Prasad, V. Environmental impact of concrete containing high volume fly ash and ground granulated blast furnace slag. *J. Clean. Prod.* **2024**, *448*, 141729. [CrossRef]
17. Saradar, A.; Nemati, P.; Paskiabi, A.S.; Moein, M.M.; Moez, H.; Vishki, E.H. Prediction of mechanical properties of lightweight basalt fiber reinforced concrete containing silica fume and fly ash: Experimental and numerical assessment. *J. Build. Eng.* **2020**, *32*, 101732. [CrossRef]
18. Stutzman, P. SEM Analysis and Computer Modelling of Hydration of Portland Cement Particles. In *Petrography of Cementitious Materials*; ASTM Special Technical Publication: Philadelphia, PA, USA, 1994.
19. Shaikh, F.U.A.; Supit, S.W.M.; Sarker, P.K. A study on the effect of nano silica on compressive strength of high volume fly ash mortars and concretes. *Mater. Des.* **2014**, *60*, 433–442. [CrossRef]
20. FAO. *World Food and Agriculture—Statistical Yearbook 2021*; FAO: Rome, Italy, 2021. [CrossRef]
21. Thomas, B.S.; Yang, J.; Mo, K.H.; Abdalla, J.A.; Hawileh, R.A.; Ariyachandra, E. Biomass ashes from agricultural wastes as supplementary cementitious materials or aggregate replacement in cement/geopolymer concrete: A comprehensive review. *J. Build. Eng.* **2021**, *40*, 102332. [CrossRef]
22. Santhosh, K.G.; Subhani, S.M.; Bahurudeen, A. Recycling of palm oil fuel ash and rice husk ash in the cleaner production of concrete. *J. Clean. Prod.* **2022**, *354*, 131736. [CrossRef]
23. Su, Q.; Xu, J. Mechanical properties of rice husk ash and glass powder concrete: Experimental and mesoscopic studies. *J. Build. Eng.* **2024**, *95*, 110278. [CrossRef]
24. Sobuz, M.H.R.; Al-Imran Datta, S.D.; Jabin, J.A.; Aditto, F.S.; Sadiqul Hasan, N.M.; Hasan, M.; Zaman, A.A.U. Assessing the influence of sugarcane bagasse ash for the production of eco-friendly concrete: Experimental and machine learning approaches. *Case Stud. Constr. Mater.* **2024**, *20*, e02839. [CrossRef]
25. Pazouki, G.; Tao, Z.; Saeed, N.; Kang, W.H. Using artificial intelligence methods to predict the compressive strength of concrete containing sugarcane bagasse ash. *Constr. Build. Mater.* **2023**, *409*, 134047. [CrossRef]
26. Althoey, F.; Zaid, O.; Martínez-García, R.; de Prado-Gil, J.; Ahmed, M.; Yosri, A.M. Ultra-high-performance fiber-reinforced sustainable concrete modified with silica fume and wheat straw ash. *J. Mater. Res. Technol.* **2023**, *24*, 6118–6139. [CrossRef]
27. Bheel, N.; Chohan, I.M.; Alwetaishi, M.; Waheeb, S.A.; Alkhattabi, L. Sustainability assessment and mechanical characteristics of high strength concrete blended with marble dust powder and wheat straw ash as cementitious materials by using RSM modelling. *Sustain. Chem. Pharm.* **2024**, *39*, 101606. [CrossRef]
28. Saraswat, A.; Kumar Parashar, A.; Bahadur, R. Effect of coconut shell ash substitute with cement on the mechanical properties of cement concrete. *Mater. Today Proc.* **2023**, 1–4. [CrossRef]
29. Varghese, A.; Unnikrishnan, S. Mechanical strength of coconut fiber reinforced concrete. *Mater. Today Proc.* **2023**. [CrossRef]
30. Sharma, N.; Sharma, P.; Parashar, A.K. Incorporation of Silica Fume and Waste Corn Cob Ash in Cement and Concrete for Sustainable Environment. *Mater. Today Proc.* **2022**, *62*, 4151–4155. [CrossRef]
31. Oriola, K.O.; Raheem, A.A.; Ogundele, A.B. Investigation of compressive strength and thermal properties of corn cob ash cement concrete. *Mater. Today Proc.* **2023**, *86*, 128–133. [CrossRef]

32. FAO. *Global Cereal Production Heading for a Record High*; FAO: Rome, Italy, 2023; Available online: https://www.fao.org/worldfoodsituation/csdb/en/ (accessed on 27 October 2024).
33. Pan, X.; Sano, Y. Fractionation of wheat straw by atmospheric acetic acid process. *Bioresour. Technol.* **2005**, *96*, 1256–1263. [CrossRef]
34. FAO. *Global Forest Resources Assessment 2010 Main Report*; FAO Forestry Paper; FAO: Rome, Italy, 2010; Volume 163, pp. 37–72. Available online: https://www.fao.org/3/i1757e/i1757e.pdf (accessed on 27 October 2024).
35. Ungerfeld, E.; Vial, M.; Jobet, C.; Mathias, M.; Peñalillo, K. Problemas de la Quema de Rastrojos y Alternativas Posibles—Mundoagro. 2021. Available online: https://mundoagro.cl/problemas-de-la-quema-de-rastrojos-y-alternativas-posibles/# (accessed on 29 June 2023).
36. Ruiz, C.; Wolff, M.; Claret, M. *Rastrojos de Cultivos Anuales y Residuos Forestales*; Instituto de Investigaciones Agropecuarias: Santiago, Chile; Centro Regional de Investigación Quilamapu: Chillán, Chile, 2015; Volume 308, pp. 10–29.
37. Muñoz, P.; Mendívil, M.A.; Letelier, V.; Morales, M.P. Thermal and mechanical properties of fired clay bricks made by using grapevine shoots as pore forming agent. Influence of particle size and percentage of replacement. *Constr. Build. Mater.* **2019**, *224*, 639–658. [CrossRef]
38. Biricik, H.; Aköz, F.; Berktay, I.; Tulgar, A.N. Study of pozzolanic properties of wheat straw ash. *Cem. Concr. Res.* **1999**, *29*, 637–643. [CrossRef]
39. Amin, M.N.; Murtaza, T.; Shahzada, K.; Khan, K.; Adil, M. Pozzolanic potential and mechanical performance of wheat straw ash incorporated sustainable concrete. *Sustainability* **2019**, *11*, 519. [CrossRef]
40. Maddalena, R.; Roberts, J.J.; Hamilton, A. Can Portland cement be replaced by low-carbon alternative materials? A study on the thermal properties and carbon emissions of innovative cements. *J. Clean. Prod.* **2018**, *186*, 933–942. [CrossRef]
41. Al-akhras, N.M.; Abu-alfoul, B.A. Effect of wheat straw ash on mechanical properties of autoclaved mortar. *Cem. Concr. Res.* **2002**, *32*, 859–863. [CrossRef]
42. Ataie, F.F.; Riding, K.A. Thermochemical Pretreatments for Agricultural Residue Ash Production for Concrete. *J. Mater. Civ. Eng.* **2013**, *25*, 1703–1711. [CrossRef]
43. Memon, S.A.; Wahid, I.; Khan, M.K.; Tanoli, M.A.; Bimaganbetova, M. Environmentally friendly utilization of wheat straw ash in cement-based composites. *Sustainability* **2018**, *10*, 1322. [CrossRef]
44. *ASTM C595-21*; Standard Specification for Blended Hydraulic Cements. ASTM: Philadelphia, PA, USA, 2021. Available online: www.astm.org (accessed on 27 October 2024).
45. González-Kunz, R.N.; Pineda, P.; Bras, A.; Morillas, L. Plant biomass ashes in cement-based building materials. Feasibility as eco-efficient structural mortars and grouts. *Sustain. Cities Soc.* **2017**, *31*, 151–172. [CrossRef]
46. Khan, K.; Ishfaq, M.; Amin, M.N.; Shahzada, K.; Wahab, N.; Faraz, M.I. Evaluation of Mechanical and Microstructural Properties and Global Warming Potential of Green Concrete with Wheat Straw Ash and Silica Fume. *Materials* **2022**, *15*, 3177. [CrossRef] [PubMed]
47. Al-Kadhim Hameed, M.A.; Alzerjawi, A.K.R.; Mahdi, Z.A. Studying the behavior of the concrete mixture with wheat straw as part of the cement. *J. Phys. Conf. Ser.* **2021**, *1973*, 012174. [CrossRef]
48. Muthukumar, M.; Mohan, D.; Rajendran, M. Optimization of mix proportions of mineral aggregates using Box Behnken design of experiments. *Cem. Concr. Compos.* **2003**, *25*, 751–758. [CrossRef]
49. Memon, S.A.; Javed, U.; Haris, M.; Khushnood, R.A.; Kim, J. Incorporation of wheat straw ash as partial sand replacement for production of eco-friendly concrete. *Materials* **2021**, *14*, 2078. [CrossRef] [PubMed]
50. Luo, T.; Hua, C.; Liu, F.; Sun, Q.; Yi, Y.; Pan, X. Effect of adding solid waste silica fume as a cement paste replacement on the properties of fresh and hardened concrete. *Case Stud. Constr. Mater.* **2022**, *16*, e01048. [CrossRef]
51. Bhanja, S.; Sengupta, B. Influence of silica fume on the tensile strength of concrete. *Cem. Concr. Res.* **2005**, *35*, 743–747. [CrossRef]
52. Das, S.K.; Mustakim, S.M.; Adesina, A.; Mishra, J.; Alomayri, T.S.; Assaedi, H.S.; Kaze, C.R. Fresh, strength and microstructure properties of geopolymer concrete incorporating lime and silica fume as replacement of fly ash. *J. Build. Eng.* **2020**, *32*, 101780. [CrossRef]
53. Box, G.E.P.; Wilson, K.B. On the Experimental Attainment of Optimum Conditions. *J. R. Stat. Soc. Ser. B* **1951**, *13*, 1–38. [CrossRef]
54. *ASTM-C305-20*; Standard Practice for Supplementation. ASTM Standard Guide. ASTM: Philadelphia, PA, USA, 2020. [CrossRef]
55. *ASTM C109-21*; Standard Test Method for Compressive Strength of Hydraulic Cement Mortars (Using 2-in. or [50 mm] Cube Specimens). American Society for Testing and Materials: West Conshohocken, PA, USA, 2021. [CrossRef]
56. *ASTM-C511-13*; Standard Specification for Mixing Rooms, Moist Cabinets, Moist Rooms, and Water. ASTM Standard Guide. ASTM: Philadelphia, PA, USA, 2015. [CrossRef]
57. Mena, J.; González, M.; Remesar, J.C.; Lopez, M. Developing a very high-strength low-CO_2 cementitious matrix based on a multi-binder approach for structural lightweight aggregate concrete. *Constr. Build. Mater.* **2020**, *234*, 117830. [CrossRef]
58. Yajun, J.; Cahyadi, J.H. Effects of densified silica fume on microstructure and compressive strength of blended cement pastes. *Cem. Concr. Res.* **2003**, *33*, 1543–1548. [CrossRef]
59. *ASTM C1679-14*; Standard Practice for Measuring Hydration Kinetics of Hydraulic Cementitious Mixtures Using Isothermal Calorimetry. ASTM: Philadelphia, PA, USA, 2014.
60. Mitchell, D.R.G.; Hinczak, I.; Day, R.A. Interaction of silica fume with calcium hydroxide solutions. *Cem. Concr. Res.* **1998**, *28*, 1571–1584. [CrossRef]

61. De Rojas, M.I.S.; Rivera, J.; Frías, M. Influence of the microsilica state on pozzolanic reaction rate. *Cem. Concr. Res.* **1999**, *29*, 945–949. [CrossRef]
62. Shar, I.A.; Memon, F.A.; Bheel, N.; Shaikh, Z.H.; Dayo, A.A. Use of Wheat Straw Ash as Cement Replacement Material in the Concrete. In Proceedings of the International Conference on Sustainable Development in Civil Engineering, MUET, Jamshoro, Pakistan, 16–18 February 2019; Volume 3, pp. 5–7.

sustainability

MDPI

Article

The Adhesion Performance in Green-Glued Finger Joints Using Different Wood Ring Orientations

Gonzalo Rodríguez-Grau [1,2], Pierre-Louis Cordonnier [3], Benjamín Navarrete [1], Claudio Montero [4,5], Claudia Alvarado [6], Régis Pommier [7], Víctor Rosales [8,9] and Carlos Galarce [1,*]

1 School of Civil Construction, Faculty of Engineering, Pontificia Universidad Católica de Chile, Santiago 7820436, Chile; grodriguezg@uc.cl (G.R.-G.)
2 Centro Nacional de Excelencia para la Industria de la Madera (CENAMAD), Pontificia Universidad Católica de Chile, Santiago 7820436, Chile
3 Ecole Nationale Supérieure d'arts et Métiers (ENSAM), 151 Boulevard de L'hôpital, 75013 Paris, France
4 Adhesives & Composites Material Laboratory, Universidad del Bío-Bío, Concepción 4051381, Chile
5 Wood Design & Technology Laboratory, Universidad del Bío-Bío, Concepción 4051381, Chile
6 Centro de Investigación y Desarrollo en Ciencias Aeroespaciales, Fuerza Aérea de Chile, Av. José Miguel Carrera 11087, Santiago 8020744, Chile
7 Institut de Mécanique et D'Ingienierie (I2M), Arts et Métiers, University Bordeaux, UMR 5295, 351 Cours de la Libération, 33400 Talence, France
8 Centro Nacional de Excelencia para la Industria de la Madera (CENAMAD), Universidad del Bío-Bío, Concepción 4051381, Chile
9 Department of Construction, Universidad del Bío-Bío, Concepción 4051381, Chile
* Correspondence: cegalarc@uc.cl

Abstract: Structural lumber is designed to meet the technical standards that ensure safety, cost-effectiveness, and sustainability. However, some tree species face limitations in their growth, which restricts their widespread use. An example of this is *Nothofagus alpina*, which has excellent mechanical properties but is not utilized much due to the challenges in extracting its timber and poor utilization, mainly because of the length of the wood. There is little information concerned with the uses and better management of small pieces using Nothofagus species, but it is still insufficient. This study investigates the adhesion performance of green-glued finger joints with varying wood ring orientations and moisture contents ranging from 21% to 25% using *Nothofagus alpina*. The primary aim is to assess how ring orientation and wet timber affect the green gluing process for creating larger wood pieces than sawn wood. The resulting products could meet the standards for wood serviceability number three for native Chilean wood. The findings indicate that finger joint performance improves with higher timber moisture levels. However, the orientation of the wood fibers did not significantly affect the performance under the tested conditions. It is important to note that this effect may become more significant near the fiber saturation point. These findings emphasize the need for a detailed protocol on the green gluing technique for *Nothofagus alpina* and the associated drying and surface processes in finger joint construction.

Keywords: *Nothofagus alpina*; green gluing; finger joints

1. Introduction

Wood is a highly competitive material due to its favorable weight-to-strength ratio and sustainability criteria [1]. However, due to its natural growth patterns, wood production faces limitations in size, consistency, and mechanical and physical properties homogeneity. Glued-Laminated Timber (GLT) or glulam is a technique developed to overcome these challenges [2,3]. This technology involves creating structurally Engineered Wood Products (EWPs) by finger-jointing and gluing wood lamellae end-to-end, resulting in a more homogeneous, shape-stable material. GLT allows the production of massive timbers, up to 30 m in length, which would be difficult or impossible to obtain from a single tree [4]. This technique is also valuable for eliminating the natural defects of plain wood, such as knots or any other defect that reduces mechanical properties.

At the same time, Laminated Veneer Lumber (LVL) and Cross-Laminated Timber (CLT) are two popular engineered wood products extensively used in modern construction due to their strength, stability, and sustainability. LVL is mainly used for linear structural elements and provides high strength and stiffness in one direction. On the other hand, CLT is used to create large panels with strength and stability in multiple directions, making it ideal for the walls, floors, and roofs of multi-story buildings. Both LVL and CLT are solid, stable, and sustainable building materials. They offer significant environmental benefits, making them a more eco-friendly choice than traditional construction materials [5,6].

In the same way, glued laminated timber, often called glulam, is an engineered wood product created by bonding layers of wood (known as laminas) with adhesives. This process enables the production of structural elements in different sizes and shapes, including long spans and unique designs that captivate interest and are often challenging to achieve with solid wood [7].

The glulam technique uses moisture above the fiber saturation point (FSP) timber [8]. On the contrary, green gluing is a technique used to bond freshly sawn lumber that has not been through a kiln or air-drying process [9]. Several authors have reported the impact of this technique in protecting and assessing native forests. For instance, Lissouck et al. [10] explained that developing the glulam industrialization of species of wood native from the Congo basin can mitigate their overexploitation, thus protecting them against rarefaction and extinction risk. Similar studies have been developed using different native species worldwide [11–13], a robust background to implement this technique in other species. However, limited information on research on native Chilean species is found [8,14].

In a previous study by Rosales et al., the feasibility of using the green gluing technique with Nothofagus Chilean species was tested [14]. The green gluing with timber near or above the fiber saturation point shows promising results. However, their potential is limited by the standard for laminated wood products, which require kiln-dried wood for manufacturing. Typically, wood products are made by machining and gluing dried timber with a moisture content of around 10%. While manufacturing GLT with green wood and letting it dry as a final product saves time and energy, the wood tends to retract while drying, which can cause shrinkage and internal strain in the material until it ruptures. Hence, the primary challenge of green wood gluing is to ensure that the product can withstand the drying cycle. At the same time, the lack of regulations for using the green gluing technique in hardwoods presents a challenge. As a result, this study intends to push to develop this technique for use in more minor structural elements like pergolas, garages, and playgrounds. If the method's focus shifts to small companies and producers, it could lead to the more sustainable management and production of these types of constructions.

Previously, Sterley and Gustafsson [15] studied the shear fracture properties of one-component polyurethane (PUR) wood-adhesive bonds. Their article examines the impact of moisture content (MC) and pressing conditions during the gluing process. This study aims to improve our understanding of how these factors affect the mechanical performance of adhesive bonds, especially in green gluing where unseasoned wood is used. The findings indicate that the precise control of moisture content and pressing conditions is essential for optimizing the performance of PUR adhesive bonds in wood applications. Additionally, the study highlights the potential of green gluing techniques to enhance wood utilization and energy efficiency in the wood processing industry, potentially leading to more sustainable wood use. These results can serve as a valuable reference for understanding potential outcomes when using the *Nothofagus alpina* species and incorporating finger joints.

The finger joint technique may provide a solution to overcome this issue. Pommier and Elbez demonstrated that finger joints made with green-glued softwood could be more potent than those made with gluing on dry wood [16]. This situation is due to the higher ductility of greenwood, which allows the finger profile to fit up to the tipping point, resulting in a more regular glue line [16]. Despite this information, it is crucial to consider the hygroscopicity of wood, which allows the wood products to endure thermal and hygroscopic variations with the seasons' changes throughout their life. This anisotropic wood swelling with humidity variations provokes internal strain that may harm the beam [4]. Green-glued wood products can suffer swelling while drifting to moisture equilibrium, but pieces with deformations are reduced.

Furthermore, other factors, such as the length and slope of the finger joint, are closely related to the strength of a specimen [17,18]. The applied pressure is also crucial for evaluating the ultimate tensile strength [2,19]. Additionally, the stiffness and thickness of the bond line are essential factors, as emphasized by Groom and Leichti [20]. Therefore, the performance of the wood product is critical in the initial cycle.

Wood's drying process and its direction-dependent shrinkage may be particularly harmful to the finger joint region depending on the relative orientations of wood rings on the two sides of the joint. This information could be crucial to overcoming the structural standard of the norm and implementing green gluing using native Chilean wood. Moreover, equipping small industries with simple tools and procedures is vital, especially considering their access to smaller woods. This action not only protects forests with these species but also reduces the need to replace them with other species (pine or eucalyptus), underlining the practicality of the proposal and its potential to empower small producers. Due to that, this research studied the adhesion performance in green glued finger joints using different wood ring orientations. The results of this research can provide helpful information for understanding the behavior of green-glued wood sections and thus use them in longer sections for construction.

2. Materials and Methods

2.1. Selection, Preparation, and Characterization of Wood

The study used *Nothofagus alpina* (Rauli), a native wood species in Chile with a 500 kg/m^3 oven-dry density. Softwood green-glued finger joints have been shown to have potential by Pommier et al. [16]. Moreover, this study used a one-component polyurethane adhesive (1C-PUR) due to its good mechanical properties and low environmental impact [2, 21]. This adhesive is certified for structural use in the traditional dry bonding process (13–18% humidity), that is, on the support of wooden structures (safety data sheet according to Chilean standard NCh2245:2015) [22]. Nonetheless, this product was used in a previous report to carry out the green gluing technique with promising results [14].

At the same time, to reduce the uncertainty and variability of wood, all specimens were made with wood from the same lot, and half of them were dried before gluing. According to the EN 408 standard, the moisture content in the specimen must be stable when tested (less than 0.1% mass variation in 6 h). The modulus of elasticity (MOE) of wood was measured using a FAKOPP Microsecond Timer© (Agfalva, Hungary), designed explicitly for wood measurements by the Hungarian company Fakopp Enterprise (Agfalva, Hungary) [23,24]. Therefore, the green-glued specimens were dried to Equilibrium Moisture Content (EMC) corresponding to 14% moisture content for the Santiago of Chile city before being tested. All specimens were tested at 6–8% moisture content measured with the RDM3 tool by the Delmhorst Instrument Company (Towaco, NJ, USA). To appreciate the effect of relative wood ring orientation, a batch of tests was made for each of these configurations (Figure 1): Tangential–Tangential (T/T): six dry-glued and six green-glued, Radial–Radial (R/R): six dry-glued and seven green-glued, Radial–Tangential (R/T): 11 dry-glued and 11 green-glued. A batch of 10 plain beams without junctions was also tested as a reference. In total, 57 tests were conducted.

Moisture content, reference density (mass and density at moisture content at 14%) and oven-dry density (mass and density at 0% moisture content) were determined for each end of the tested pieces according to Chilean standard NCh 176/1 [25] and NCh 176/2 [26], respectively.

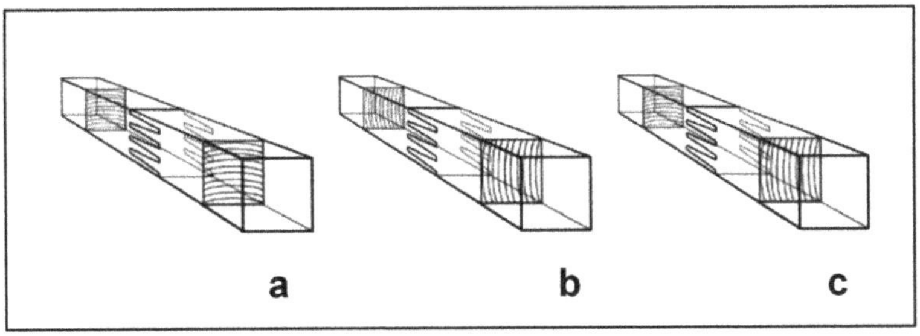

Figure 1. Different orientation ring combinations to build the finger joint samples: (**a**) Tangential–Tangential (T/T); (**b**) Radial–Radial (R/R); and (**c**) Radial–Tangential (R/T).

2.2. Measurement of Moisture

Upon receipt, the sawn board samples measuring 50 mm × 50 mm × 920 mm were observed and separated based on the orientation of the rings concerning the lumber's faces. Only treads with rings tangent to one side (around 20 degrees) were selected for the study. Each lumber's humidity was measured using the RDM3 tool from Delmhorst Instrument Company. To prevent the lumbers from drying, they were kept in plastic bags when not being manipulated for cutting or gluing. All specimens were prepared within a minimum period of approximately two weeks to maintain their humidity at environmental temperature (20 ± 3 °C). These efforts were successful as the moisture content was re-measured for the green-glued specimens and had not changed. The moisture content measures varied between 21% and 25%. The moisture measurements were conducted using the sensor mentioned, which is faster and less intrusive for determining the moisture content. Furthermore, small companies or producers will be able to utilize this method.

2.3. Drying Protocol

When received, the reference lumber and dry-glued specimens were placed in a dry oven at 40 degrees Celsius and kept there for nine days. Afterward, they were moved to laboratory conditions for an additional week before being cut down in a controlled environment of 20 ± 3 °C and less than 60 ± 5% relative moisture. In Chilean standards (NCh 176/1 [25] and NCh 176/2 [26]), it is mandatory to work within the specified range of temperature and moisture to certify the data's value. This protocol helped to achieve a measured moisture content of 6–8%, with very few cracks on the lumber. Most of the cracks were located at the ends of the lumber, which were subsequently removed.

2.4. Finger Joint Profile Cutting Protocol

The green-glued lumbers squared to 46 mm × 46 mm × 920 mm, while the others were left to dry. Then, as shown in Figure 2, a 3-axis CNC machine and a finger joint cutting bit were used to cut the finger joints off. The machine corresponds to Shopbot, the PRSalpha model. The depth of the finger cutting was determined through tip gap adjustment tests to find the optimal depth of 11 mm. The ends of the beams were planned to ensure consistency in the depth of the finger joints, as illustrated in Figure 2. The fingers were cut in four passes, each being 3 mm, except for the last one, which was 2 mm to limit defects during the cutting process.

Support was built to cut the finger profile in half and specimens 12 by 12, using the same CNC machine as the finger profile to guarantee orthogonality between the fingers and the specimen's axis. Slots were machined in plywood boards, and a system blocking the specimens was mounted on top of this board. The slots ensure the position of the specimen on two axes, and the first pass provides the precision on the third axis. The CNC has a 0.01 mm precision on each axis. The cutting speed was set to 18,000 rpm.

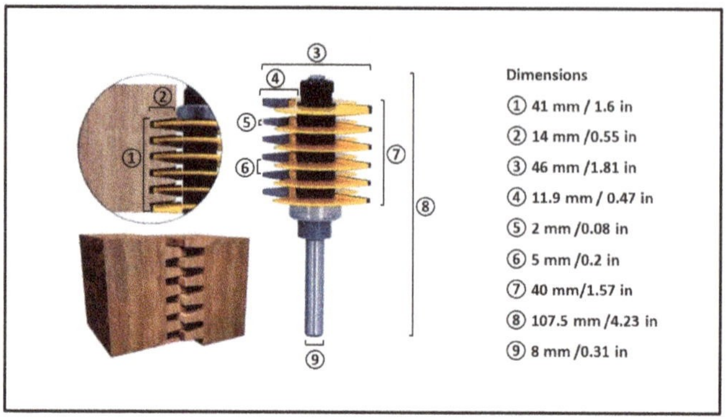

Figure 2. Details of finger joint construction. Each number indicates the position and the length used.

Lastly, the 92 cm lumbers were cut in half just before cutting the finger profile to prevent the finger from drying. The lumbers were immediately covered with plastic film to keep it from drying out.

Forty-eight hours after cutting the parts, they were glued together using Jowapur 685.12 adhesive and pressed at 6 N/mm^2, following Pommier's and Elbez's instructions [10]. The pressure was kept for a few seconds, as recommended by He et al. [27]. The pressure was applied end-to-end without guidance, resulting in a maximum 1-degree misalignment, which was corrected during the following squaring process. After two days for the glue to polymerize, the specimens were dried using the same protocol (nine days in a dry-40 degrees Celsius oven followed by one week at lab conditions).

The reference lumbers were immediately squared and cut to the correct dimensions for the EN 408 four-point bending tests. The dimensions were 36 mm × 36 mm × 800 mm (Figure 3) since the lab's test bench had a 700 mm span. On the other hand, the dry-glued specimens were squared down to 43 mm × 43 mm × 920 mm lumbers. These pieces were cut in half, and the fingers were cut. They were then glued and pressed at 6 N/mm^2. After two days of polymerization, they were cut to 36 mm × 36 mm × 800 mm and made ready for testing. The green-glued specimens were squared again once dry, cut down to the 36 mm × 36 mm × 800 mm dimensions, and prepared for testing. Slices showing the glued zone were kept as much as possible when cutting the specimen to its ultimate size to observe the glue joint and compare it to the mechanical test results.

The green gluing technique described in this research was developed by Rosales et al. (2023) [14]. In this procedure, a spatula is used to apply 150 g/m^2 of adhesive to each contact surface. A clamping pressure of approximately 0.6 MPa is maintained for two hours. The pressing operations were carried out in a controlled environment at 20 ± 3 °C. The NCh 2148 standard [28] sets out the operating range for pressing conditions.

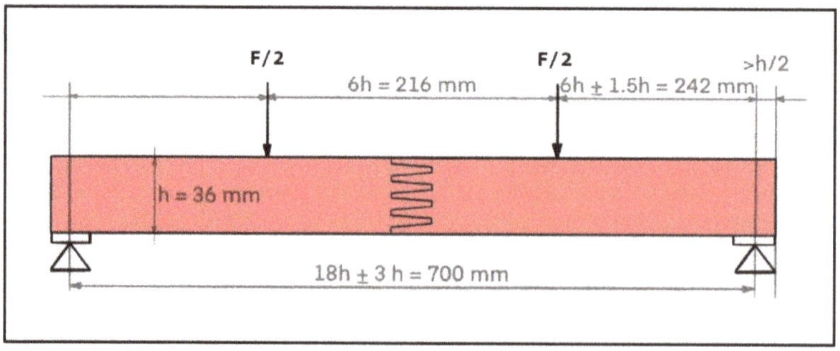

Figure 3. Dimensions of the specimens following the EN 408 standard.

2.5. Mechanical Tests

The bending tests were conducted following the instructions of the EN 408 standard [29]. Although the norm required more specimens, a clear tendency was observed with the specimens used in the study. The study determined each specimen's bending strength and rupture mode (MOR). The determination of the local modulus of elasticity (MOE) in bending for each specimen was subjected to a symmetrical load by being bent at two points throughout 18 times its depth, as illustrated in Figure 3. The lateral restraint was applied as needed to prevent lateral torsional buckling. This action allowed the specimen to deflect with minimal frictional resistance. All the procedures followed the EN 408 standard [29]. The press model was TC10001, and a local manufacturer made it [14].

The two following equations are used to calculate the MOR and MOE of finger wood [30].

$$\text{MOR} = \frac{3 \cdot P \cdot a}{b \cdot h^2} \tag{1}$$

$$\text{MOE} = \alpha \cdot \frac{6 \cdot P \cdot a \cdot L^2}{\delta \cdot b \cdot h^3} \tag{2}$$

where:

$$\alpha = \frac{1}{8} - \frac{1}{6} \cdot \left[\frac{a}{L}\right]^2 \tag{3}$$

P: applied load (N);
a: distance from one of the supports to the applied load (mm). In this case, 242 mm;
L: span of the tested element (mm). In this case, 700 mm;
b: width of the section of the part (mm);
h: height of the section of the part (mm);
α: coefficient of elastic deformation;
δ: deformation at span center.

In summary, MOR relates to the stress at yield during bending, while breaking strain quantifies the material's deformation capacity before fracture occurs.

2.6. Micro-Computed Tomography (Micro-CT)

A high-resolution Bruker SkyScan 1272 microtomography (Kontich, Belgium) was used to scan the samples. The scanner operated a rotation of 0.4 at 80 kV, 125 mA, and used an aluminum filter of 0.25 mm. The resolution of the analysis was 12 µm. The images were visualized with Nrecon reconstruction software (Version 3.1.1), and three-dimensional scanned images were obtained. The images were repositioned with Data Viewer software v1.5.1.9 to standardize the sample positioning. Quantitative assessments were performed using micro-CT and software analysis in a volume of interest (VOI) of approximately 2 cm^3 in the transversal plane. Images of each sample were obtained using a visualization software, with a pixel size of 24 µm. Voxels with grayscale values between 25 and 255 were considered solid material, while voxels between 24 and 0 were considered void space [14].

The image analysis was similar to that of the research group in [14]. This analysis consisted of processing and analysis with Image J software version 154e (The National Institutes of Health, Bethesda, MD, USA). Images obtained from micro-CT were corrected by removing the background. The corrected image was then converted to a binary black-and-white format for post-processing and parameter estimation. The thresholding procedure used was the iterative selection method. Finally, the percent porosity was calculated by dividing the number of black pixels by the total number of pixels and multiplying by 100 [31].

2.7. Roughness Measurement

The average roughness (Ra) was carefully measured for each piece. Duplicate measurements were taken on each side of the five finger joints of every piece. The fingers on each end were identified and cut with a carpenter's saw. The pieces were then conditioned at 20 ± 3 °C and 55 ± 5% RH for 48 h. Validating the procedure is a critical step. Before registering the value and realizing the measurement, it is essential that the parameters are in the specified range. The average roughness (Ra) was measured using a PCE-RT 1200 roughness tester. This thorough approach involved taking duplicate measurements on both faces of the five finger joints of 30 pieces, resulting in 1200 roughness measurements. This meticulous process ensured a comprehensive and accurate analysis.

2.8. Statistical Analysis

A one-way ANOVA was conducted to compare species under each condition. Significant differences were noted on each graph. An unpaired *t*-test with Welch's correction compared the same orientation wood under different moisture levels. The analysis and the graphics were carried out using the GraphPad Prism 6 program.

3. Results

3.1. Wood Characterizations

Various wood parameters can affect the outcome of the green gluing technique, and density is one such parameter. Table 1 below summarizes this parameter for samples with varying moisture content in finger jointing using the green gluing technique. The values of samples extracted from the finger joint and the opposite side of the finger were obtained.

Table 1. Physical parameters of wood and finger joint tested.

Physical Parameter	Statistical Tests	Finger Joint of Green Gluing	Wet Wood	Finger Joint of Dry Wood	Dry Wood
Oven-dry density (kg/m^3)	Average	0.511	0.534	0.541	0.548
	Standard deviation	0.046	0.045	0.039	0.037
	Size sample	14	14	16	16

The results suggest that the moisture had a low impact on the density using Nothofagus alpina. However, the water gradient into the wood could affect the gluing efficiency. On the other hand, the roughness of the wood (Table 2) showed significative statistical differences in the variance test among the groups (Table 3), which gives evidence of the influence of the roughness on the performance of the green gluing technique.

Table 2. Roughness (μm) values for finger-jointed wood under different treatments were assessed.

Statistical Tests	Finger Joint of Green Gluing	Wet Wood	Finger Joint of Dry Wood	Dry Wood
Average	12.345	8.707	9.650	10.497
Standard deviation	2.900	1.727	2.475	2.709
Size sample	69	70	80	80

Table 3. Analysis of variance for the roughness value.

Factors	Degrees of Freedom	Sum of Squares	Mean Square	F-Value	Probability	F Critical
Between groups	3	502.765	167.588	26.847	2.2169×10^{-15} *	2.635
Within groups	295	1841.500	6.242			
Total	298	2344.265				

* significant difference.

In Table 4, Tukey's mean comparison test identified significant differences in the roughness of the finger joint used for the green gluing technique compared to the other conditions. Additionally, there were significant differences in the roughness of dry wood compared to wet wood and the finger joint area used for the green gluing technique.

Table 4. Tukey test of roughness for the conditions tested.

	Finger Joint of Green Gluing	Wet Wood	Finger Joint of Dry Wood	Dry Wood
Finger-joint of green gluing	-	<0.001 *	<0.001 *	<0.001 *
Wet wood		-	0.0988 NS	<0.001 *
Finger-joint of dry wood			-	0.1421 NS
Dry wood				-

NS: non-significative. * significant difference.

3.2. Mechanical Test

The study's results that focused on the orientation effect of wood rings on the green gluing technique are presented in Table 5. The table includes the mean modulus of rupture (MOR), MOR coefficient of variation, mean modulus of elasticity (MOE), MOE coefficient of variation, and the type of failure. The types of failure indicated in Table 5 are presented in Figure 4.

Table 5. Mechanical results of the experimental design.

Condition	Orientation					
	R/T		R/R		T/T	
Dry	Mean MOR (MPa)	23.09	Mean MOR (MPa)	20.10	Mean MOR (MPa)	18.02
	COV	0.175	COV	0.396	COV	0.398
	Type of failure	c (100%)	Type of failure	b (20%) c (80%)	Type of failure	c (100%)
	Mean MOE (Mpa)	10,665.4	Mean MOE (Mpa)	10,319.97	Mean MOE (Mpa)	9842.91
	COV	0.99	COV	0.056	COV	0.148
Green	Mean MOR (MPa)	51.32	Mean MOR (MPa)	50.42	Mean MOR (MPa)	45.80
	COV	0.158	COV	0.074	COV	0.109
	Type of failure	a (25%) b (60%) c (15%)	Type of failure	a (30%) b (40%) c (30%)	Type of failure	a (50%); b (50%)
	Mean MOE (Mpa)	11,151.88	Mean MOE (Mpa)	11,813.25	Mean MOE (Mpa)	11,771.55
	COV	0.126	COV	0.092	COV	0.182

Different combinations of wood fiber orientation: Radial–Tangential (R/T); Radial–Radial (R/R); Tangential–Tangential (T/T).

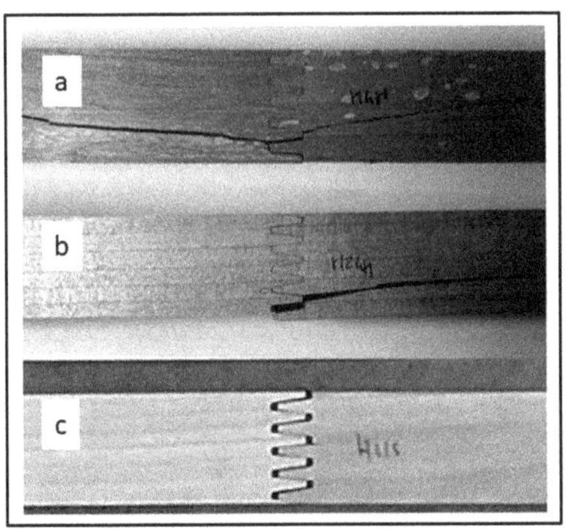

Figure 4. Types of failures observed: (**a**) wood failure crossing the joint zone; (**b**) failure initiated by the fingertip gap; and (**c**) separation of the fingers.

The comparison of the average MOR values supports the fact that the different combinations of orientation did not influence the results. This trend was observed in dry wood specimens and those made with green wood.

However, the ANOVA test noted a significant difference (p-value < 0.0001) when comparing dry and green wood samples. These results suggest that using green wood in the finger joint resulted in superior performance compared to using dry wood. The same conclusion was drawn when comparing the results of each orientation combination using wood with different moisture levels, as determined by an unpaired t-test with Welch's correction. On the other hand, the MOR value of the reference samples (wood of total length of finger joints) showed a much higher value, almost four times more (approximately 84.23 ± 17.25 MPa). This value confirms the excellent mechanical properties of the wood.

The modulus of elasticity was calculated based on each specimen's ultimate load and strain. This modulus was adjusted for each specimen's linear elastic behavior range. The results showed that the modulus of elasticity of the wet-glued specimens was similar to that of the standard specimen. However, the modulus of elasticity in bending for the wet-glued samples was higher than that of the dry-glued samples, although this difference was not statistically significant.

The variability index of the results indicated low values for both the wet and dry-glued samples, with 10.1% for the dry-glued samples and 13.3% for the wet-glued samples. It was also observed that the stiffness of wooden parts was higher when the gluing was realized using wet timbers, but the difference in stiffness between damp and dry wood was not as significant as the breaking strain.

Furthermore, the analysis of deformations revealed that the tension was proportional to the loads and higher in wet specimens. The tension was significant when comparing wet conditions in the same ring orientations. Additionally, the load that caused a deformation in all specimens exhibited a linear behavior without a range of inelastic behavior, suggesting an extremely brittle behavior.

Regarding the type of failure, the dry-glued specimens unexpectedly broke utterly at the joint (failure type "a") when using the R/T timber orientation with dry timber. This type of failure suggests a possible issue with the adhesion process. On the other hand, the green-glued specimens showed a mix of wood failure crossing the joint zone (B) and failure initiated by the finger gap (C). Some specimens had partial separation of the fingers, but never entirely, and only when other rupture modes occurred. Strain concentration at the tip gap of the fingers caused a rupture that followed the glue joint on the side of the fingers. However, a defective glue joint for the green-glued specimens did not initiate the rupture. The shift in the roughness in the zone of a finger could be responsible for these results.

3.3. Characterization by Microcomputed Tomography (Micro-CT)

Micro-CT was used to examine the internal morphology of finger joint samples made from native wood with varying moisture levels. The analysis revealed a wide range of wood porosity in dry and moist samples, ranging from 50% to 70%, despite being sourced from the same batch of wood (Figure 5). This situation could potentially impact the specimen performance, especially regarding adhesive penetration. However, no significant difference was observed in the thickness of the glued line between the samples with the same orientation but different humidity levels. This comparison was made in the end finger, allowing for better results. The adhesive covered a large surface area in this union zone, using between 75% and 97% of the surface area. All the relevant data are presented in Table 6.

The data on the porosity of the glue line and its thickness indicate that the fiber orientation does not affect the performance of the finger joints made with *Nothofagus alpina*. However, it is important to investigate why finger joints with green wood have better MOR and MOE values.

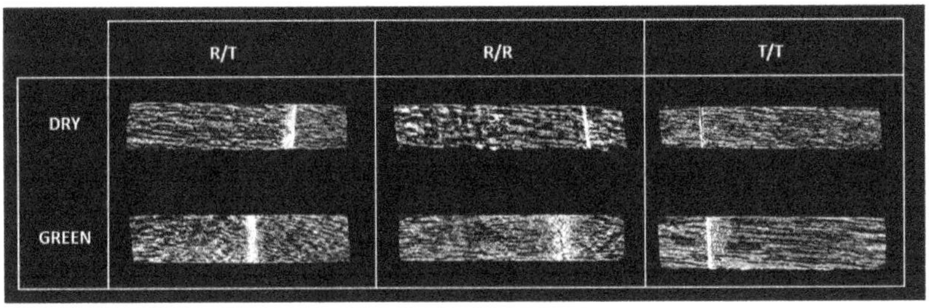

Figure 5. Porosity distribution with different timber orientations in finger joints under different moisture levels.

Table 6. Summarize the influence of different fiber orientations and moisture levels on the average porosity and thickness of the glue line in various types of finger joints.

Sample	Thickness of the Glued Line (μm)	Average Porosity (%)	Average Porosity in Glue Line (%)
Dry R/T	340 ± 72.1	56.7 ± 0.79	13.7
Green R/T	341.3 ± 120.7	59.6 ± 0.33	11.3
Dry R/R	229 ± 63.7	70.7 ± 0.29	9.4
Green R/R	350 ± 55.7	50.2 ± 0.48	1.49
Dry T/T	187.7 ± 51.9	66.3 ± 0.23	6.07
Green T/T	247 ± 61.4	52.5 ± 1.41	7.03

Different combinations of wood fiber orientation: Radial–Tangential (R/T); Radial–Radial (R/R); Tangential/Tangential (T/T).

4. Discussion

This study promotes the use and conservation of *Nothofagus alpina*, a species known for its excellent mechanical properties. However, its potential use is restricted due to the limited availability of specimens in the right size for different applications. The study aims to identify the best techniques to extend the length of the pieces and improve their mechanical response under green gluing conditions.

The wood's characterization showed an MOE and MOR value almost four times higher than the samples with finger joints, indicating good timber quality. However, unexpected results were observed in the finger joints sample.

Using the green gluing technique mixed with finger joints showed better behavior than the samples made with dry timbers (Table 5). According to the research, the green gluing technique effectively produces long-lasting pieces with good mechanical behavior. The data gathered show a significant difference in the MOR and tension values between the samples made with dry wood and those made with wet timber. The use of timber with high moisture content seems to improve the adhesion capacity of glue compared to other factors, as there was almost no difference in the line thickness and porosity between the samples with the same treatment but using timber with a different moisture content. The difference in the average porosity line was more significant in the samples with R/R orientation. Probably, the R/R orientation had a more substantial influence on the glue polymerization than the T/T orientation, where the difference was not noticeable (Table 6).

In a previous study, the samples made with dry wood performed better than those made with higher moisture contents using *Nothofagus alpina* [14]. However, the results of this research are contrary. The finger joint construction probably creates a localized drying effect and a gradient throughout the finger. This assumption is based on the thickness of the glued line between the specimens, which had similar values between dry and wet samples, without the influence of the ring orientations (Table 6). Moreover, previous backgrounds supported the hypothesis since it was demonstrated that the thickness of the glued line was almost double in wet samples compared to dry samples [14].

The study by Jakob et al. highlights the crucial role of fiber orientation in determining the mechanical properties of wood products [32]. The research indicates that unidirectional fiber orientation is necessary for high composite strength and stiffness. However, the present study found that fiber orientation did not significantly affect the MOR and MOE values. In the analysis of the mechanical behavior of the finger joint, we observe that when pure bending is applied, there are no shear stresses in the studied section. However, due to the shape of the connection, tensile stresses appear perpendicular to the fiber. This observation confirms that the bending resistance should be lower when the piece is arranged in the T/T plane than in the R/R arrangement. The tensile strength normal to the signature is higher in the radial plane than in the tangential plane. This finding suggests that fiber orientation would play a more significant role in the mechanical performance of composite materials than solid wood products. In practical applications, a structure such as cross-laminated timber (CTL) could distribute forces over a wide surface area and the wood fibers.

In contrast, finger joints could create rupture points that may concentrate forces and reduce mechanical performance. The fibers' orientation had a more significant impact on how well the adhesive penetrated the body of the finger, where the wood structure could differ from the end of the finger. This situation was shown in a study by Shirmohammadli et al. [33], which demonstrated how the cellular structure of softwoods and hardwoods can affect the penetration of 1C PUR adhesive.

Choosing the correct type of glue is essential for achieving the best results with green gluing. The selection can be a difficult decision because the performance of the glue depends on many factors, including the type of wood, wood density, application process, surface treatment, and the final timber product [8,14,34,35]. In this research, the glue selected had previously been tested with good performance [14], whereby the possibility of a wrong selection was discarded. Despite this, moisture can significantly affect the performance of adhesives in wood bonding, impacting both short-term and long-term bond integrity. In this study, a high moisture content in wood may inhibit adhesive penetration. For example, adhesives like urea-formaldehyde (UF) [36] and polyvinyl acetate (PVA) [37] may not penetrate as deeply into wet wood fibers, resulting in weaker mechanical interlocking and reduced bond strength. Wood's moisture content can affect its mechanical properties, including tensile strength. It can also impact the bonding strength of adhesives and the overall performance of finger-jointed lumber. Generally, wood with higher moisture content tends to have lower mechanical strength due to reduced bonding efficiency and potential for dimensional changes. Therefore, the drying process of the wood influences the bonding and could be relevant to ensure consistent and reliable results in terms of tensile strength under high-temperature conditions [38].

However, the mechanical results may be attributed to changes in roughness in the finger joint area and shifts in pore size, which may have influenced the green gluing performance more than the type of glue used and the moisture content in this study.

To confirm this suggestion, a more thorough study should be carried out that includes various microscopy techniques and adhesives. These considerations will help to determine whether the changes observed could be due to the formation of intra-chain hydrogen bonds at dehydrated crystallite surfaces caused by drying cycles. These bonds may cause conformational changes within the cellulose chains and increase packing density. A similar phenomenon was previously reported by other authors [39,40].

The construction of finger joints was the most relevant action for this research. This step is influenced by the finger's characteristics and the wood's drying process. In this sense, the shape and size of the finger probably helped in the local drying process, triggering a better performance in the wet samples. This behavior was reflected in a higher MOR value compared to the samples made with the dry wood. In the same way, the results in the glue line could support this suggestion because the thickness of the glue line was similar between the samples with the same orientations but with different moisture levels (Table 6, Figure 5). The orientation probably limited the adhesive penetration, but the union capacity was given by the moisture localized in the body of the finger.

The study by González-Prieto et al. [41] reported similar findings. They made green-glued finger joints using Eucalyptus globulus wood and a single-component polyurethane adhesive. The study evaluated the mechanical behavior of the joints and found that the moisture state was a significant factor. The finger joints made in a moist environment showed higher elastic modulus and resistance compared to those made in a dry environment. The results suggest that these joints had resistance to inflection, similarly to solid pieces without finger joints for this species. The authors concluded that using finger joints in green castings is feasible and can improve the quality of the joint for hardwood species.

It should be noted that finger-jointed pieces have reduced bending strength compared to a solid piece of wood. In this case, wet-glued pieces showed an average drop in strength of 41.6%, while dry-glued pieces showed a drop in bending strength of 75.8%. This behavior can be attributed to the discontinuity of the piece, which makes it structurally weaker. However, it should be noted that the green gluing technique combined with finger joints still requires significant improvement. The MOR and MOE values need to be higher than 90% of solid wood in order to be used for structural applications. Creating smaller structures that meet the service class 3 wood standards is more feasible.

Although the wet specimens showed the best results in the mechanical test, there is still a long way to go. The green gluing technique involves various areas of knowledge [8]. One important consideration is the characteristics of wood, where density plays a significant role. Previous research by Alia-Syahirah et al. [42] reported that higher density can harm adhesion in wood products. This finding is consistent with the results of Ahmad et al. [13], who found that denser wood species have worse bonding characteristics than less dense ones. However, the density was not a relevant factor in the finger joint construction using *Eucalyptus globulus* wood with a high moisture content [41]. These findings suggest that the gluing procedure must be adapted depending on the species studied.

This study used a preliminary approach to identify the factors that influence surface roughness. The results of the multiple linear regression model will help control these factors better, thus improving the performance and application of green gluing using finger joints in future projects. It has been reported by Yu et al. [43] that surface roughness can significantly impact the ability of glue to wet wood surfaces. This effect should be evaluated more accurately to determine the best way to apply the green gluing technique to Nothofagus species.

In future studies, it is essential to investigate how moisture levels affect the bending modulus of elasticity and compressive strength of wood. Research has shown that the pliability of treated wood differs significantly depending on its moisture content since the most favorable pliability is achieved when bending the wood at or close to its fiber saturation point [44].

5. Conclusions

Green gluing is a method that could increase the value of various wood species to increase the size of wood pieces. In this study, it was found that the finger joint samples performed best when made with timber that had high moisture levels. Future studies must examine the influence of moisture close to the fiber saturation point on the finger joint, the effect of surface roughness, its mechanical response, and the drying process.

Research and development in this technology will enable us to produce higher-quality components, maximize our use of local wood resources, and encourage improved forest management. This method could create laminated beams and decorative and construction panels. It also has applications in the production of indoor and outdoor furniture.

Author Contributions: Conceptualization, G.R.-G. and R.P.; methodology, G.R.-G., V.R. and R.P.; formal analysis, P.-L.C., B.N., C.M. and C.G.; investigation, C.G., C.A. and C.M.; resources, G.R.-G., V.R. and R.P.; writing—original draft preparation, P.-L.C. and C.G. All authors contributed to the review and editing of the manuscript. All authors have read and agreed to the published version of the manuscript.

Funding: The authors want to acknowledge funding from Centro Nacional de Excelencia para la Industria de la Madera (CENAMAD, ANID BASAL FB210015). In particular, Víctor Rosales, Gonzalo Rodríguez-Grau, Claudio Montero, Claudia Alvarado, Carlos Galarce, and Régis Pommier thank the financial support of ANID + FONDEF/Concurso IDeA I+D, FONDEF/ANID 2021 ID 21 I 10241.

Institutional Review Board Statement: Not applicable.

Informed Consent Statement: Not applicable.

Data Availability Statement: The data presented in this study are available upon request from the corresponding author.

Acknowledgments: The authors would like to acknowledge the technical support of Maria Jose Salas (FONDEQUIP project EQM 130028) in the micro-CT measurements. Also, the authors thank FabLab of Centro de innovación UC Anacleto Angelini for permitting the use of their facilities.

Conflicts of Interest: The authors declare no conflicts of interest.

References

1. Cuadrado, J.; Zubizarreta, M.; Pelaz, B.; Marcos, I. Methodology to Assess the Environmental Sustainability of Timber Structures. *Constr. Build. Mater.* **2015**, *86*, 149–158. [CrossRef]
2. Gilbert, B.P.; Davies, T.E.; McGavin, R.L.; Dowse, C.J. Towards Reducing the Capital Cost of Manufacturing Laminated Veneer Lumbers: Investigating Finger Jointing Solutions. *Constr. Build. Mater.* **2024**, *411*, 134158. [CrossRef]
3. Hou, J.; Taoum, A.; Kotlarewski, N.; Nolan, G. Study on the Effect of Finger-Joints on the Strengths of Laminations from Fiber-Managed Eucalyptus Nitens. *Forests* **2023**, *14*, 1192. [CrossRef]
4. Clouet, B. Comportement Hydromécanique d'assemblages Bois Collés à l'état Vert: Approches Expérimentale et Numérique. Ph.D. Thesis, Université de Bordeaux, Bordeaux, France, 2014.
5. Romero, A.; Odenbreit, C. Experimental Investigation on Strength and Stiffness Properties of Laminated Veneer Lumber (LVL). *Materials* **2023**, *16*, 7194. [CrossRef] [PubMed]
6. De Araujo, V.; Aguiar, F.; Jardim, P.; Mascarenhas, F.; Marini, L.; Aquino, V.; Santos, H.; Panzera, T.; Lahr, F.; Christoforo, A. Is Cross-Laminated Timber (CLT) a Wood Panel, a Building, or a Construction System? A Systematic Review on Its Functions, Characteristics, Performances, and Applications. *Forests* **2023**, *14*, 264. [CrossRef]
7. Hermawan, A.; Mohammad Sofi, A.Z.; Roszalli, M.N. Performance of Glued Laminated Timber (Glulam) Made from Rubberwood with Different Lamina Assembly Patterns and Adhesive Spreads Rates. In *Proceedings of the IOP Conference Series: Earth and Environmental Science*; Institute of Physics: London, UK, 2023; Volume 1145.
8. Rodríguez-Grau, G.; Marín-Uribe, C.; Cortés-Rodríguez, P.; Montero, C.; Rosales, V.; Galarce, C. Bibliometric Analysis of the Green Gluing Technique (2000–2020): Trends and Perspectives. *Forests* **2022**, *13*, 1714. [CrossRef]
9. Sterley, M. *Green Gluing of Wood*; KTH—Royal Institute of Technology: Stockolm, Sweden, 2004.
10. Lissouck, R.O.; Pommier, R.; Castéra, P.; Ayina Ohandja, M.L. Timber Engineering as a Tool for Species in Tropical Rain Forests: The Case of Congo Basin Forest. In Proceedings of the World Conference on Timber Engineering, Auckland, New Zealand, 15–19 July 2012.
11. Bidzo, C.H.N.; Nziengui, C.F.P.; Ikogou, S.; Kaiser, B.; Pitti, R.M. Mechanical Behavior of Tropical Glued Laminated Timber Beams with Fingers Joints. *Procedia Struct. Integr.* **2022**, *37*, 447–452. [CrossRef]
12. Morin-Bernard, A.; Blanchet, P.; Dagenais, C.; Achim, A. Glued-Laminated Timber from Northern Hardwoods: Effect of Finger-Joint Profile on Lamellae Tensile Strength. *Constr. Build. Mater.* **2021**, *271*, 121591. [CrossRef]
13. Ahmad, Z.; Lum, W.C.; Lee, S.H.; Razlan, M.A.; Mohamad, W.H.W. Mechanical Properties of Finger Jointed Beams Fabricated from Eight Malaysian Hardwood Species. *Constr. Build. Mater.* **2017**, *145*, 464–473. [CrossRef]
14. Rosales, V.; Rodríguez-Grau, G.; Galarce, C.; Montero, C.; Alvarado, C.; Muñoz, L.; Pommier, R. Feasibility of Bonding High-Moisture-Content Wood Using Nothofagus Chilean Species. *Forests* **2023**, *14*, 2386. [CrossRef]
15. Sterley, M.; Gustafsson, P.J. Shear Fracture Characterization of Green-Glued Polyurethane Wood Adhesive Bonds at Various Moisture and Gluing Conditions. *Wood Mater. Sci. Eng.* **2012**, *7*, 93–100. [CrossRef]
16. Pommier, R.; Elbez, G. Finger-Jointing Green Softwood: Evaluation of the Interaction between Polyurethane Adhesive and Wood. *Wood Mater. Sci. Eng.* **2006**, *1*, 127–137. [CrossRef]
17. Ayarkwa, J.; Hirashima, Y.; Sasaki, Y.; Yamasaki, M. Influence of Finger-Joint Geometry and End Pressure on Tensile Properties of Three Finger-Jointed Tropical African Hardwoods. *S. Afr. For. J.* **2000**, *188*, 37–49. [CrossRef]
18. Tsalagkas, D.; Börcsök, Z.; Pásztory, Z. The Effect of Finger Length and Adhesive Type on the Curing Time of Finger-Jointed Black Pine (*Pinus nigra* L.) Lumber. *Wood Mater. Sci. Eng.* **2021**, *16*, 312–320. [CrossRef]
19. Bustos, C.; Hernández, R.E.; Beauregard, R.; Mohammad, M. Effects of End-Pressure on the Finger-Joint Quality of Black Spruce Lumber: A Microscopic Analysis. *Maderas. Cienc. Tecnol.* **2011**, *13*, 319–328. [CrossRef]

20. Groom, L.H.; Leichti, R.J. Effect of Adhesive Stiffness and Thickness on Stress Distributions in Structural Finger Joints. *J. Adhes.* **1994**, *44*, 69–83. [CrossRef]
21. Faircloth, A.; Kumar, C.; McGavin, R.; Gilbert, B.; Leggate, W. Mechanical Performance and Bond Integrity of Finger Jointed High-Density Sub-Tropical Hardwoods for Residential Decking. *Forests* **2023**, *14*, 956. [CrossRef]
22. *NCh2245:2015*; Hoja de Datos de Seguridad para Productos Químicos—Contenido y Orden de las Secciones. Instituto Nacional de Normalización: Santiago, Chile, 2015.
23. Grabianowski, M.; Manley, B.; Walker, J.C.F. Acoustic Measurements on Standing Trees, Logs and Green Lumber. *Wood Sci. Technol.* **2006**, *40*, 205–216. [CrossRef]
24. Balmori Roiz, J.A.; Acuña Rello, L.B.; Otero, L.A. Estudio de La Influencia de La Dirección de La Fibra En La Velocidad de Propagación de Ultrasonidos (FAKOPP) En Madera Estructural de "*Pinus sylvestris* L." y "*Pinus radiata* D. Don.". In Proceedings of the 6th Rehabend Congress, Burgos, Spain, 24–27 May 2016.
25. *NCh176/1*; Madera-Parte 1: Determinación Del Contenido de Humedad. Instituto Nacional de Normalización: Santiago, Chile, 2019.
26. *NCh176/2*; Madera-Parte 2: Determinación de Densidad. Instituto Nacional de Normalización: Santiago, Chile, 1988.
27. He, S.; Lin, L.; Wu, Z.; Chen, Z. Application of Finite Element Analysis in Properties Test of Finger-Jointed Lumber. *J. Bioresour. Bioprod.* **2020**, *5*, 124–133. [CrossRef]
28. *NCh 2148*; Madera Laminada Encolada Estructural—Requisitos, Métodos de Muestreo e Inspección. Instituto Nacional de Normalización: Santiago, Chile, 2013.
29. *EN 408:2010+A1:2012*; Timber Structures—Structural Timber and Glued Laminated Timber—Determination of Some Physical and Mechanical Properties. European Committee for Standardization: Brussels, Belgium, 2010.
30. Meseguer, A.G. *Hormigon Armado (Vol. II): Calculos en Estados Limite*; Fundación Escuela de la Edificación: Madrid, Spain, 2001.
31. Kazup, Á.; Fegyverneki, G.; Gácsi, Z. Evaluation of the Applicability of Computer-Aided Porosity Testing Methods for Different Pore Structures. *Metallogr. Microstruct. Anal.* **2022**, *11*, 774–789. [CrossRef]
32. Jakob, M.; Mahendran, A.R.; Gindl-Altmutter, W.; Bliem, P.; Konnerth, J.; Müller, U.; Veigel, S. The Strength and Stiffness of Oriented Wood and Cellulose-Fibre Materials: A Review. *Prog. Mater. Sci.* **2022**, *125*, 100916. [CrossRef]
33. Shirmohammadli, Y.; Pizzi, A.; Raftery, G.M.; Hashemi, A. One-Component Polyurethane Adhesives in Timber Engineering Applications: A Review. *Int. J. Adhes. Adhes.* **2023**, *123*, 103358. [CrossRef]
34. Clerc, G.; Brülisauer, M.; Affolter, S.; Volkmer, T.; Pichelin, F.; Niemz, P. Characterization of the Ageing Process of One-Component Polyurethane Moisture Curing Wood Adhesive. *Int. J. Adhes. Adhes.* **2017**, *72*, 130–138. [CrossRef]
35. Yang, G.; Gong, Z.; Luo, X.; Chen, L.; Shuai, L. Bonding Wood with Uncondensed Lignins as Adhesives. *Nature* **2023**, *621*, 511–515. [CrossRef] [PubMed]
36. Park, S.; Jeong, B.; Park, B.-D. A Comparison of Adhesion Behavior of Urea-Formaldehyde Resins with Melamine-Urea-Formaldehyde Resins in Bonding Wood. *Forests* **2021**, *12*, 1037. [CrossRef]
37. Iždinský, J.; Reinprecht, L.; Sedliačik, J.; Kúdela, J.; Kučerová, V. Bonding of Selected Hardwoods with PVAc Adhesive. *Appl. Sci.* **2020**, *11*, 67. [CrossRef]
38. Yue, K.; Wang, F.; Lu, W.; Tang, Z.; Chen, Z.; Liu, W. Tensile Properties of Finger-Jointed Lumber under High-Temperature and Oxygen-Free Conditions. *Holzforschung* **2021**, *75*, 838–846. [CrossRef]
39. Hill, S.J.; Kirby, N.M.; Mudie, S.T.; Hawley, A.M.; Ingham, B.; Franich, R.A.; Newman, R.H. Effect of Drying and Rewetting of Wood on Cellulose Molecular Packing. *Holzforschung* **2010**, *64*, 421–427. [CrossRef]
40. Tomad, J.; Leelatanon, S.; Jantawee, S.; Srisuchart, K.; Matan, N. Internal Stress Development within Wood during Drying: Regime and Kinetics. *Dry. Technol.* **2023**, *41*, 77–88. [CrossRef]
41. González-Prieto, O.; Casas Mirás, J.M.; Torres, L.O. Finger-Jointing of Green Eucalyptus Globulus L. Wood with One-Component Polyurethane Adhesives. *Eur. J. Wood Wood Prod.* **2022**, *80*, 429–437. [CrossRef]
42. Alia-Syahirah, Y.; Paridah, M.T.; Hamdan, H.; Anwar, U.M.K.; Nordahlia, A.S.; Lee, S.H. Effects of Anatomical Characteristics and Wood Density on Surface Roughness and Their Relation to Surface Wettability of Hardwood. *J. Trop. For. Sci.* **2019**, *31*, 269–277. [CrossRef]
43. Yu, Q.; Pan, X.; Yang, Z.; Zhang, L.; Cao, J. Effects of the Surface Roughness of Six Wood Species for Furniture Production on the Wettability and Bonding Quality of Coating. *Forests* **2023**, *14*, 996. [CrossRef]
44. Báder, M.; Németh, R. Moisture-Dependent Mechanical Properties of Longitudinally Compressed Wood. *Eur. J. Wood Prod.* **2019**, *77*, 1009–1019. [CrossRef]

sustainability

MDPI

Review

Bibliometric Review of Prefabricated and Modular Timber Construction from 1990 to 2023: Evolution, Trends, and Current Challenges

Nohelia Gutiérrez [1,*], João Negrão [2], Alfredo Dias [1] and Pablo Guindos [3]

[1] Department of Civil Engineering, ISISE, Universidade de Coimbra, 3030-788 Coimbra, Portugal
[2] Department of Civil Engineering, Universidade de Coimbra, 3030-788 Coimbra, Portugal; jhnegrao@dec.uc.pt
[3] Centro Nacional de Excelencia para la Industria de la Madera (CENAMAD), Pontificia Universidad Católica de Chile, Santiago 7820436, Chile
* Correspondence: nohelia.gutierrez@uc.pt

Abstract: Due to its inherent characteristics, such as a light weight and ease of workability, timber is ideal for prefabricated and modular construction. However, since the research in this field covers a wide range of niches such as structural engineering, building physics, design for assembly and disassembly, and life cycle analysis, among others, and since there has been considerable development of the field in past years—boosted by new mass timber products and tall timber construction—it is difficult to critically analyze the current state of the art, current trends, and research challenges. Therefore, this research aimed to cover a systematic review of 409 articles to assess the field of prefabricated and modular timber construction critically. The methodology comprised a co-word network approach using the Science Mapping Analysis Software Tool (SciMAT, Version 3) to illustrate their evolution from 1990 to 2023. The findings show that the circular economy and digital technologies significantly impact the development of these technologies, which can potentially provide practical solutions for designing buildings with a circular approach and improving productivity and efficiency in the construction process. However, it is essential to acknowledge a notable deficiency in the research and understanding of these subjects. Therefore, various sectors must take the lead in conducting a thorough reassessment to enhance research and development in the field. Finally, the findings from this research can significantly contribute to existing knowledge and serve as a comprehensive platform for the further exploration of prefabricated and modular timber construction.

Keywords: prefabrication; modular timber construction; circular economy; digital technologies

1. Introduction

Off-site construction, also commonly referred to as prefabrication, involves manufacturing various structural components within a controlled factory environment, subsequently facilitating the transportation of complete assemblies or subassemblies to the designated construction site where the intended structure is to be established [1]. Modular construction is widely recognized as a fundamental approach within offsite construction [2]. This technology can achieve a remarkable level of completion, reaching upwards of 95% before its transportation to the designated location [1]. While some researchers state that modular construction is constituted by a dimensionally unified spatial unit or a 3D volumetric section [3–6], others describe it in the form of 2D planar components (pre-assembled elements) or fully volumetric units [7–9]. From the viewpoint of these authors, modules can cover a diverse range of prefabricated elements or systems, including component groups and complete units of significant size and volume.

These two approaches will be appropriate depending on the demands of each project. The assembly work involved in a 2D panelized system is intricate due to its requirement for additional internal finishing. Nevertheless, it provides superior flexibility. In contrast, volumetric units require adaptation to building specifications, which demands early involvement in the design process to ensure the desired result. Still, it has the potential for the most favorable efficiency and time savings [10,11]. If existing criteria are no longer satisfied, prefabricated elements or systems can be removed individually for expansion and reused to produce new buildings, providing the opportunity for adaptability.

Even though the materials utilized in modular buildings are generally identical to those employed in on-site installations, they exhibit enhanced quality since the components are fabricated within regulated conditions [12]. Timber constructions possess a notable competitive edge in this aspect due to their substantial degree of prefabrication (quick and precise assembly), lightweight nature [13,14], and simplicity of connections [15]. Furthermore, timber structures "*can be renewed, manipulated, and laminated*" [1]. These are favorable conditions for the effective recycling and repurposing of complete architectural elements, specific sections of buildings, and individual components [16], reducing expenses and shortening installation schedules [17]. In addition, the waste produced can be reused in the manufacturing process for various applications, which ensures a significantly cleaner and more regulated environment. Simultaneously, it enhances health and safety because the operational process is greatly minimized [15]. Given all these benefits, timber may be considered an optimal modularization material. Therefore, the utilization of timber as an off-site material has been extensively researched and promoted in the industry.

Recently, there has been a comprehensive evaluation of prefabricated/modular structures, examining various aspects related to management strategies [18–20], structural systems in high-rise buildings [7], building design [8], environmental and economic performance [21], as well as the features of modular construction utilizing exclusively cross-laminated timber [22]. However, the specific advancements and recent innovations in prefabricated/modular timber structures, which include light-frame, mass timber, and prefabricated hybrid elements, systems, or structures have not been reviewed and critically analyzed to identify current trends and challenges in the field.

Therefore, the objective of this research entails the assessment of the current state of the art regarding modular timber construction, analyzing its evolution as a research field and critically assessing the current trends and challenges. This contribution aims to assist researchers in understanding the current state of technology and orient them towards researching the most impactful needs. To achieve this, the applied methodology included the following: (i) to quantitatively evaluate the existing literature; (ii) to develop a scientific mapping analysis utilizing a co-word network approach; (iii) to establish the evolution of the underlying themes within this domain; (iv) to identify future trends and gaps in the current literature; and (v) to provide valuable insights and directions for future research endeavors. The organization of the article encloses a comprehensive methodology section (Section 2), a scientometric analysis and presentation of the results (Section 3), a critical analysis of the results (Section 4), and a conclusion section (Section 5), which encapsulates the essential findings and potential approaches to further investigation.

2. Methodology

2.1. Bibliographic Search

The initial phase involved acquiring data and selecting the primary source of the Web of Science (WOS) collection database. This database was chosen due to its reputation for including many valuable publications, such as journals, book chapters, conference proceedings, and other scholarly works [23]. A comprehensive set of keywords was developed by drawing upon the relevant literature and diverse reviews [7,18,24,25]. These keywords were then categorized into two themes delineated in Table 1: (a) prefabricated construction/modular construction and (b) wood. In the case of the former, terms associated with prefabrication and modularization were also considered to ensure the inclusion of all pertinent information. WOS includes multiple search criteria, and the following were taken into consideration: (a) the topic, which encompasses various parameters such as the title, abstract, author keywords, and Keywords Plus (additional keywords automatically generated by the WOS website concerning the research topic); (b) the abstract, which provides a summary of the existing knowledge about the subject matter under investigation; and (c) the title, which encapsulates keywords that reflect the main issues addressed in the study [26]. As for the relationship between the keywords, the *"OR"* connector was utilized for the terms prefabricated construction/modular construction and their associated terms, while the *"AND"* connector was employed to incorporate the term wood. Therefore, the selected literature must contain at least one keyword from each theme.

Table 1. Selected keywords, connectors, and associated terms applied to the search criteria.

Theme 1	Connector	Theme 2
Topic: prefab* pre-fab* modular* *assembl* *OR* Abstract: "offsite construction" "off-site construction" "offsite manufacture" "off-site manufacture" "modular building" "modular construction" "modular technology" "prefab building" "prefab construction" "pre-fab building" "pre-fab construction" "industrialized construction" "industrialized building" "panelized construction" "tilt-up construction" "volumetric construction" "prefinished construction" "preassembly building" "preassembly construction" "plug & play modular"	*AND*	Title: timber* OR wood*

By defining the search domain based on the parameters above (topic, abstract, and title), a total of 3573 publications were retrieved. However, the search was restricted to journal articles, as these documents are considered high-quality research studies [21]. Moreover, the search was limited to specific fields, namely, civil engineering, construction building technology, architecture, green sustainable science technology, engineering manufacturing, management, and industrial engineering. The selected information was limited to publications written in English. Finally, the analysis period was set from 1990 to November/2023 to elucidate the development and evolution of prefabrication with wood and its connection to future trends, obtaining 623 articles. Subsequently, these articles underwent analysis, resulting in the exclusion of papers focusing on timber shells, timber grid shells, folded structures, wooden domes, or timber textile modules, as these elements or structures did not align with the definition of prefabrication levels (linear, panelized systems, or volumetric) [4,15]. Furthermore, the study did not embrace research on elements or structures based on other materials, even if it included relevant parts fabricated from wood or cellulose-based materials, plastic, or biocarbon, which are increasingly attracting attention for their utilization in the sustainable advancement of functional materials [27,28].

It is important to emphasize that the compiled publications must examine several aspects of the performance of prefabricated/modular timber elements or structures. These aspects included the structural, acoustic, hygrothermal, and thermal characteristics. Furthermore, the scope of the analysis covered single-level and multi-level constructions made of light-frame and mass timber, including cross-laminated timber (CLT), nail-laminated timber (NLT), dowel-laminated timber (DLT), glue-laminated timber (GLT), and laminated veneer lumber (LVL). The investigation also explored hybrid elements combining mass timber with steel or concrete. In addition to these technical aspects, the review focused on the design, construction, production, management, and life cycle assessment (LCA) of prefabricated and modular timber structures. Finally, a comprehensive final assessment was carried out based on a thorough examination, leading to the identification of 11 papers that lacked the author's keywords, which were vital to developing the scientometric analysis. Therefore, after the selection criteria, a total of 409 articles were considered.

2.2. Scientometric Analysis

The subsequent step involved executing a scientific mapping analysis based on a co-word network using Scientific Mapping Analysis Software Tool (SciMAT), software developed by the University of Granada (Spain) in 2011. According to [29], this tool incorporates a range of essential functionalities such as data retrieval, data preparation, network extraction, network normalization, mapping, analysis, and visualization. In this investigation, the organization of co-word analysis facilitated an in-depth exploration of the conceptual framework of the topic through the associations between keywords. This powerful technique allowed for obtaining insights into the principal investigation concepts while also enabling distinguishing the trajectory of their evolution and the interaction between scientific research themes [30]. These research themes represent different research fields and can be visualized on the two-dimensional strategic diagram in Figure 1.

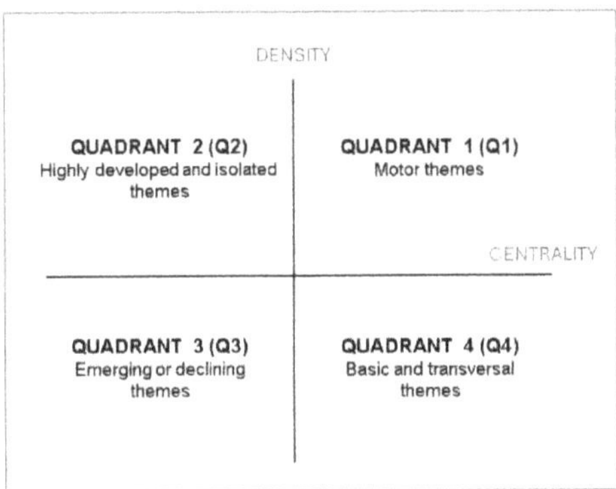

Figure 1. Strategic diagram adapted from [31].

3. Results

This section presents an interpretation and examination of the outcomes originating from the co-word network of the scientific mapping analysis. This outline is structured into two distinct subsections: (i) a comprehensive thematic analysis conducted by periods with the relationship between the diverse themes, and (ii) an assessment of the thematic conceptual evolution.

3.1. Thematic Analysis Conducted by Periods

As stated before, the investigation involved the analysis of articles published between 1990 and 2023. The research was divided into three subperiods: (i) 1990–2004, (ii) 2005–2019, and (iii) 2020–2023 (see Table 2). Particular attention was given to CLT's emergence and subsequent development as an industrial product, which commenced in 2005 [32]. Furthermore, to conclude the same subperiod, constructing the tallest all-timber Mjøstårnet building in 2019, located in Norway, marked another crucial milestone. It is important to note that, as defined by the Council on Tall Buildings and Urban Habitat (CTBUH), an all-timber building is characterized by having its main vertical and lateral structural elements, as well as its floor spanning systems, constructed primarily from timber while other materials, such as concrete elements, play a secondary role in the structure [33]. The three subperiods provide a comprehensive understanding of the topic and illustrate the progression of various factors such as the number of documents, number of citations, h-index, and average citations attained.

Table 2. Subperiods from 1990 to 2023 and the number of publications developed during the subperiods.

Subperiod Name	Number of Documents
1990–2004	11
2005–2019	198
2020–2023	200

3.1.1. Subperiod 1990–2004

1. Quantitative measures and strategic diagram.

The 11 documents of this period (about 2.7% of all published papers) were linked with specific themes; thus, it can be observed that this subperiod exhibits a relatively low number of keywords associated with each document. Following the strategic diagram illustrating the characteristics of this initial period (see Figure 2), it becomes evident that there are no motor themes (1st quadrant) or highly developed themes (2nd quadrant). Additionally, there is an absence of emerging or declining themes (3rd quadrant) and basic themes (4th quadrant) that guide the field of prefabricated timber construction. However, this period is focused on two distinct clusters. The first cluster, denoted as TIMBER-STRUCTURES (within the 1st and 2nd quadrants of the strategic diagram), can be considered the motor topic of the research. The second cluster, DESIGN, appears to possess characteristics of a motor or basic theme (within the 1st and 4th quadrants of the strategic diagram). Upon closer examination of the quantitative measures presented in Table 3, the first theme has the highest number of citations, amounting to 38, while the second theme has 27 related publications. As previously mentioned, due to the limited number of papers associated with this specific period's themes, both exhibit relatively low h-index scores, further emphasizing the scarcity of scholarly contributions in this domain.

Table 3. Performance measures for the themes of the subperiod 1990–2004.

Theme Name	Number of Documents	Number of Citations	Average of Citations	h-Index
TIMBER-STRUCTURES	3	38	12.66	3
DESIGN	3	27	9	3

2. Thematic network
    - Themes
    The network graph analysis reveals that the clusters of TIMBER-STRUCTURES and DESIGN (Figure 3) are highly interconnected (in other words, each cluster focuses on a specific topic, and all their keywords are related between them). The first cluster is related to various aspects and components involved in constructing and evaluating timber structures, including *steel*, *fiber-plaster-boards*, *trusses*, *walls*, *experimental-test*, *fem-analysis*, *structural-performance*, and *joints*. On the other hand, the second cluster, centered around the efficient performance of structures and design processes, delves into concepts such as *resistance*, *reliability*, *construction*, *civil-engineering*, *behavior*, *loads*, and *connections*.

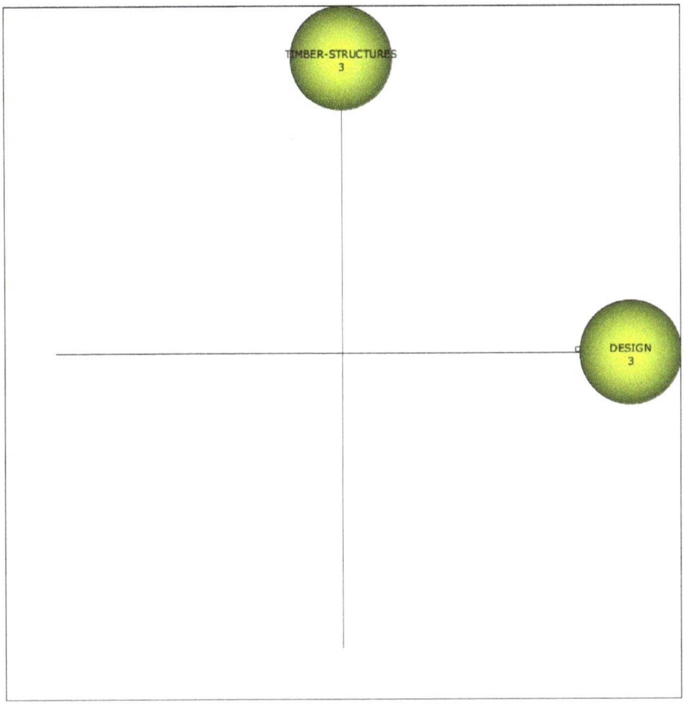

Figure 2. Strategic diagram for the 1st subperiod with two distinct clusters (1990–2004).

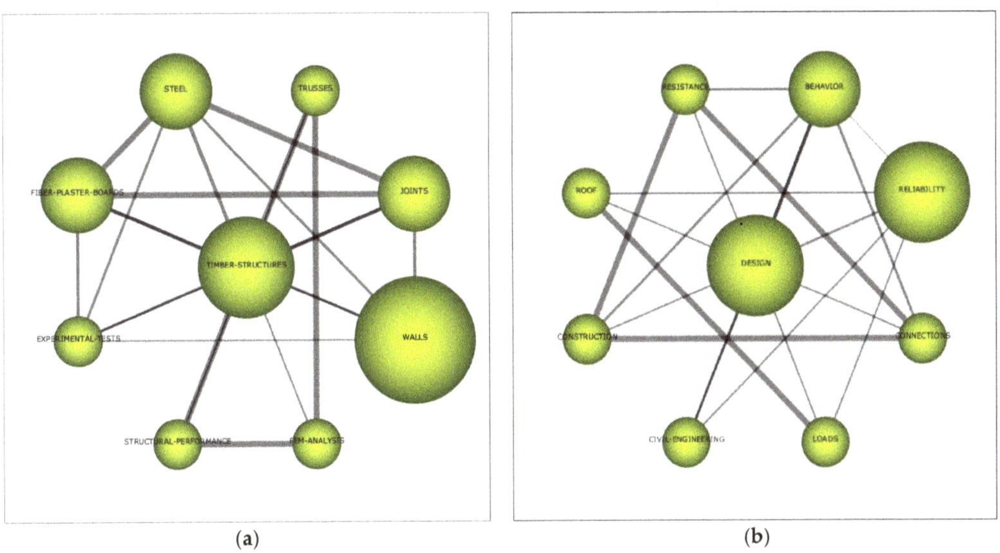

Figure 3. Cluster networks of the 1st period: (**a**) TIMBER-STRUCTURES; (**b**) DESIGN.

- Relationship between themes with most-cited articles
  The collection of keywords provides valuable insights into the research focus during this period. For example, Dobrila and Premrov [34] discussed the im-

portance of the composite timber frame–fiberboard panel wall design. Their study demonstrated that inserting diagonal steel strips attached to the wooden structure represents the most effective method to enhance the resistance and flexibility of the elements. Among the investigations that concentrate on finite element method (FEM) analysis in timber structures, the work of [35] stood out due to its introduction of an analytical technique for examining the behavior of light-frame wood structures when subjected to static loads. The study primarily focused on the behavior of substructures and system components that have not been previously included in a comprehensive structural model. The findings indicate remarkable similarities between the experimental and analytical outcomes. Similarly, Tarabiate and Itani [36] described a three-dimensional computer model for evaluating light-frame wood constructions. Their study showcased the model's capability to explore various structural configurations and generate design requirements for wood structures.

3.1.2. Subperiod 2005–2019

1. Quantitative measures and strategic diagram.

One hundred and ninety-eight (198) documents were published within the research focused on prefabricated timber construction in the specified subperiod, representing a significant increase compared to the initial period. The comprehensive analysis of these publications revealed the presence of nine distinct research themes, which are subsequently described through the quantitative measurements presented in Table 4 and strategic diagrams illustrated in Figure 4. Within the 1st quadrant of the strategic diagram, the theme CONNECTIONS emerged as a highly relevant and influential area of study (motor theme). This theme achieved an impressive number of citations, amounting to 2028, making it the most frequently referenced theme. Furthermore, it had the highest number of published articles, with a total of 76, thus solidifying its substantial impact on the research field, as evidenced by its high h-index score of 26. Another motor theme within this subperiod is LIFE-CYCLE, which, despite being ranked fourth in the number of citations (517), exhibited a low number of published articles related to the topic (13). For this reason, it presented a relatively low impact within the research field, as indicated by its h-index score of 11.

Table 4. Performance measures for the themes of the subperiod 2005–2019.

Theme Name	Number of Documents	Number of Citations	Average of Citations	h-Index
CONNECTIONS	76	2028	26.684	26
MODELING	26	521	20.038	14
WALLS	24	454	18.917	13
INDUSTRIALIZED-CONSTRUCTION	21	723	34.429	14
NUMERICAL-MODELING	16	215	13.438	10
HYGROTHERMAL-PERFORMANCE	14	339	24.214	11
LIFE-CYCLE	13	517	39.769	11
THERMAL-PERFORMANCE	11	225	20.455	9
CONCRETE	10	353	35.3	8

In the second quadrant of the strategic diagram, there are highly developed and isolated themes, HYGROTHERMAL-PERFORMANCE, MODELING, and THERMAL-PERFORMANCE. HYGROTHERMAL-PERFORMANCE ranks in the seventh position regarding the number of citations (339). Similarly, MODELING and THERMAL-PERFORMA NCE rank in the eighth and ninth positions, respectively, with 225 and 215 citations. The theme WALL, situated within the 3rd quadrant of the strategic diagram, can be perceived as either an emerging or declining theme despite occupying the fifth position regarding the number of citations, with a count of 454. It does not make a substantial impact, as evidenced by its relatively low h-index score of 13. Finally, in the 4th quadrant, one notes themes deemed relevant to the research field (basic and transversal themes). INDUSTRIALIZED-CONSTRUCTION emerges as the central and most basic theme, with 723 citations. The next theme is MODELING, with 521 citations, while CONCRETE ranks sixth in the number of citations, with a count of 353. It is crucial to emphasize that the themes positioned within the last three quadrants (2nd, 3rd, and 4th) exhibit the lowest impact or h-index scores, predominantly due to the limited number of published papers dedicated to the themes.

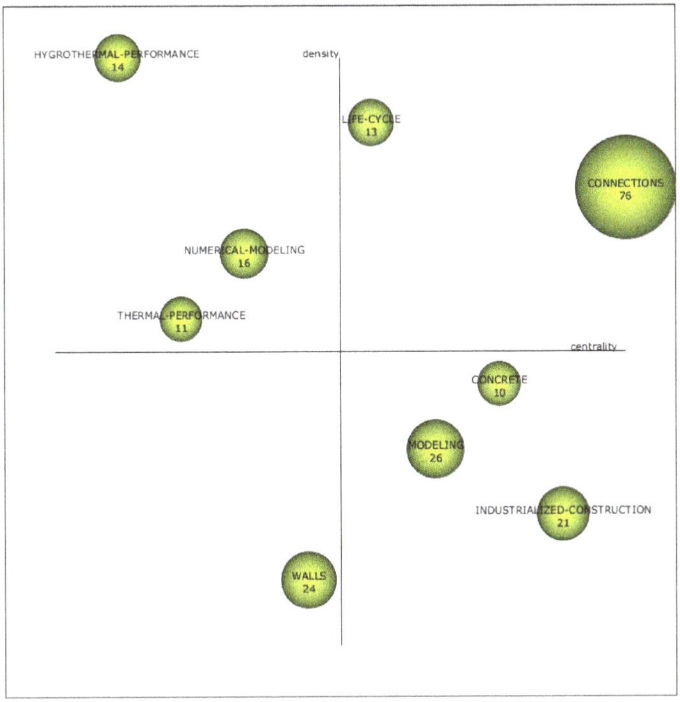

Figure 4. Strategic diagram for the 2nd subperiod (2005–2019).

2. Thematic network
    - Motor themes

      Analyzing the network graph (Figure 5a), it can be observed that the principal motor theme, CONNECTIONS, forms a highly interconnected cluster that explores various structure-related aspects. These aspects include *timber-structures, behavior, experimental-tests*, and structural elements such as *beams*. It is also linked to engineered timber materials like *clt* and the combination of timber with other materials, such as *hybrid* and *composites*. It is worth noting that during this period, certain studies assessed the performance of connectors in prefabricated timber-framed walls [37] or prefabricated timber composite systems [38].

      Moving on to the second motor theme, LYFE-CYCLE (Figure 5b), this cluster covers many aspects related to the Life Cycle Assessment (LCA) methodology, such as *energy-performance, environmental-impact, emissions, carbon, impacts*, and *climate*. Additionally, it includes other terms associated with construction processes, such as *civil-engineering* and *production-planning*. However, it is essential to note that despite being a motor theme, it does not have an extensive collection of published articles. This can be attributed to the lack of specific rules governing the LCA of whole buildings and wooden building products before 2011. Consequently, this led to a vast dissimilarity in approaches, results, and interpretations among various studies, as highlighted by [39]. For instance, the study conducted by [40] emphasizes prefabricated techniques, specifically wood-frame systems, as an alternative approach to reducing a building's carbon footprint.

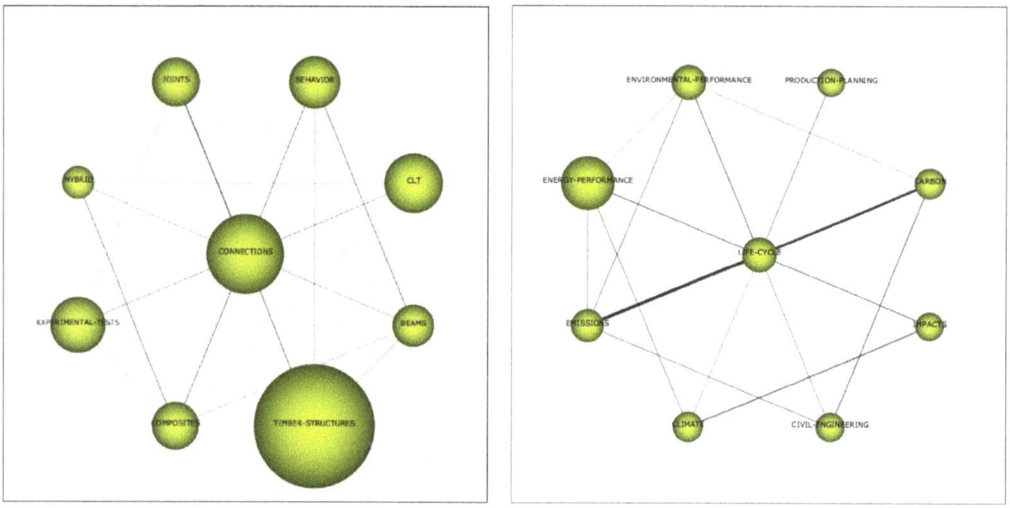

(**a**) (**b**)

Figure 5. Cluster networks of the motor themes: (**a**) CONNECTIONS; (**b**) LYFE-CYCLE.

- Highly developed and isolated themes
  In the second quadrant, the theme NUMERICAL-MODELING (Figure 6a) delves into various aspects related to the seismic behavior of structures. These aspects include *analysis*, *seismic-performance*, *cyclic-testing*, materials, and structural elements such as *steel*, *timber-framed-walls*, and *timber-shear-walls*. Furthermore, this theme considers components in the design of facades, such as *openings* and *sheating*. The research undertaken by [41] investigated a hybrid structural system comprising a steel moment-resisting frame and prefabricated infill light-frame timber shearwalls. The author stated that quasi-static cyclic and monotonic loading tests were carried out to analyze such structures' seismic performance accurately, and the results were evaluated using a numerical model.
  HYGROTHERMAL-PERFORMANCE (Figure 6b) focuses on factors leading to potential issues, such as *mold* and *moisture*. It covers the parameter measurement of insulation performance such as *u-values*, evaluation forms such as *simulations*, construction elements such as *assemblies*, and construction elements such as *highly-insulated-wood-frame-walls*. Also, it is related to other factors that impact hygrothermal performance, such as *air-leakage* and *risks*. Moving forward, THERMAL-PERFORMANCE (Figure 6c) is linked to energy transfer aspects, such as *heat*, materials, and construction typologies, such as *insulation*, *phase-change-materials*, *systems*, *prefabricated*, and *residential-buildings*, and processes such as *retrofitting*. Additionally, it includes other keywords, such as *interviews*. As the demand for energy-efficient buildings continues to rise, an urgent need is to improve building envelope insulation [42]. Glass et al. [43] focused on analyzing the moisture characteristics of highly insulated wood-frame walls in mixed-humid climates. The authors of [44] employed hygrothermal simulations and field measurements to analyze the moisture content in the surface layers of cross-laminated timber (CLT) panels. Other studies have attempted to opti-

mize the structural and thermal performance of light wood-frame structures to enhance thermal efficiency and structural capacity [45].

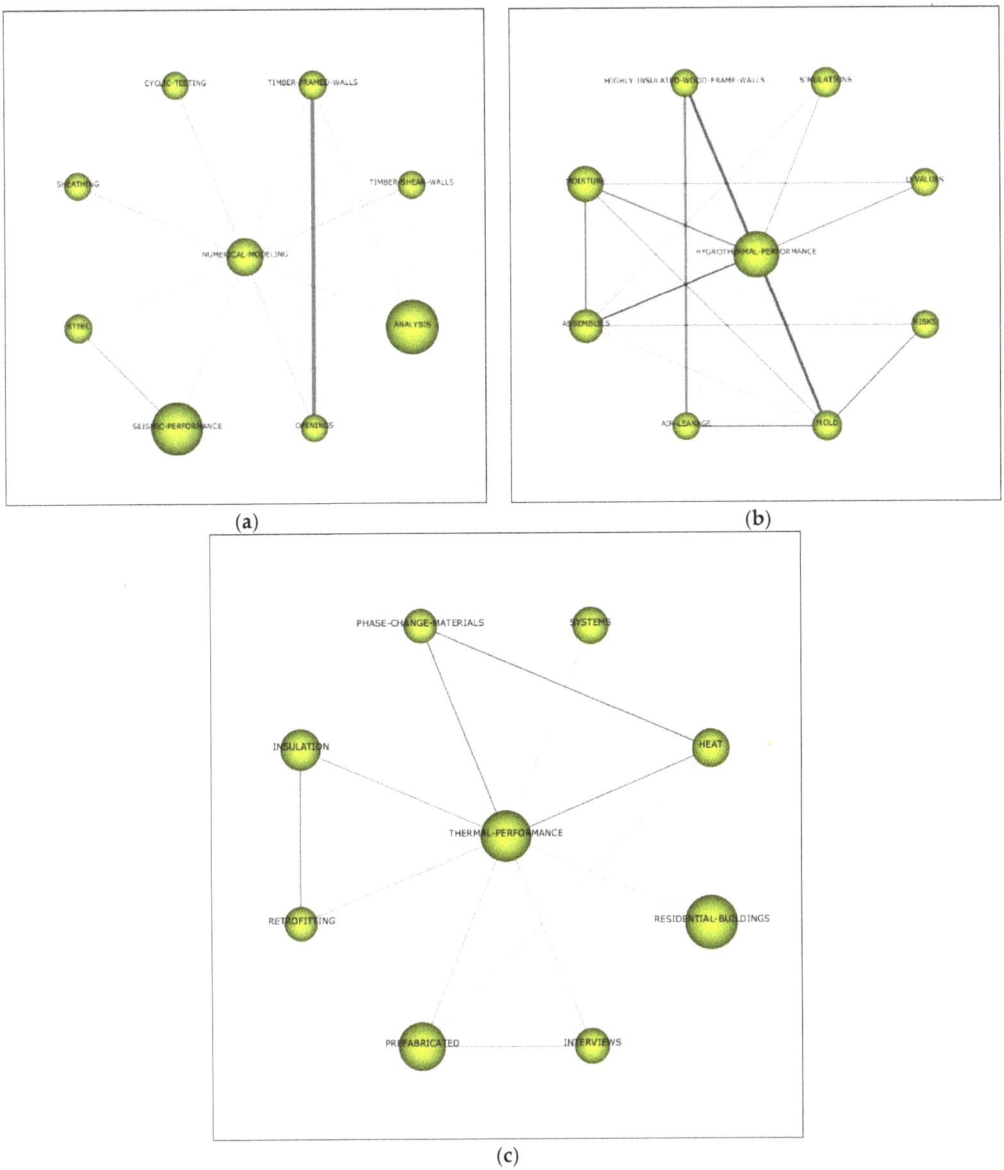

Figure 6. Cluster networks of the highly developed and isolated themes: (**a**) NUMERICAL-MODELING; (**b**) HYGROTHERMAL-PERFORMANCEM; (**c**) THERMAL-PERFORMANCE.

- Emerging or declining themes
  The theme WALLS (Figure 7) focused on structural forces, characteristics that influence the behavior of structures, and methodologies employed to measure them, such as *loads, ductility, failures, full-scale-tests*, and *nonlinear*. It also includes architectural typologies and structural elements, such *as buildings, frames*, and

timber-framed-structures. For instance, Van De Lindt et al. [46] focused on examining the damage incurred by lateral load-carrying assemblies within wood frame structures, specifically shearwalls. The researchers also explored the potential for developing a holistic predictive damage model for the entire structure and integrating it into seismic design practices.

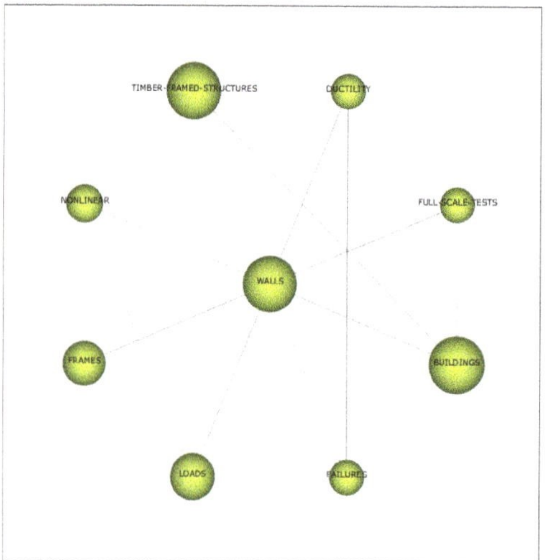

Figure 7. Cluster networks of the emerging or declining themes: WALLS.

- Basic and transversal themes
  Among the themes that hold significant relevance to the field but have not yet been extensively explored, we found the theme MODELING (Figure 8a), which was linked to various aspects of timber structures, encompassing *structural-performance, design, strength,* and *fire-behavior*; different types of timber structures, like *light-frame-structures*; and structural support elements, such as *shear-walls*. It was also associated with hurricanes and wind. The authors of [47] presented a three-dimensional model of a two-story house design utilizing a prefabricated timber-frame structural system. Using parametric analysis, the study offered a single-variable parameter to find the suitable glazing area size value for all modern prefabricated timber buildings.
  INDUSTRIALIZED-CONSTRUCTION (Figure 8b) is a highly interconnected cluster that incorporates various topics related to the construction process and structural design, such as *modular-construction, timber-construction, innovation, shrinkage*, and *performances*, as well as the structural components of buildings, such as *floors* and *concrete-composite-structures*. Novel modular structures that could facilitate the rapid fabrication of new residential complexes were proposed in this period. As demonstrated by [48], a versatile construction system in which the prominent structural members, made by combining steel–timber hybrid shearwall and floor components, present a low environmental impact, given their recyclability and quick replaceability. Similarly, Ref. [49] proposed an innovative composite steel–timber prefabricated floor component that showcased the system's potential in bearing capacity, stiffness, and construction methods.
  Finally, the theme CONCRETE (Figure 8c) primarily focuses on exploring structural timber elements, such as *mass-timber* and *timber-concrete-composite-structures*. Furthermore, it mainly delves into materials that bond these elements, including *adhesives* and *gluing*. It also covers structural topics like *shear, balances,* and *reinforcement*, and numerical methods like *fem-analysis*. To illustrate this point, some

researchers have dedicated their efforts to conducting experimental tests that examine the structural performance of prefabricated timber–concrete composite floors [50,51].

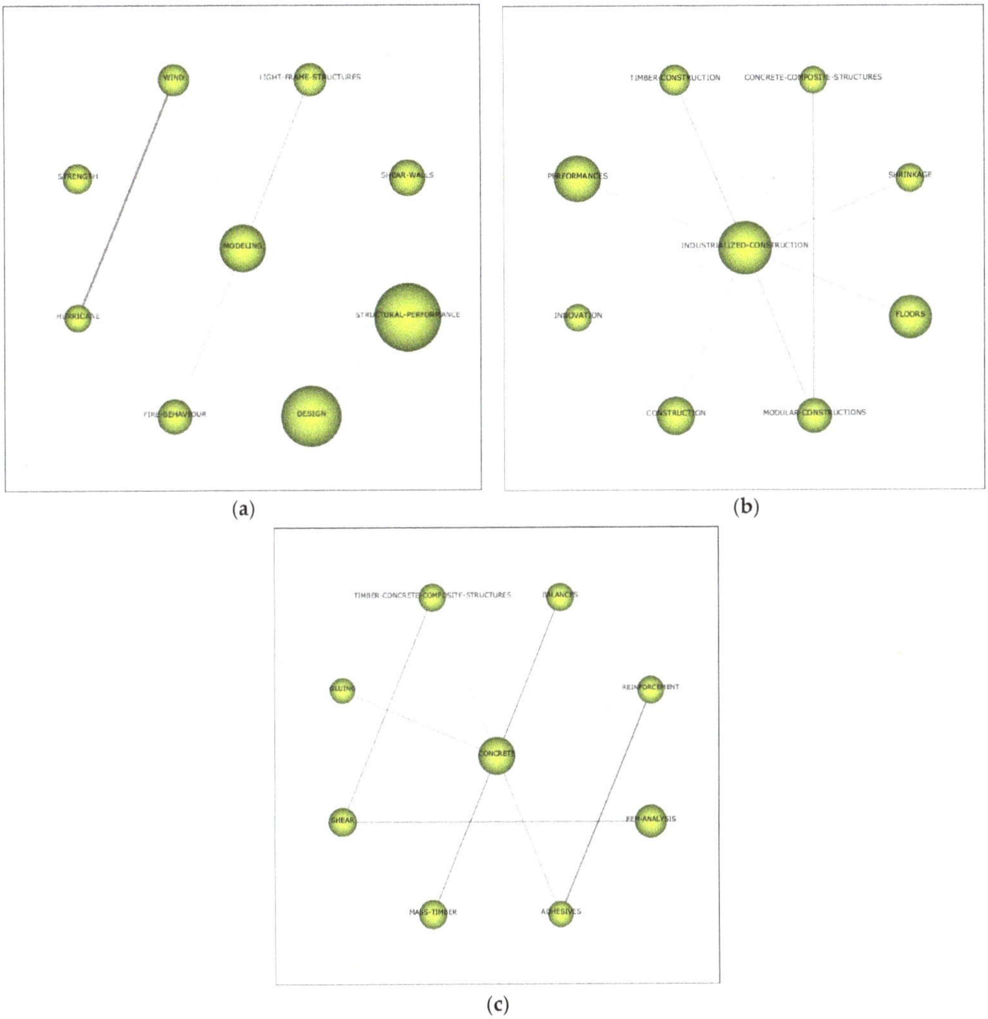

Figure 8. Cluster networks of the of the basic and transversal themes: (**a**) MODELING; (**b**) INDUSTRIALIZED-CONSTRUCTION; (**c**) CONCRETE.

- Relationship between motor and basic themes with most cited articles
  In this period, numerous investigations persisted to determine the structural behavior of light-framed prefabricated timber buildings [52–55]. Nonetheless, the comprehensive examination of the structural performance of mass timber construction methods has not been fully addressed. Among the most crucial aspects of the design process in mass timber structures lies the design of connections, as they impact the joining of members and the load capacity, stiffness, and ductility of the entire system [56].
  Regarding the most frequently mentioned subjects (related to the motor or fundamental themes and their networks), a noteworthy instance of connections and

their strong relationship with CLT was the complete-scale seven-story framework composed of CLT panels following the European seismic standard Eurocode 8. It was subjected to earthquake-induced loading on a three-dimensional shaking table. The seven-story structure was part of the research project "SOFIE-Sistema Costruttivo Fiemme", whose goal was to exhibit the capability to self-center, possess high stiffness, and possess sufficient flexibility to avert brittle failures. Additionally, tensile joint assemblies were developed, and their importance in dissipating energy in structures with rigid CLT panels became apparent. Lastly, the authors noted that the seven-story structure experienced relatively strong accelerations at higher levels, which could result in secondary damage and should be addressed in future investigations [57]. Another approach to the same topic was the work of [58]. The study examined alternative mechanical "dry-dry" connections previously integrated into a prefabricated concrete slab. Direct shear tests were conducted on small blocks constructed from glulam segments affixed to a prefabricated concrete slab. The experiment measured the relationships between shear force and relative slip and calculated mechanical properties, such as slip moduli and shear strengths. The findings indicated that innovative connection technologies for precast concrete slabs can perform comparably to cast-in situ slabs while being more cost-effective.

Lastly, one of the most frequently addressed topics was the energy efficiency of timber buildings. The authors of [59] utilized a consequential-based lifecycle approach to examine the carbon implications of conventional (Swedish building regulations) and low-energy (Swedish passive home standards) versions of three timber multi-story building systems. The building systems were constructed using mass timber (CLT components), beam-and-column structures (glulam and LVL components), and prefabricated modules of light-frame volume elements. The analysis encompassed the entire resource chain throughout the lifecycle of the building, tracking carbon flows from fossil energy, industrial processes, changes in carbon stocks in materials, and potential avoidance of fossil emissions through the substitution of fossil energy with woody residues. In conclusion, the study emphasized the potentially significant role of multi-story timber structures in carbon-efficient construction by demonstrating that the low-energy variant of the CLT building emits the least amount of carbon over its lifespan. In comparison, the standard beam-and-column construction exhibits the highest lifespan emission.

3.1.3. Subperiod 2020–2023

1. Quantitative measures and strategic diagram.

   CONNECTION emerged as the dominant motor theme (Figure 9). It accumulated the highest number of citations, totaling 518, and had the most significant number of published documents, reaching 80. However, it had a relatively low impact on the field of research, as indicated by its h-index score of 12. The following motor theme was ENERGY-PERFORMANCE, which ranked third in the number of citations with a count of 183. However, the number of published papers was comparatively low (36), resulting in a correspondingly low h-index score of 9. Additionally, within the first quadrant of the strategic diagram, other motor themes emerged: CIRCULAR-ECONOMY and COMPOSITES. The former had a number citation count of 127, while the latter had 123 citations. Both themes shared the same number of publications, 16, and achieved an h-index score of 7, as can be seen in Table 5.

Table 5. Performance measures for the themes of the subperiod 2020–2023.

Theme Name	Number of Documents	Number of Citations	Average of Citations	h-Index
CONNECTIONS	80	518	6.475	12
ENERGY-PERFORMANCE	36	183	5.083	9
CONSTRUCTION	32	232	7.25	8
BUILDING-ENVELOPE	19	127	6.684	8
CIRCULAR-ECONOMY	16	127	7.938	7
COMPOSITES	16	123	7.688	7
FLOORS	14	84	6	5
MODELING	13	98	7.538	5
SEISMIC-PERFORMANCE	13	74	5.692	6

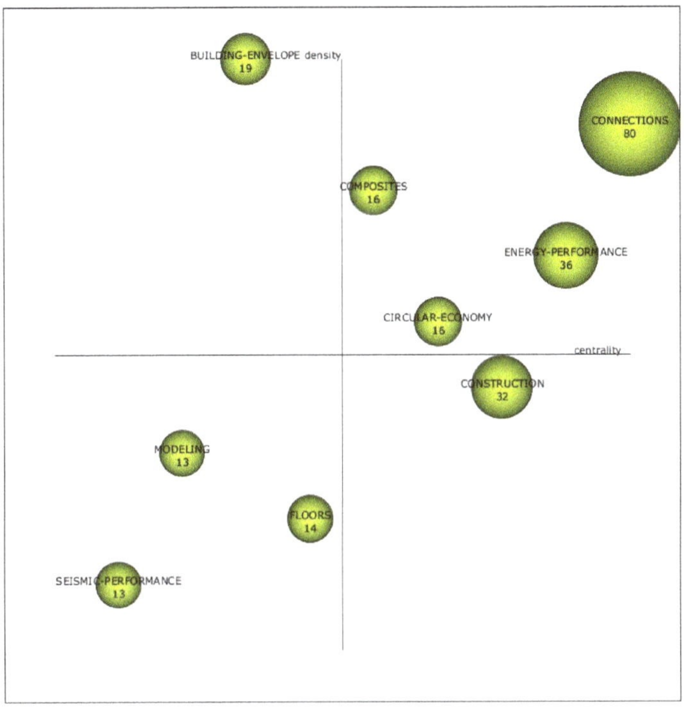

Figure 9. Strategic diagram of the 3rd subperiod (2020–2023).

Moving on to the second quadrant, the highly developed and isolated theme was BUILDING-ENVELOPE. This theme was characterized by 19 published documents, 127 citations, and an h-index score of 8. In the third quadrant, we observed emerging or disappearing themes such as FLOORS, MODELING, and SEISMIC-PERFORMANCE. FLOORS was represented by 14 published documents and achieved 84 citations. The h-index score was the lowest among all the themes analyzed, standing at 5. On the other hand, MODELING and SEISMIC-PERFORMANCE boasted citation counts of 98 and 74, respectively. The h-index score for MODELING was 5, while for SEISMIC-PERFORMANCE, it was 7. Finally, the fourth quadrant was occupied by the basic and transversal theme, CONSTRUCTION. This theme ranked second in the number of citations, with a count of 232. It was accompanied by 32 published documents and an h-index score of 8.

Upon comparing this subperiod to the previous one, the number of themes remained relatively consistent. However, there was a noticeable increase in the quantity of motor themes within the quadrant. Additionally, the number of published documents in this subperiod, totaling 200, was almost on par with the previous subperiod. Similarly, we found that the motor themes (first quadrant) and basic themes (fourth quadrant) continued to be the most highly cited and were associated with the highest number of documents. It is worth noting that the analysis only covered three years, yet it was evident that researchers have shown a growing interest in the motor and basic themes.

2. Thematic network

- Motor themes

As illustrated in the thematic networks, the motor themes CONNECTIONS (Figure 10a) and ENERGY-PERFORMANCE (Figure 10b) are closely connected clusters (each cluster is focused on a specific topic). The first motor theme is mainly focused on the conditions of structures under loading and construction materials, which include *structural-performance, experimental-tests, shear, performances, behavior, clt*, and *timber-concrete-composite-structures*. Additionally, this cluster incorporates analytical models such as *fem-analysis*. On the other hand, the second motor theme is associated with crucial measures for assessing the sustainability of a structure, including *life-cycle, environmental-performance, thermal-performance, carbon*, and *heat*. In addition, it contains structural and construction topics such as *residential-buildings, timber-structures*, and *blast*. Although CLT is now being utilized for previously considered improbable structures, the technology for connecting these elements is still evolving [60], and it is discussed in various papers published during this period [61,62]. Furthermore, there is a specific focus on enhancing the energy dissipation capacities of mass timber building assemblies by utilizing steel energy-absorbing connections [63].

Another motor theme that exhibits significant connectivity is CIRCULAR-ECONOMY (Figure 10c), which delves into construction and structural concepts, including *architecture, design-for-disassembly, business-models, systems, challenges*, and *buildings*. It also covers construction materials such as *mass-timber* and *concrete*. The circular economy concept is extensively discussed in the publication by [64], which presents a systematic literature review highlighting significant knowledge gaps and urgent research needs for developing holistic approaches to circularity potentials for panelized mass timber structures.

In contrast to the previous ones, the theme COMPOSITES (Figure 10d) does not possess a well-connected cluster. This theme focuses on the structural behavior of the structures, construction elements, and materials such as *loads, collapse, strength, resistance, part-I, validation, beams*, and *steel*. In the scope of composites, several investigations concentrate on the structural behavior of prefabricated timber-concrete composite elements with innovative connections [65,66].

- Highly developed and isolated themes

  The theme BUILDING-ENVELOPE (Figure 11) is related to the external behavior and properties of the building's envelope. This covers various aspects, such as *hydrothermal-performance* and *airtightness*. Moreover, it covers construction elements, potential issues, and the materials utilized in the building envelope, such as *wood-frame-walls, walls, mold, moisture, air-barrier*, and *oriented-strandboard*. Research efforts have been ongoing to study timber-framed and mass-timber external wall assemblies, as timber proves to be effective in reducing thermal bridges due to its low thermal conductivity, which enhances a building envelope's overall energy efficiency [67]. For instance, [68] demonstrated the correlation between the hygrothermal simulation tool (WUFI) model output and the measured values of OSB moisture content (MC), cavity temperature (T), and cavity relative humidity (RH) for wood-framed walls with various exterior insulations in cold climates.

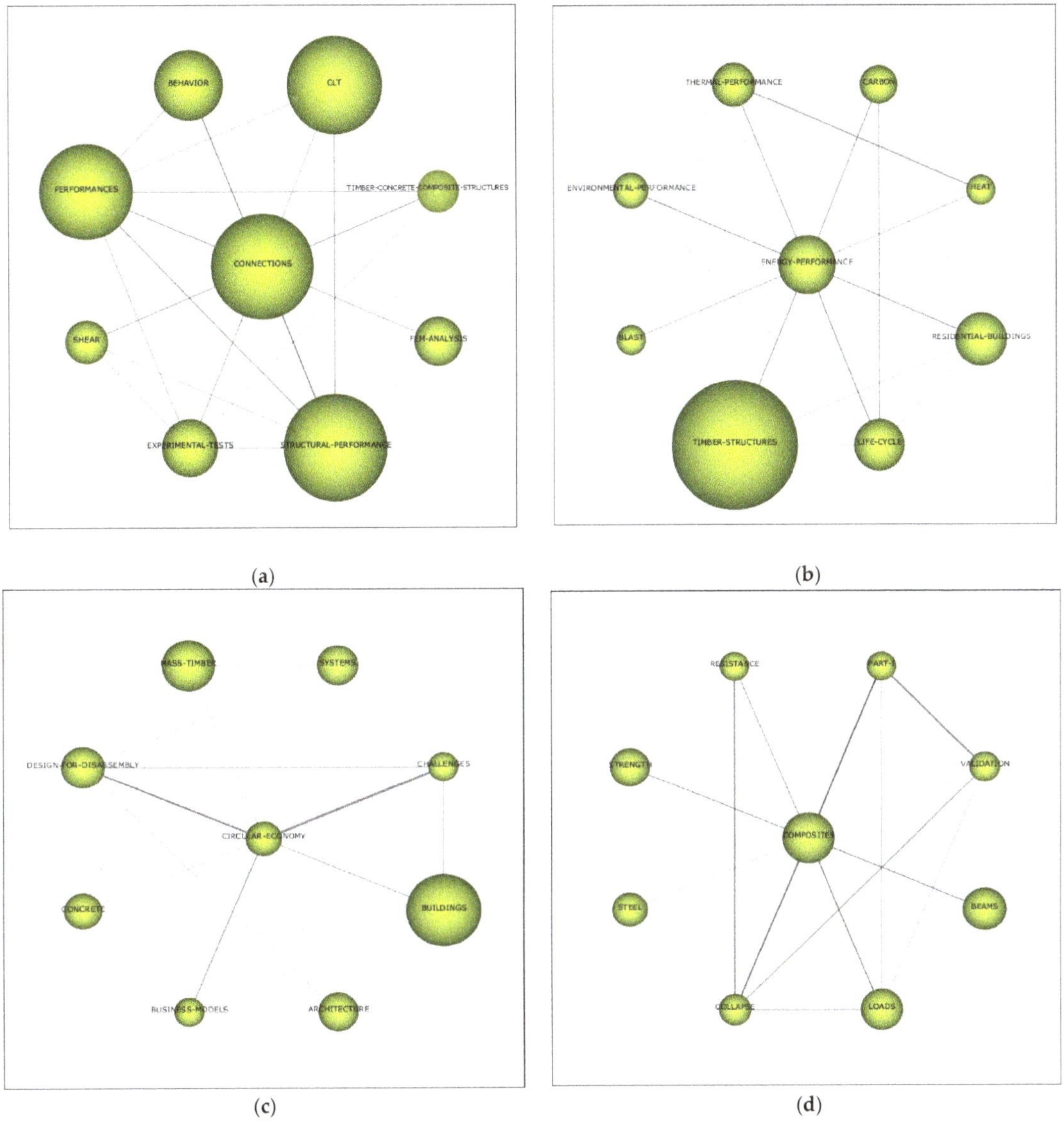

Figure 10. Cluster networks of the basic and transversal themes: (**a**) CONNECTIONS; (**b**) ENERGY-PERFORMANCE; (**c**) CIRCULAR-ECONOMY; (**d**) COMPOSITES.

Figure 11. Cluster networks of the of the basic and transversal themes: BUILDING-ENVELOPE.

- Emerging or declining themes
  FLOORS (Figure 12a) covers structural material properties and processes to improve the structural material properties such as *flexural-behavior* and *prestress*, and types of connections including *shear-connections, joints*, and *screws*. It also covers the parts and features of the structures, including *elements* and *prefabricated*. The analysis presented by [69] shed light on the acoustic performance of timber buildings and complex composite floor systems in the Australian construction industry. The study indicated that prefabricated construction methods showcased improved performance compared to on-site construction. However, direct comparisons are challenging due to the varying timber products and construction techniques. Thus, establishing well-defined measurement methods, predictive models for sound insulation properties, and numerical simulations for airborne and impact sound transmission becomes imperative.
  The theme MODELING (Figure 12b) focuses on the structural behavior of systems, structural elements, and the evaluation of processes. This includes *fragility, failures, post-tensioned, stiffened-engineered-timber-wall-systems, timber-framed-structures*, and *damage-assessment*. In addition, it covers technology types and visual media such as *digital-design* and *virtual-design*. For example, Wei et al. [70] proposed implementing a Digital Twin model, a collection of technologies to establish seamless physical and virtual connections. This model was designed to enhance the productivity of a wood-panel construction company. The company could foresee significant waste reduction and improve overall efficiency by utilizing this model. Additionally, it facilitates the widespread application of technology by providing a holistic framework.
  SEISMIC-PERFORMANCE (Figure 12c) is related to the seismic characteristics and evaluations of the behavior of a structure, covering various factors such as *ductility, lateral-performance, damage-free*, and *analysis*. Also, it includes types of connections, *dowel*, and construction systems, such as *frames*. In the study conducted by [71], the focus was on the redesign of a nine-story residential building constructed with reinforced concrete (RC) using a hybrid disassembly building approach involving cross-laminated timber floors, glue-laminated timber frames, and concentrically braced steel frames. This research shed light on the fact that although the lateral load capacity of the hybrid building is lower than that of

the RC building, its deformation capacity is significantly higher. Moreover, the LCA results indicated that the carbon emissions from the RC building were twice as much as those emitted by the hybrid building. Therefore, material selection and construction technologies are critical for implementing circular projects in seismically active areas.

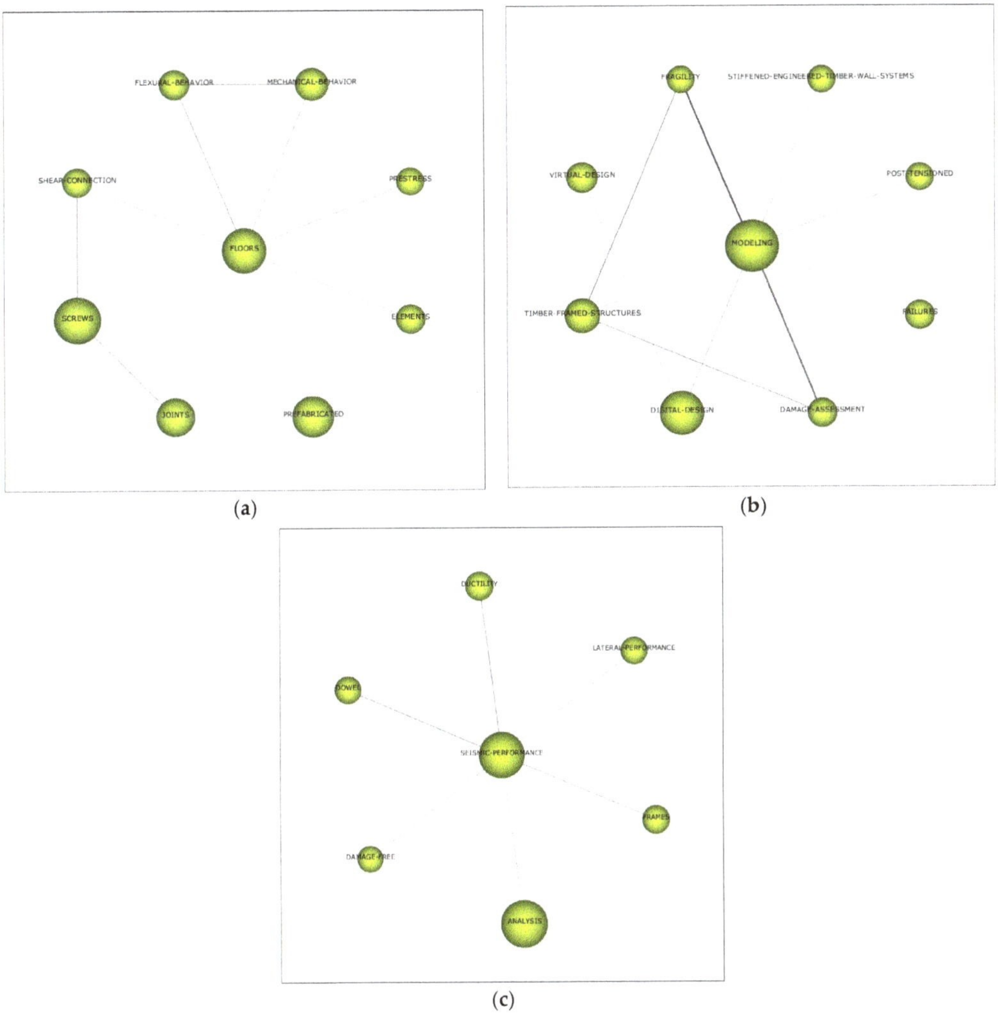

Figure 12. Cluster networks of the emerging or declining themes: (**a**) FLOORS; (**b**) MODELING; (**c**) SEISMIC-PERFORMANCE.

- Basic and transversal themes
  CONSTRUCTION (Figure 13), a relevant theme in the field, presents a highly interconnected cluster. It mainly focuses on structural processes, including *design, management, manufacturing, climate, opportunities*, and *production-planning*. It also covers the application of technology to improve them, such as *automation* and *industrialized-construction*. Understanding the multitude of variables that influence the adoption of prefabricated timber construction enhances the

implementation of the construction industrialization process. In their study, Du et al. [72] examined modern timber structures (MTSs) in China from the stakeholders' perspective. This analysis encompassed critical elements, including government policy, public awareness and acceptability, cost considerations, and the connections between designers, prefabricated component producers, and construction companies. Finally, the research provided recommendations to enhance existing perceptions, bridge the knowledge gap, and accelerate current development efforts to expand the prefabricated timber construction or MTS building sector.

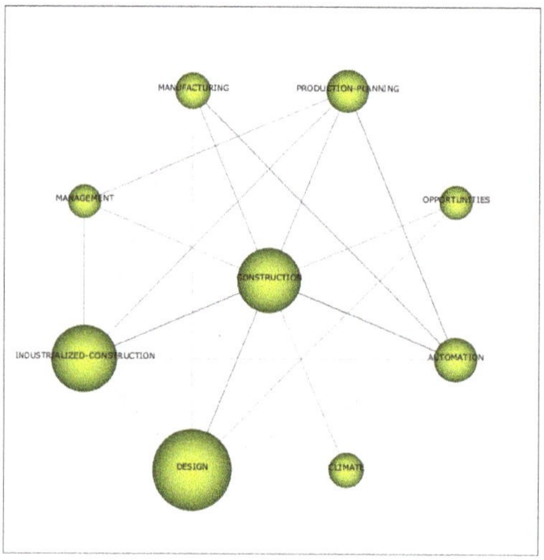

Figure 13. Cluster networks of basic and transversal themes: CONSTRUCTION.

- Relationship between motor and basic themes with most cited articles
  As shown below, the motor and basic themes' networks are the predominant subjects of reference in the publications developed in this subperiod. It is highlighted that there is a growing interest in automation in timber manufacturing with advancements in digital fabrication technology (e.g., robotic timber manufacturing). This interest stems from the potential for the fully automated production of non-standard timber structures [73] and prefabricated timber buildings. A pioneering approach to robotic manufacturing within the construction industry was introduced by developing, designing, and testing a large-scale, portable, and adaptable automated wood construction platform called TIM [74]. This mechanical system can be swiftly incorporated into existing manufacturing setups in standard carpentry projects, thereby facilitating the exploitation of synergies between traditional craftsmanship and specialized automation technology. As a result, the quality and productivity of the trade can be significantly enhanced. Additionally, the study underscored an interdisciplinary co-design approach that emphasizes cyber-physical design and production methods for multi-story timber buildings. The researchers emphasized the need to address several significant challenges in future studies before the widespread adoption of robotic timber-building platforms can be realized in the industry.
  Considering the current focus on energy performance and the circular economy, Margani et al. [75] proposed an innovative solution for rehabilitating reinforced concrete (RC)-framed buildings considering energy, seismic, and architectural factors. This solution involves the utilization of prefabricated cross-laminated timber (CLT) panels in conjunction with wood-framed panels on the exterior

walls. The combination of these panels incorporates insulation and cladding materials, enhancing the energy performance and aesthetics of the renovated building. Furthermore, innovative seismic energy dissipation devices connect the CLT panels to the existing RC frame, reducing structural damage and repair costs during seismic events. In conclusion, retrofitting significantly enhances the energy performance of multi-story residential buildings in Italy, resulting in a reduction of up to 60% in global energy demand.

Projections indicate that global demand for mass timber will continue rising upward, particularly in Australia, Japan, and New Zealand [76]. Responding to this trend, Li et al. [77] studied the out-of-plane bending behavior of Radiata Pine XLam CLT panels subjected to four-point loading. The experiments revealed brittle failure modes when the samples were loaded along the central and minor axes. Various analytical approaches were employed to evaluate the bending characteristics of the test materials, with the shear analogy proving to be the most accurate predictor. Other theoretical techniques yielded satisfactory results in determining effective bending stiffness and bending and shear strengths. The bending characteristics of CLT were further assessed through finite element analysis, with the thickness of the panels being quantitatively evaluated concerning their influence on bending behavior. The authors concluded that further parametric analysis was required to comprehensively understand these effects and providing data support for future practical applications.

3.2. Thematic Conceptual Evolution

The evolution map (Figure 14) showcases the changes in themes over time within each subperiod. According to [31], each column in the map represents the themes established during a specific subperiod. Additionally, themes from successive subperiods are connected if they share common keywords. Solid lines indicate that the connected themes have the same name or that one theme's name is part of the other theme, while a dotted line signifies that the themes share different keywords that are not their names [29].

When examining the evolution of the number of themes, we can observe a significant increase during the first and second subperiods, with four times as many themes as in the first subperiod. However, for the last subperiod, the number of themes remains the same as in the second subperiod, indicating a continuous evolution of themes from the second subperiod until the final one studied. Furthermore, most thematic areas exhibit increasing evolution regarding the number of documents (dimensions of the spheres). This signifies a growth in the number of documents compared to the previous period, indicating the sustained interest of the research community in these themes.

Regarding the thematic composition of each thematic area, the following observations can be made:

1. The themes TIMBER-STRUCTURES and DESIGN (undefined themes) evolve into motor, basic, highly developed, and emerging or disappearing themes in the second subperiod.
2. TIMBER-STRUCTURES and DESIGN are the origins of the theme CONNECTIONS, a motor theme in the second and third subperiods.
3. Two thematic areas consist of motor or basic themes throughout all the subperiods:
    - LYFE-CYCLE (motor theme) evolves into ENERGY-PERFORMANCE (motor theme) and CONSTRUCTION (basic theme).
    - CONCRETE (basic theme) evolves into CIRCULAR-ECONOMY (motor theme).
4. Only one thematic area comprises a highly developed theme in all the subperiods: HYGROTHERMAL-PERFORMANCE evolves into BUILDING-ENVELOPE.
5. Four thematic areas consist of motor or basic themes and an emerging or disappearing theme:

- CONNECTIONS (motor theme) evolves into CONNECTIONS (motor theme), ENERGY-PERFORMANCE (motor theme), COMPOSITES (motor theme), and FLOORS (emerging or disappearing theme).
- NUMERICAL-MODELING (motor theme) evolves into SEISMIC-PERFORMANCE (emerging or disappearing theme) and COMPOSITES (motor theme).
- INDUSTRIALIZED-CONSTRUCTION (basic theme) evolves into CONSTRUCTION (basic theme) and FLOORS (emerging or disappearing theme).
- MODELING (basic theme) evolves into MODELING (emerging or disappearing theme), COMPOSITES (motor theme), CONSTRUCTION (basic theme), and CONNECTIONS (motor theme).

6. Only one thematic area comprises a highly developed theme, two motor themes, and an emerging or disappearing theme: THERMAL-PERFORMANCE (highly developed theme) evolves into ENERGY-PERFORMANCE (motor theme), CIRCULAR-ECONOMY (motor theme), and FLOORS (emerging or disappearing theme).
7. Only one thematic area consists of three emerging or disappearing themes and two motor themes: WALLS (emerging or disappearing theme) evolves into BUILDING-ENVELOPE (highly developed theme), COMPOSITES (motor theme), CIRCULAR-ECONOMY (motor theme), MODELING (emerging or disappearing theme), and SEISMIC PERFORMANCE (emerging or disappearing theme).

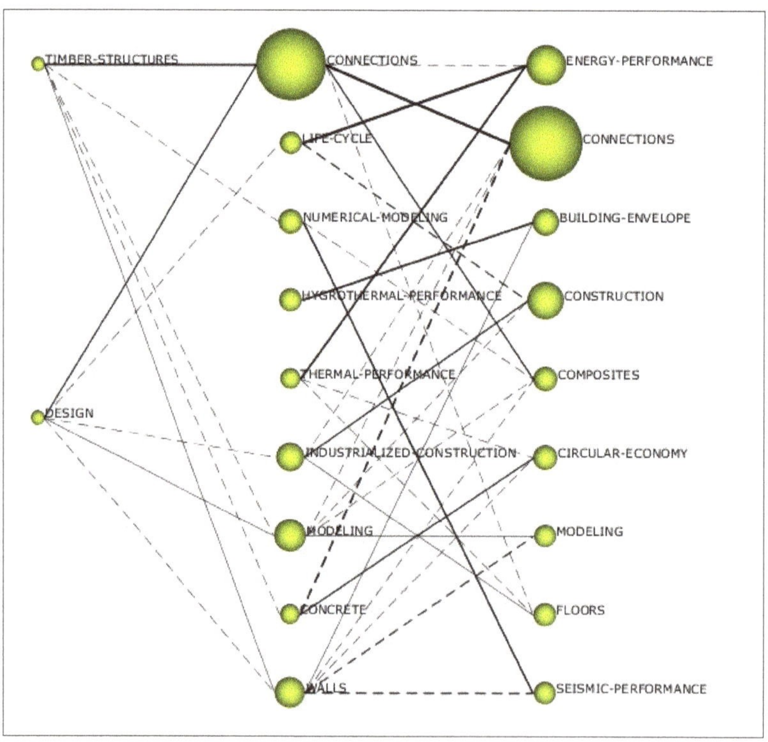

Figure 14. Thematic evolution network of the research field from 1990 until 2023.

4. Analytical Description by Subperiod

According to the insights provided by the bibliometric analysis, an analytical description of the leading research topics and current challenges is outlined as follows:

4.1. Superiod 1990–2004

It demonstrates a notable inclination toward studying timber-framed panel walls' design and structural behavior. These walls are commonly employed as the primary bearing capacity elements in constructing prefabricated timber structures [78]. Furthermore, there is a noticeable growth in the development and utilization of simulation tools, such as FEM, to analyze the behavior of timber structures. Overall, this period denotes a significant advancement in the research and exploration of prefabricated timber structures, beginning the way for further improvements and innovations.

4.2. Superiod 2005–2019

The utilization of CLT as a prefabricated component in medium- and high-rise buildings became critical, and it worked hand-in-hand with connections. While concrete is still used, it assumes a different role in this context as a material that complements emerging prefabricated timber structures, such as hybrid systems. It is also clear that model approaches remain an essential tool for comprehending the behavior of timber structures. Two fundamental concepts emerge within the construction industry: life cycle and industrialized construction. Unquestionably, one of the most significant advantages associated with employing prefabricated timber systems in construction lies in reducing costs, labor, and time, and the subsequent decrease in carbon emissions and their impact on the structure's overall life cycle.

4.3. Superiod 2020–2023

During this subperiod, the researchers kept attracting attention by exploring the structural behavior of prefabricated and modular timber buildings (mass timber and composite elements). Vital aspects involved examining new wood species that exhibited potential for being employed in the manufacturing process of CLT and innovative proposals for developing connections. Furthermore, attention was being given to improving the energy efficiency of timber buildings by enhancing building envelope systems. There were ongoing research studies on prefabricated timber walls (light timber-framed and mass-timber external assemblies) due to wood's susceptibility to humidity and temperature changes. Apart from these matters, the construction industry was also witnessing the emergence of the circular economy concept as a prominent topic. Prefabricated timber technologies possess various qualities that make them suitable for implementing a circular construction approach. According to the existing literature, these qualities include a faster construction process, reduced pollution, and the potential for material reuse and recycling.

Finally, another significant change observed in the timber construction industry during this period was the application of new digital technologies into industrialized construction practices. There was a clear indication of interest in the extensive application of technology to provide a holistic framework for designing, managing, manufacturing, and producing prefabricated/modular timber buildings.

5. Conclusions

The primary contribution of this study is the identification of relevant topics, emerging trends, and prospective pathways in the research of prefabricated and modular timber structures through the analysis of strategic diagrams and cluster networks with the aid of the SciMAT software.

The thematic evolution revealed an initial interest in understanding the structural behavior of timber buildings with light structures. This interest was achieved using numerical factors, analytical models, and experimental tests. The rationale behind this focus was based on the fact that this system constituted a substantial percentage of the houses constructed in North America, Australia, New Zealand, Japan, and Europe [36]. Subsequently, with the introduction of mass timber products and composite elements, the design and construction methods demanded increased concern in detailing critical elements and new planning methods. Consequently, research efforts were primarily concentrated on the study of two critical elements, namely, connections (which impact the structural functionality, manufacturing, and construction of the entire building system) and the building envelope systems (which safeguard the timber building structure from undesirable weather conditions during the construction phase and enhance the energy efficiency of timber buildings during their operational lifespan). Despite the increasing demands and challenges, the number of CLT modular projects achieved a remarkable milestone in 2017, reaching its peak with the completion of 26 buildings [22].

Furthermore, the analysis results indicated a strong correlation between design and prefabricated and modular timber structures. This correlation was observed from the first subperiod and represents potential opportunities for future research. It was found that prefabrication and modularization offer numerous advantages, such as flexibility in design and the ability to customize and standardize constructive elements. However, Ribeiro et al. [79] identified specific barriers associated with prefabrication. One such barrier is that architectural design often fails to consider the modular construction aspect, primarily due to aesthetic limitations. Similarly, Svatoš-Ražnjević et al. [80] pointed out that only some modular timber projects differ from the traditional rectangular and orthogonal footprints. From the publications analyzed, the modular design approach has predominantly been applied to houses rather than large-scale projects, potentially due to the limited exposure to modular technology in the popular literature and timber databases.

In addition, the analysis of thematic clusters and trends revealed that two factors influence the development of prefabricated and modular timber structures:

- Circular economy. Researchers have shifted their focus towards the reusability and efficiency of prefabricated elements. Prefabricated timber construction is characterized by utilizing large-format building elements and the associated logic of joining them together [16]. This characteristic makes it suitable for disassembly, reuse, and recycling. Therefore, adopting strategies and measures that reduce the carbon footprint, minimize waste, and increase the life cycle of structures aligns with the principles of circular construction. To this end, various initiatives have been developed, exemplified by the MAI-Ivalsa Modular House. This environmentally sustainable system of prefabricated housing modules employs reused CLT for the load-bearing structure, allowing for the prefabrication of components at one site and their successful transportation to another [81]. Another example is the Collegium Academicum IBA, the first multi-story modular timber student residence that offers significant spatial flexibility through simple modular pieces that can be easily disassembled and recycled [82,83]. Additionally, a ten-square-meter prototype utilizing a fully modular precast lightweight engineered timber structural system, as developed by [84], has demonstrated the potential of building-scale additive manufacturing with cost-effective materials to enable deconstruction and material recovery. Studies focusing on hybrid structures have recently begun incorporating assessment strategies such as design for disassembly or deconstruction (DfD). For instance, Grüter et al. [85] presented two case studies of modern residential timber hybrid buildings in Switzerland, employing digital tools to evaluate strategies that facilitate a circular design process for timber elements. These strategies specifically address the elements' start and end of life, thus serving as a solid foundation for future research. Similarly, Derikvand and Fink [86] proposed directions for future developments and advocated for DfD in timber–concrete composite (TCC) floors, emphasizing deconstructable connectors that enable material recovery and reuse as the preferred end-of-life scenario. These projects underscore the importance of integrating sustainable practices into the design and construction processes. Despite the recent release of a standard by the International Organization for Standardization (ISO), which provides general rules for integrating design for disassembly and adaptability (principles of circular economy) into the service life of structures [87], few studies have, thus far, presented innovative solutions and designed buildings that fully address these strategies. From the literature content analysis focusing on connections and building envelope systems (as was established before, they are critical elements in the design and construction of prefabricated and modular timber structures), only one study stands out as it introduced a novel reversible mass timber connector for prefabricated systems [88] while another explored the design and construction of fully prefabricated façade components to reduce materials and production energy [89].

- Digital technologies. Digital architecture, computer science, engineering informatics, virtual reality (VR), and building information modeling (BIM) are critical in prefabricated and modular timber construction. Integrating these parameters into industrialized construction practices necessitates the implementation of automation in timber manufacturing. For instance, in Canada, there is a strong focus on integrating BIM information into the construction phase, specifically in the context of automated fabrication. A study by [90] delved into a BIM-based framework that aims to automate the machine capability evaluation of timber frame assemblies. This proposal highlights the system's ability to accurately determine whether a user-selected machine can effectively manufacture a construction product that has been pre-designed using BIM software. In the United Kingdom, the research developed by [91] adopted BIM models in the game environment (gamification) to facilitate the implementation of small- and medium-sized architectural and construction practices dedicated to visualization creation. The collected evidence showcases that combining a game-like platform and BIM can lead to simplified data delivery to clients, ultimately resulting in customer satisfaction, confidence, and increased sales of timber frame houses. According to [92], *"the level of automation is high in the initial stages of prefabrication but relatively low in the assembly stages, which require flexibility and human knowledge."* To address this, some research groups are focused on developing robotic systems. Robotic fabrication, directly connected to a precisely planned virtual model, significantly reduces the risk of construction mistakes and ensures high global precision. This results in a cost-effective and efficient construction system [93]. The literature reveals that robotic systems have enabled the construction of unconventional structures, such as the BUGA Wood Pavilion. Furthermore, companies like Weinmann and Randek have gained recognition for adopting these technologies in their processes [94]. Similarly, researchers in the United States have devised a simulation-based methodology to evaluate the automated assembly of wood frames through BIM, utilizing various robotic systems [95]. By employing simulation techniques, researchers can assess the effectiveness of different robotic systems in automating the assembly of timber frames.

However, among the selected papers, only a few studies delved into the practical implementation of these technologies. This can be attributed to the persistent obstacles that hinder the adoption of digital technology, such as the fragmented and disorganized nature of reverse supply chains, resulting in a lack of practical information sharing among stakeholders. Moreover, these supply chains' intricate and diverse nature renders the recreation of accurate (BIM) models representing the existing conditions often expensive or time-consuming. Additionally, actors involved may lack the necessary knowledge or skills to embrace certain technologies, further impeding their application [96]. Furthermore, establishing and improving manufacturing environments, which typically necessitate substantial investments, also pose a significant challenge in adopting these technologies.

Drawing from the results obtained in the investigation, it becomes evident that the market ratio of these construction technologies continues to evolve. This development highlights the interrelation of digital technologies in effectively supporting and facilitating advanced circular economy strategies in the design and construction processes. These strategies include supporting forestry, reducing expenses, shortening installation schedules, and ensuring a significantly cleaner and more regulated environment, which enhances workers' health and safety, as the operational processes are greatly minimized. However, it is essential to acknowledge a significant gap in research and knowledge surrounding these topics, which necessitates further exploration and understanding. Studies conducted in Austria and Australia focused on developing perception surveys and interviews of relevant industry stakeholders to gain insight into the barriers to the widespread application of digital technologies and circular strategies on prefabricated and modular timber construction. For instance, Santana-Sosa et al. [97] highlighted a lack of knowledge and experience, limited implementation of digital tools such as BIM, inadequate capacity of current infrastructure and procedures, and absence of standardization in design, planning, optimized construction management, and manufacturing. Similarly, in Australia, the study conducted by [98] identified a lack of regulations, performance concerns, limited application experience among local manufacturers and suppliers, inadequate regulatory and insurance policies, financing challenges, difficulty in incentivizing clients, and a shortage of a skilled workforce.

Considering the significance of these findings, it becomes crucial for multiple sectors to act:

- Reevaluating the regulatory framework is necessary to encourage research collaborations between countries, facilitating the transfer of specialized technology knowledge to the labor market.
- Open new research areas in the construction sector, contributing to the growth of existing scientific systems and advancing these fields.
- Implementing professional training programs in industries can promote competitiveness and employment opportunities.
- Integrating timber expertise, including designers, producers, and construction companies, at different stages of the construction process can further enhance the adoption of prefabricated and modular timber construction.
- Raising society's awareness and knowledge transfer of timber's potential, such as its safety, cost-effectiveness, adequate structural performance, and environmental aspects, can foster the acceptance of these technologies.

Ultimately, the findings of this research significantly contribute to the existing body of knowledge and serve as an initial comprehensive platform that supports further investigations of these construction technologies. Future research will explore the relationship between circular economy strategies concerning prefabricated and modular timber construction, such as design for disassembly/deconstruction (DfD) and design of adaptability (DfA).

Author Contributions: Conceptualization, N.G. and P.G.; methodology, N.G. and P.G.; software, N.G.; formal analysis, N.G.; investigation, N.G.; resources, A.D. and P.G.; writing—original draft preparation, N.G.; writing—review and editing, N.G., J.N., A.D. and P.G.; visualization, N.G. All authors have read and agreed to the published version of the manuscript.

Funding: This research was funded by CENTRO-08-5864-FSE-000031 with the title Forest for Future (F4F) project, financed through co-financing by the European Social Fund through the Regional Operational Program for the Centre (Centro2020) of Portugal2020 and the Institute for Sustainability and Innovation in Structural Engineering (ISISE). In addition, the authors would like to recognize the financial support provided by the Timber Innovation Center UC, ANID BASAL FB210015 (CENAMAD).

Institutional Review Board Statement: Not applicable.

Informed Consent Statement: Not applicable.

Data Availability Statement: No data were used for the research described in the article.

Acknowledgments: Technical support of the SciMat software: Alexander Wenzel. Centro Nacional de Excelencia para la Industria de la Madera (CENAMAD), Pontificia Universidad Católica de Chile.

Conflicts of Interest: The authors declare no conflicts of interest.

References

1. Smith, R. *Prefab Architecture: A Guide to Modular Design and Construction*; John Wiley & Sons, Inc.: Hoboken, NJ, USA, 2010. Available online: https://ia801906.us.archive.org/23/items/Prefab_Architecture_A_Guide_to_Modular_Design_and_Construction_R._Smith_Wiley_20/Prefab_Architecture_A_Guide_to_Modular_Design_and_Construction_R._Smith_Wiley_2010_BBS.pdf (accessed on 10 December 2023).
2. Kamali, M.; Hewage, K. Development of performance criteria for sustainability evaluation of modular versus conventional construction methods. *J. Clean. Prod.* **2017**, *142*, 3592–3606. [CrossRef]
3. Hořínková, D. Advantages and Disadvantages of Modular Construction, including Environmental Impacts. *IOP Conf. Ser. Mater. Sci. Eng.* **2021**, *1203*, 032002. [CrossRef]
4. Boafo, F.E.; Kim, J.H.; Kim, J.T. Performance of modular prefabricated architecture: Case study-based review and future pathways. *Sustainability* **2016**, *8*, 558. [CrossRef]
5. Iacovidou, E.; Purnell, P.; Tsavdaridis, K.D.; Poologanathan, K. Digitally enabled modular construction for promoting modular components reuse: A UK view. *J. Build. Eng.* **2021**, *42*, 102820. [CrossRef]
6. Gunawardena, T.; Mendis, P. Prefabricated Building Systems—Design and Construction. *Encyclopedia* **2022**, *2*, 70–95. [CrossRef]
7. Thai, H.T.; Ngo, T.; Uy, B. A review on modular construction for high-rise buildings. *Structures* **2020**, *28*, 1265–1290. [CrossRef]
8. Abdelmageed, S.; Zayed, T. A study of literature in modular integrated construction—Critical review and future directions. *J. Clean. Prod.* **2020**, *277*, 124044. [CrossRef]
9. Schnell, P. Correlation between modular construction and sustainability in the building life cycle. *E3S Web Conf. EDP Sci.* **2022**, *349*, 04006. [CrossRef]
10. Bertram, N.; Fuchs, S.; Mischke, J.; Palter, R.; Strube, G.; Woetzel, J. Modular construction: From Projects to Products. 2019. Available online: https://www.mckinsey.com/~/media/mckinsey/business%20functions/operations/our%20insights/modular%20construction%20from%20projects%20to%20products%20new/modular-construction-from-projects-to-products-full-report-new.pdf (accessed on 31 October 2023).
11. Brege, S.; Stehn, L.; Nord, T. Business models in industrialized building of multi-storey houses. *Constr. Manag. Econ.* **2014**, *32*, 208–226. [CrossRef]
12. Ferdous, W.; Bai, Y.; Ngo, T.D.; Manalo, A.; Mendis, P. New advancements, challenges and opportunities of multi-storey modular buildings—A state-of-the-art review. *Eng. Struct.* **2019**, *183*, 883–893. [CrossRef]
13. Palma, P.; Fink, G. Design for Robustness, Adaptability, Disassembly and Reuse, and Repairability of Taller Timber Buildings: A State of the Art Report, COST Action CA 20139 Holistic Design of Taller Timber Buildings (HELEN). 2022. Available online: https://cahelen.eu/wp-content/uploads/2022/12/CA20139_STAR_WG1.pdf (accessed on 3 May 2023).
14. Lattke, F.; Lehmann, S. Multistorey residential timber construction: Current developments in Europe. *J. Green Build.* **2023**, *2*, 119–129. Available online: https://meridian.allenpress.com/jgb/article/2/1/119/200068/Multi-Storey-Residential-Timber-Construction (accessed on 27 April 2023). [CrossRef]
15. Carvalho, L.F.; Jorge, L.F.C.; Jerónimo, R. Plug-and-Play Multistory Mass Timber Buildings: Achievements and Potentials. *J. Archit. Eng.* **2020**, *26*, 04020011. [CrossRef]
16. Schuster, S.; Geier, S. CircularWOOD—Towards Circularity in Timber Construction in the German Context. *IOP Conf. Ser. Earth Env. Sci. Inst. Phys.* **2022**, *1078*, 012030. [CrossRef]
17. Ramage, M.; Burridge, H.; Busse-Wicher, M.; Fereday, G.; Reynolds, T.; Shah, D.; Wu, G.; Yu, L.; Fleming, P.; Densley-Tingley, D.; et al. The wood from the trees: The use of timber in construction. *Renew. Sustain. Energy Rev.* **2017**, *68*, 333–359. [CrossRef]
18. Li, C.Z.; Li, S.; Li, X.; Wu, H.; Xiao, B.; Tam, V.W.Y.; Asiedu-Kwakyewa, C. A Scientometric Review of Management of Prefabricated Construction from 2011–2021. *Buildings* **2022**, *12*, 1515. [CrossRef]
19. Koppelhuber, J.; Bauer, B.; Wall, J.; Heck, D. Industrialized Timber Building Systems for an Increased Market Share—A Holistic Approach Targeting Construction Management and Building Economics. *Procedia Eng.* **2017**, *171*, 333–340. [CrossRef]
20. Duncheva, T.; Bradley, F.F. Multifaceted Productivity Comparison of Off-Site Timber Manufacturing Strategies in Mainland Europe and the United Kingdom. *J. Constr. Eng. Manag.* **2019**, *145*, 04019043. [CrossRef]
21. Aghasizadeh, S.; Tabadkani, A.; Hajirasouli, A.; Banihashemi, S. Environmental and economic performance of prefabricated construction: A review. *Env. Impact Assess. Rev.* **2022**, *97*, 106897. [CrossRef]
22. Bhandari, S.; Riggio, M.; Jahedi, S.; Fischer, E.C.; Muszynski, L.; Luo, Z. A review of modular cross laminated timber construction: Implications for temporary housing in seismic areas. *J. Build. Eng.* **2023**, *63*, 105485. [CrossRef]
23. Hou, L.; Tan, Y.; Luo, W.; Xu, S.; Mao, C.; Moon, S. Towards a more extensive application of off-site construction: A technological review. *Int. J. Constr. Manag.* **2022**, *22*, 2154–2165. [CrossRef]
24. Liu, W.; Zhang, H.; Wang, Q.; Hua, T.; Xue, H. A Review and Scientometric Analysis of Global Research on Prefabricated Buildings. *Adv. Civ. Eng.* **2021**, *2021*, 8869315. [CrossRef]
25. Yin, X.; Liu, H.; Chen, Y.; Al-Hussein, M. Building information modelling for off-site construction: Review and future directions. *Autom. Constr.* **2019**, *101*, 72–91. [CrossRef]
26. Ecarnot, F.; Seronde, M.F.; Chopard, R.; Schiele, F.; Meneveau, N. Writing a scientific article: A step-by-step guide for beginners. *Eur. Geriatr. Med.* **2015**, *6*, 573–579. [CrossRef]
27. Gurunathan, T.; Mohanty, S.; Nayak, S.K. A review of the recent developments in biocomposites based on natural fibres and their application perspectives. *Compos. Part A Appl. Sci. Manuf.* **2015**, *77*, 1–25. [CrossRef]

28. Codou, A.; Pin, J.M.; Misra, M.; Mohanty, A.K. Impact of temperature andin situFeCo catalysis on the architecture and Young's modulus of model wood-based biocarbon. *Green Chem.* **2021**, *23*, 3015–3027. [CrossRef]
29. Cobo, M.J.; Lõpez-Herrera, A.G.; Herrera-Viedma, E.; Herrera, F. SciMAT: A new science mapping analysis software tool. *J. Am. Soc. Inf. Sci. Technol.* **2012**, *63*, 1609–1630. [CrossRef]
30. Cobo, M.J.; López-Herrera, A.G.; Herrera-Viedma, E.; Herrera, F. An approach for detecting, quantifying, and visualizing the evolution of a research field: A practical application to the Fuzzy Sets Theory field. *J. Inf.* **2011**, *5*, 146–166. [CrossRef]
31. López-Robles, J.R.; Cobo, M.J.; Gutiérrez-Salcedo, M.; Martínez-Sánchez, M.A.; Gamboa-Rosales, N.K.; Herrera-Viedma, E. 30th Anniversary of Applied Intelligence: A combination of bibliometrics and thematic analysis using SciMAT. *Appl. Intell.* **2021**, *51*, 6547–6568. [CrossRef]
32. Salvadori, V. Multi-Storey Timber-Based Buildings: An International Survey of Case-Studies with Five or More Storeys Over the Last Twenty Years. Ph.D. Thesis, Technische Universität Wien, Vienna, Austria, 2021.
33. Safarik, D.; Elbrecht, J.; Miranda, W. State of Tall Timber 2022. *CTBUH Res.* **2022**, *2021*, 22–31. Available online: https://global.ctbuh.org/resources/papers/4530-Journal2022_IssueI_StateofTallTimber+TBIN.pdf (accessed on 10 December 2023).
34. Dobrila, P.; Premrov, M. Reinforcing methods for composite timber frame-fiberboard wall panels. *Eng. Struct.* **2003**, *25*, 1369–1376. [CrossRef]
35. Kasal, B.; Leichti, R.; Itani, R. Nonlinear Finite-Element Model of Complete Light-Frame Wood Structures. *J. Struct. Eng.* **1994**, *120*, 100–119. Available online: https://ascelibrary.org/doi/10.1061/(ASCE)0733-9445(1994)120:1(100) (accessed on 4 January 2024). [CrossRef]
36. Tarabiate, A.; Itani, R. Static and dynamic modeling of light-frame wood buildings. *Comput. Struct.* **1997**, *63*, 319–334. [CrossRef]
37. Premrov, M.; Dobrila, P. Modelling of fastener flexibility in CFRP strengthened timber-framed walls using modified γ-method. *Eng. Struct.* **2008**, *30*, 368–375. [CrossRef]
38. Negrão, J.; De Oliveira, F.; De Oliveira, C.; Cachim, P. Glued Composite Timber-Concrete Beams. II: Analysis and Tests of Beam Specimens. *J. Struct. Eng.* **2010**, *136*, 1246–1254. [CrossRef]
39. Achenbach, H.; Wenker, J.L.; Rüter, S. Life cycle assessment of product- and construction stage of prefabricated timber houses: A sector representative approach for Germany according to EN 15804, EN 15978 and EN 16485. *Eur. J. Wood Wood Prod.* **2018**, *76*, 711–729. [CrossRef]
40. Padilla-Rivera, A.; Amor, B.; Blanchet, P. Evaluating the link between low carbon reductions strategies and its performance in the context of climate Change: A carbon footprint of awood-frame residential building in Quebec, Canada. *Sustainability* **2018**, *10*, 2715. [CrossRef]
41. Li, Z.; Asce, M.; Wang, X.; He, M.; Dong, W.; Dong, H. Seismic Performance of Timber-Steel Hybrid Structures. I: Subassembly Testing and Numerical Modeling. *J. Struct. Eng.* **2019**, *145*, 04019113. [CrossRef]
42. Wang, L.; Ge, H. Stochastic modelling of hygrothermal performance of highly insulated wood framed walls. *Build Env.* **2018**, *146*, 12–28. [CrossRef]
43. Glass, S.V.; Kochkin, V.; Drumheller, S.C.; Barta, L. Moisture performance of energy-efficient and conventional wood-frame wall assemblies in a mixed-humid climate. *Buildings* **2015**, *5*, 759–782. [CrossRef]
44. McClung, R.; Ge, H.; Straube, J.; Wang, J. Hygrothermal performance of cross-laminated timber wall assemblies with built-in moisture: Field measurements and simulations. *Build. Environ.* **2014**, *71*, 95–110. [CrossRef]
45. Awad, H.; Gül, M.; Zaman, H.; Yu, H.; Al-Hussein, M. Evaluation of the thermal and structural performance of potential energy efficient wall systems for mid-rise wood-frame buildings. *Energy Build* **2014**, *82*, 416–427. [CrossRef]
46. Van De Lindt, J.W.; Asce, M.; Gupta, R. Damage and Damage Prediction for Wood Shearwalls Subjected to Simulated Earthquake Loads. *J. Perform. Constr. Facil.* **2006**, *20*, 176–184. [CrossRef]
47. Leskovar, V.Ž.; Premrov, M. Design approach for the optimal model of an energy-efficient timber building with enlarged glazing surface on the south façade. *J. Asian Archit. Build. Eng.* **2012**, *11*, 71–78. [CrossRef]
48. Loss, C.; Piazza, M.; Zandonini, R. Connections for steel–timber hybrid prefabricated buildings. Part II: Innovative modular structures. *Constr. Build Mater.* **2016**, *122*, 796–808. [CrossRef]
49. Loss, C.; Davison, B. Innovative composite steel-timber floors with prefabricated modular components. *Eng. Struct.* **2017**, *132*, 695–713. [CrossRef]
50. Daňková, J.; Mec, P.; Šafrata, J. Experimental investigation and performance of timber-concrete composite floor structure with non-metallic connection system. *Eng. Struct.* **2019**, *193*, 207–218. [CrossRef]
51. Lukaszewska, E.; Fragiacomo, M.; Johnsson, H. Laboratory Tests and Numerical Analyses of Prefabricated Timber-Concrete Composite Floors. *J. Struct. Eng.* **2010**, *136*, 46–55. [CrossRef]
52. Doudak, G.; Mcclure, G.; Smith, I.; Hu, L.; Stathopoulos, T. Monitoring Structural Response of a Wooden Light-Frame Industrial Shed Building to Environmental Loads. *J. Struct. Eng.* **2005**, *131*, 794–805. [CrossRef]
53. Kim, J.H.; Rosowsky, D.V.; Asce, M. Fragility Analysis for Performance-Based Seismic Design of Engineered Wood Shearwalls. *J. Struct. Eng.* **2005**, *131*, 1764–1773. [CrossRef]
54. Pei, S.; van de Lindt, J.W. Influence of structural properties and hazard level on seismic loss estimation for light-frame wood structures. *Eng. Struct.* **2010**, *32*, 2183–2191. [CrossRef]
55. Pfretzschner, K.S.; Gupta, R.; Miller, T.H. Practical Modeling for Wind Load Paths in a Realistic Light-Frame Wood House. *J. Perform. Constr. Facil.* **2014**, *28*, 430–439. [CrossRef]

56. Ottenhaus, L.M.; Jockwer, R.; van Drimmelen, D.; Crews, K. Designing timber connections for ductility—A review and discussion. *Constr. Build Mater.* **2021**, *304*, 124621. [CrossRef]
57. Ceccotti, A.; Sandhaas, C.; Okabe, M.; Yasumura, M.; Minowa, C.; Kawai, N. SOFIE project—3D shaking table test on a seven-storey full-scale cross-laminated timber building. *Earthq. Eng. Struct. Dyn.* **2013**, *42*, 2003–2021. [CrossRef]
58. Lukaszewska, E.; Johnsson, H.; Fragiacomo, M. Performance of connections for prefabricated timber-concrete composite floors. *Mater. Struct. Mater. Et Constr.* **2008**, *41*, 1533–1550. [CrossRef]
59. Dodoo, A.; Gustavsson, L.; Sathre, R. Lifecycle carbon implications of conventional and low-energy multi-storey timber building systems. *Energy Build* **2014**, *82*, 194–210. [CrossRef]
60. Ringhofer, A.; Brandner, R.; Blaß, H.J. Cross laminated timber (CLT): Design approaches for dowel-type fasteners and connections. *Eng. Struct.* **2018**, *171*, 849–861. [CrossRef]
61. Udele, K.E.; Sinha, A.; Morrell, J.J. Effects of Re-drying on properties of cross laminated timber (CLT) connections. *J. Build. Eng.* **2023**, *76*, 107298. [CrossRef]
62. Li, Z.; Wang, X.; He, M.; Shu, Z.; Huang, Y.; Wu, A.; Ma, Z. Mechanical performance of pre-fabricated metal dovetail connections for Cross-Laminated Timber (CLT) structures. *Constr. Build Mater.* **2021**, *303*, 124468. [CrossRef]
63. Bérubé, A.; Doudak, G. Investigation of energy-absorbing connections for application in mass timber structures. *Eng. Struct.* **2023**, *297*, 117018. [CrossRef]
64. Ahn, N.; Dodoo, A.; Riggio, M.; Muszynski, L.; Schimleck, L.; Puettmann, M. Circular economy in mass timber construction: State-of-the-art, gaps and pressing research needs. *J. Build. Eng.* **2022**, *53*, 104562. [CrossRef]
65. Shi, B.; Liu, W.; Yang, H. Experimental investigation on the long-term behaviour of prefabricated timber-concrete composite beams with steel plate connections. *Constr. Build Mater.* **2021**, *266*, 120892. [CrossRef]
66. Tao, H.; Shi, B.; Yang, H.; Wang, C.; Ling, X.; Xu, J. Experimental and finite element studies of prefabricated timber-concrete composite structures with glued perforated steel plate connections. *Eng. Struct.* **2022**, *268*, 114778. [CrossRef]
67. Brambilla, A.; Gasparri, E. Hygrothermal behaviour of emerging timber-based envelope technologies in Australia: A preliminary investigation on condensation and mould growth risk. *J. Clean. Prod.* **2020**, *276*, 124129. [CrossRef]
68. Boardman, C.R.; Glass, S.V. Improving the accuracy of a hygrothermal model for wood-frame walls: A cold-climate study. *Buildings* **2020**, *10*, 236. [CrossRef]
69. Jayalath, A.; Navaratnam, S.; Gunawardena, T.; Mendis, P.; Aye, L. Airborne and impact sound performance of modern lightweight timber buildings in the Australian construction industry. *Case Stud. Constr. Mater.* **2021**, *15*, e00632. [CrossRef]
70. Wei, Y.; Lei, Z.; Altaf, S. An Off-Site Construction Digital Twin Assessment Framework Using Wood Panelized Construction as a Case Study. *Buildings* **2022**, *12*, 566. [CrossRef]
71. Morales-Beltran, M.; Engür, P.; Şişman, Ö.A.; Aykar, G.N. Redesigning for Disassembly and Carbon Footprint Reduction: Shifting from Reinforced Concrete to Hybrid Timber–Steel Multi-Story Building. *Sustainability* **2023**, *15*, 7273. [CrossRef]
72. Du, Q.; Zhang, R.; Cai, C.; Jin, L. Factors influencing modern timber structure building development in China. *Sustainability* **2021**, *13*, 7936. [CrossRef]
73. Willmann, J.; Knauss, M.; Bonwetsch, T.; Apolinarska, A.A.; Gramazio, F.; Kohler, M. Robotic timber construction—Expanding additive fabrication to new dimensions. *Autom. Constr.* **2016**, *61*, 16–23. [CrossRef]
74. Wagner, H.J.; Alvarez, M.; Kyjanek, O.; Bhiri, Z.; Buck, M.; Menges, A. Flexible and transportable robotic timber construction platform—TIM. *Autom. Constr.* **2020**, *120*, 103400. [CrossRef]
75. Margani, G.; Evola, G.; Tardo, C.; Marino, E.M. Energy, seismic, and architectural renovation of RC framed buildings with prefabricated timber panels. *Sustainability* **2020**, *12*, 4845. [CrossRef]
76. Udele, K.; Morrel, J.; Cappellazi, J.; Sinha, A. Characterizing properties of fungal-decayed cross laminated timber (CLT) connection assemblies. *Constr. Build Mater.* **2023**, *409*, 134080. [CrossRef]
77. Li, X.; Ashraf, M.; Subhani, M.; Kremer, P.; Kafle, B.; Ghabraie, K. Experimental and numerical study on bending properties of heterogeneous lamella layups in cross laminated timber using Australian Radiata Pine. *Constr. Build Mater.* **2020**, *247*, 118525. [CrossRef]
78. Premrov, M.; Dobrila, P.; Bedenik, B.S. Approximate analytical solutions for diagonal reinforced timber-framed walls with fibre-plaster coating material. *Constr. Build Mater.* **2004**, *18*, 727–735. [CrossRef]
79. Ribeiro, A.M.; Arantes, A.; Cruz, C.O. Barriers to the Adoption of Modular Construction in Portugal: An Interpretive Structural Modeling Approach. *Buildings* **2022**, *12*, 1509. [CrossRef]
80. Svatoš-Ražnjević, H.; Orozco, L.; Menges, A. Advanced Timber Construction Industry: A Review of 350 Multi-Storey Timber Projects from 2000–2021. *Buildings* **2022**, *12*, 404. [CrossRef]
81. Briani, A.; Simeone, P.; Ceccotti, A. MAI, IVALSA modular House. In Proceedings of the World Conference on Timber Engineering, Auckland, New Zealand, 16–19 July 2012; pp. 29–36.
82. Academicum, C. International Building Exhibitions. 2020. Available online: https://iba.heidelberg.de/en/projects/collegium-academicum (accessed on 26 June 2023).
83. Collegium Academicum IBA. DGJ Architektur. 2015. Available online: https://dgj.eu/en/portfolio/dgj223-iba-collegium-academicum/ (accessed on 5 September 2023).

84. Finch, G.; Marriage, G.; Gjerde, M.; Pelosi, A.; Patel, Y. Understanding the Challenges of Circular Economy Construction through Full-Scale Prototyping. 2020. Available online: https://anzasca.net/wp-content/uploads/2021/03/131-Understanding-the-challenges-of-circular-economy-construction-through-full-scale-prototyping.pdf (accessed on 20 July 2023).
85. Grüter, C.; Gordon, M.; Muster, M.; Kastner, F.; Grönquist, P.; Frangi, A.; Langenberg, S.; De Wolf, C. Design for and from disassembly with timber elements: Strategies based on two case studies from Switzerland. *Front. Built Env.* **2023**, *9*, 1307632. [CrossRef]
86. Derikvand, M.; Fink, G.; Deconstruction, D.F. Challenges, and Outlook for Timber–Concrete Composite Floors. *Buildings* **2023**, *13*, 1754. [CrossRef]
87. *ISO 20887*; Sustainability in Buildings and Civil Engineering Works—Design for Disassembly and Adaptability—Principles, Requirements and Guidance. International Organization for Standardization: Geneva, Switzerland, 2020.
88. Yan, Z.; Ottenhaus, L.M.; Leardini, P.; Jockwer, R. Performance of reversible timber connections in Australian light timber framed panelised construction. *J. Build. Eng.* **2022**, *61*, 105244. [CrossRef]
89. Bucklin, O.; Di Bari, R.; Amtsberg, F.; Menges, A. Environmental Impact of a Mono-Material Timber Building Envelope with Enhanced Energy Performance. *Sustainability* **2023**, *15*, 556. [CrossRef]
90. An, S.; Martinez, P.; Al-Hussein, M.; Ahmad, R. BIM-based decision support system for automated manufacturability check of wood frame assemblies. *Autom. Constr.* **2020**, *111*, 103065. [CrossRef]
91. Potseluyko, L.; Rahimian, F.P.; Dawood, N.; Elghaish, F.; Hajirasouli, A. Game-like interactive environment using BIM-based virtual reality for the timber frame self-build housing sector. *Autom. Constr.* **2022**, *142*, 104496. [CrossRef]
92. Kyjanek, O.; Al Bahar, B.; Vasey, L.; Menges, A. Implementation of an Augmented Reality AR workflow for Human Robot Collaboration in Timber Prefabrication. In Proceedings of the 36th International Symposium on Automation and Robotics in Construction (ISARC), Banff, AB, Canada, 21–24 May 2019; pp. 1223–1230.
93. Eversmann, P.; Gramazio, F.; Kohler, M. Robotic prefabrication of timber structures: Towards automated large-scale spatial assembly. *Constr. Robot.* **2017**, *1*, 49–60. [CrossRef]
94. Lachance, E.; Lehoux, N.; Blanchet, P. Automated and robotized processes in the timber-frame prefabrication construction industry: A state of the art. In Proceedings of the 2022 IEEE 6th International Conference on Logistics Operations Management, GOL, Strasbourg, France, 29 June–1 July 2022. [CrossRef]
95. Chong, O.W.; Zhang, J.; Voyles, R.M.; Min, B.C. BIM-based simulation of construction robotics in the assembly process of wood frames. *Autom. Constr.* **2022**, *137*, 104194. [CrossRef]
96. De Wolf, C.; Bocken, N.; Çetin, S. A Circular Built Environment in the Digital Age, Springer Cham, Switzerland. 2024. Available online: https://link.springer.com/book/10.1007/978-3-031-39675-5 (accessed on 16 February 2024).
97. Santana-Sosa, A.; Kovacic, I. Barriers, Opportunities and Recommendations to Enhance the Adoption of Timber within Multi-Storey Buildings in Austria. *Buildings* **2022**, *12*, 1416. [CrossRef]
98. Zaman, A.; Chan, Y.Q.; Jonescu, E.; Stewart, I. Critical Challenges and Potential for Widespread Adoption of Mass Timber Construction in Australia—An Analysis of Industry Perceptions. *Buildings* **2022**, *12*, 1405. [CrossRef]

sustainability

MDPI

Article

Using Timber in Mid-Rise and Tall Buildings to Construct Our Cities: A Science Mapping Study

Alexander Wenzel [1], Pablo Guindos [1,2,3] and Manuel Carpio [1,3,4,*]

1 Centro Nacional de Excelencia para la Industria de la Madera (CENAMAD), Pontificia Universidad Católica de Chile, 4860 Vicuña Mackenna, Santiago 7820436, Chile; aawenzel@uc.cl (A.W.); pablo.guindos@udc.es (P.G.)
2 Department of Structural and Geotechnical Engineering, Pontificia Universidad Católica de Chile, 4860 Vicuña Mackenna, Santiago 7820436, Chile
3 Department of Construction Engineering and Management, Pontificia Universidad Católica de Chile, 4860 Vicuña Mackenna, Santiago 7820436, Chile
4 Department of Construction Engineering and Project Management, University of Granada, Campus Fuentenueva, 18071 Granada, Spain
* Correspondence: manuel.carpio@uc.cl or carpio@ugr.es

Abstract: The increase in population and urban migration has incentivized the construction of mid-rise and tall buildings. Despite the incremental rise in vertical construction, there are still investigation gaps related to high-rise buildings, such as carbon emissions and the use of low-carbon materials in tall structures. Timber presents a potential sustainable solution for mid-rise and tall buildings. The history of topics in timber building investigations began with the material characterization of innovation in construction technologies such as cross-laminated timber (CLT) and practical topics like construction collaboration, sustainability, engineering, and construction science. To identify potential topics and understand the research history of mid- and high-rise timber buildings, a bibliometric analysis is proposed. Therefore, this article aims to perform a bibliometric analysis with a science mapping technique to categorize and analyze the evolution of mid- and high-rise timber building research topics and identify the most relevant trends and current challenges. A co-occurrence keyword analysis was performed with the software SciMAT to analyze the evolution and actual trends of mid-rise and tall timber buildings. The results show an evolution in the investigation topics from timber frame elements to mass timber and CLT for high-rise buildings, which was expected due to the higher structural capacity of the mass timber product. Surprisingly, sustainability topics such as carbon emission and life-cycle analysis (LCA) were transversal in all periods with concrete as a recurrent keyword in the analysis. More specialized topics such as robustness, disproportioned collapse, perceptions, and attitude were observed in the final periods. Research projections indicate that for mid-rise and tall timber buildings, the environmental potential has to be aligned with the structural feasibility and perception of the construction's actors and society to improve the carbon emissions reduction and support the increment of the population in an urban context.

Keywords: bibliometric analysis; tall building; timber; multi-storey building; science mapping

1. Introduction

An increase in population and migration trends from rural to urban contexts means an increase in the urban population influx and, considering the limitations of urban sprawl, vertical dimension should be taken into consideration [1]. For example, due to the expected increase in the population of China, the construction of high-rise buildings is a viable

 https://doi.org/10.3390/su17051928

solution to accommodate the increase in population in an urban context [2]. Despite tall buildings being a potential solution to population increase, research on high-rise buildings displays potential gaps related to energy simulation, thermal and visual comfort, renewable energies, and carbon emission, specifically with regard to the reduction in embodied carbon with the use of low carbon materials [3].

Timber, as a biobased material, presents the capability to capture carbon, which can contribute to decreasing the total embodied carbon of mass timber buildings and obtain a reduced amount of emissions in contrast to concrete buildings [4]. Constructing our cities with timber that comes from sustainably managed forests has often been argued as the most promising solution to reducing the embodied carbon of the great mass of resources used in construction, both from a scientific perspective [5] and from the perspective of the authorities [6].

Mid-rise and tall buildings represent a promising solution for urban sprawl. In addition, the application of timber in the structures can potentially achieve low emissions. However, three decades prior, it would have been unthinkable that timber could be a feasible solution to construct our cities, due to the societal need for taller buildings. However, in the 2000s, new timber products and technologies became popular, such as cross-laminated timber (CLT) and other mass timber products, which cemented timber's reputation as a feasible construction material for mid- and even high-rise buildings [7]. Early studies in multi-storey timber buildings began with fundamental subjects and topics, such as the characterization of timber properties as a building material [8]; the refinement of the structural codes such as the wind-load provisions uniform building code [9]; and seismic behaviour under the application of damped resonant appendages [10]. However, today's interest and research in timber buildings is greatly focused on critical aspects such as sustainability, engineering, construction science, and collaborative design [11]. To identify the potential topics, a clear comprehension of the research history of mid- and high-rise timber building lines of research is needed. Due to the large amount of data related to multi-storey tall timber buildings, research analytical methods such as bibliometric analysis can be used [12].

Bibliometrics analysis originated as a scientific computer-assisted review methodology that covers all the publications related to a given topic or field and can identify core research authors and their relationships [13]. Bibliometric analysis can be classified into two techniques: performance analysis and science mapping. Performance analysis focuses on contributions, while science mapping engages the relations between research variables [12]. A few bibliometric analyses on timber or wood have recently been made [14–17] mostly focused on specific structural components, materials, or forest level.

The increase in population and the necessity to develop low-emission solutions have propelled the use of timber as a building solution. Different topics in the research history of mid- and high-rise timber building research lines have evolved and emerged.

To address these issues, this study is structured around the following research questions: (1) What have been the main research lines in mid- and high-rise timber buildings over time? (2) What are the current research trends in this field? (3) What challenges and opportunities arise from implementing timber-based solutions in urban contexts? These questions guide our bibliometric analysis to provide clarity on the evolution of knowledge in this discipline.

Therefore, this article aimed to perform a bibliometric analysis with a science mapping technique to categorize and analyze the evolution of mid- and high-rise timber buildings research topics and identify the most relevant trends and current challenges. This is expected to be a relevant contribution in this incipient research field, for both established and new researchers to understand how the appearance of mid-rise and tall timber buildings is shaping the needs and trends of the entire discipline. The manuscript is organized as follows: the methodology section explains the theoretical elements, inputs, and outputs of the science-mapping analysis; the results section presents findings categorized by science mapping outputs and research periods; and finally, the conclusions summarize key insights and final reflections on mid-rise and tall timber buildings.

2. Materials and Methods

2.1. Setting the Analysis Database

2.1.1. Collecting the Publications of Mid-Rise and Tall Timber Buildings

According to Donthu et al. [12], the first step required to perform a bibliometric analysis is establishing the aim and the scope of the analysis. This bibliometric study aims to analyze retrospectively the evolution of research topics related to mid-rise and tall timber buildings and identify the actual research challenges. A flowchart of the bibliometric analysis is presented in Figure 1.

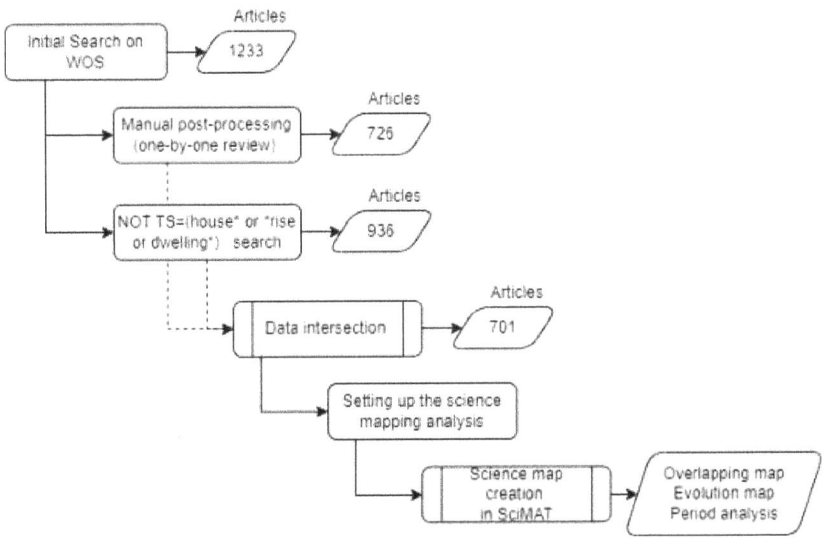

Figure 1. Flowchart of the bibliometric analysis from the data acquisition to the mapping and period analysis. (*) truncation operator that allows for the retrieval of different word variations by replacing multiple characters.

The bibliometric analysis was limited to the Web of Science (WOS) database due to the large number of articles, avoiding highly possible duplicated files, and using only one standardized file format [12]. Additionally, WOS possesses several citation and publication data for well-founded discovery and bibliometric analyses, exhibits a large collection of scientific publications and highly rated journals, and features a notable multidisciplinary nature useful to collect State-of-the-Art information [18].

The search was performed with no year restriction, but the final data obtained were until the end of 2022. The search formula used in this investigation is described in Equation (1). The format of Equation (1) can be used in the advanced search on the WOS database.

$$TS = ((\textit{timber* or wood*}) \textit{ and } (\textit{building*}) \textit{ and } (\textit{"mid" or "tall" or "high" or "multi-story" or "multi story" or "multi-storey"})) \textit{ and } (AK = (\textit{building*}) \textit{ or } TI = (\textit{building*})) \quad (1)$$

The term selection was chosen to cover timber or wood buildings with the additional characteristic of being mid-rise or tall height. The addition of the AND of building in the title or keyword was selected to filter non-related topics such as *microbiology* and *marine freshwater biology*, for example.

F0 represents the initial data set of articles matching with the above search criterion, which in this case consisted of a total of 1233 articles. The search criterion was conducted in the abstract, title, keyword, and keyword plus with an additional filter on the keywords and title, which consisted of a constraint related to the word building presented in the keyword or title. All resulting articles were selected at this first instance.

After the standardized search, a manual post-processing (one-by-one review) of the data was made. The data resulting from the manual review was called *F1* and consisted of 726 articles. This manual filtering aimed to check the contents of the articles and avoid themes not focused on mid- and high-rise timber buildings. For example, it was discarded if wood was used as fuel for building climatization. Another criterion applied to the search was to reject articles focused on the forestry level or previous processes of a timber building. In the case of window frames or louvres articles, they were not selected if the wall was made using a building material other than timber. On the other hand, elements like floors, walls, roofs, or envelopes related to timber elements applied in mid- and high-rise buildings were selected. Investigations based on timber products but without an application on timber mid- and high-rise buildings were also discarded. In the case of timber-based articles that did not mention the case of the study, the investigation's scalability was reviewed to search for the potential applicability of mid- and high-rise buildings.

In parallel to the manual post-processing, the original search formula was modified to apply an additional constraint related to low-height buildings, low-rise buildings, houses, and dwellings. The filter comprised an exclusion formula, *NOT TS = (house* or *rise or dwelling*) and* the data set obtained was called *F2*. The modified formula found a total of 936 articles instead of the 1.233 of the first search.

A data intersection was then performed between *F1* and *F2*. The objectives of this search were twofold: first, to check the accuracy of the manual filter results *F1* concerning the more specific formula *F2*, reviewing the excluded articles in *F1* but added in *F2*. If the article was rejected in the *F1* manual filter but added to *F2*, this could be an error in the manual filter, so the intersection allowed for a double check of articles manually discarded for reasons other than not being low-rise construction.

The second objective was to check the cases where an article was added to *F1* but not to *F2*; this allowed us to double-check if, after manual filtering, any low-rise research was mistakenly included in the analysis. Figure 2 shows the different data filters applied and how they intersected with the number of resultant articles.

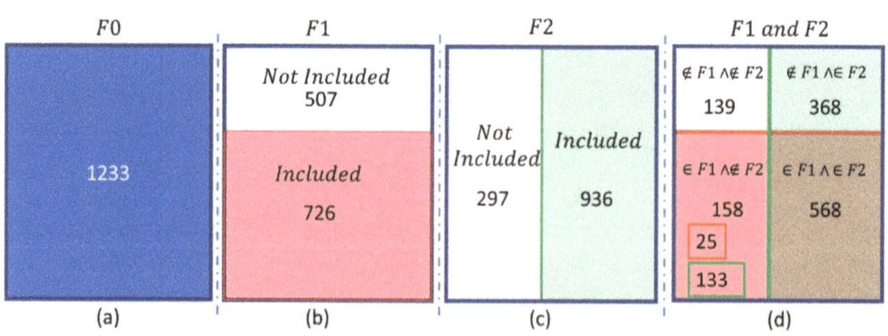

Figure 2. Research database acquisition, including the initial data set F0 (**a**), the manual filtering data set F1 (**b**), the low rise excluded filter data set F2 (**c**), and the intersections of F1 and F2 (**d**).

Regarding the first objective, by the intersection, it was found that 368 articles were excluded in *F*1 but added to *F*2. After manually reviewing the articles again, it was found that the *F*1 manual filter was correctly applied. Regarding the second objective, the number of articles included in *F*1 but not in *F*2 were 158, which were manually reviewed. Of these 158, 25 articles agreed that the focus was low-height, low-rise; houses and dwellings, therefore, were discarded. However, the remaining 133 articles were correctly added in *F*1 because the contributions could still be of relevance for tall timber construction. This resulted in a final database analysis set of 701 articles, 568 articles belonging to both manual filtering *F*1 and *F*2, plus the 133 articles that belong to *F*1 but not *F*2 and that were deemed as relevant for the investigation. The final data set resulted in the 568 articles belonging to *F*1 and *F*2, plus the 133 articles that belong to *F*1 but not *F*2 and were deemed as relevant.

2.1.2. Enriching the Analysis Database with KeyWord Plus

The tool KeyWord Plus of WoS permits the consideration of terms cited in references that do not appear in the authors' keywords of the article itself, thus, allowing for an enriched search and analysis due to multiple keywords derived from the metadata. However, the accuracy of the autogenerated words cannot always be precise. In this article, an additional review of the final 701 documents was performed, in which only 420 presented added words from KeyWord Plus. The additional review was made to ensure the added words' relationship to the articles' topics and contents. For example, in multiple articles, the word *green house gas* (*GHG*) was added to the KeyWord Plus category because several articles started their introduction to research problems by introducing the positive impact of timber in the reduction in GHG emissions [19,20]. Another example of misleading by KeyWord Plus was the use of the term *cyclic analysis*, which was added in a structural article referring to laboratory tests of elements under cyclic load [21]. The article presented a methodology to determine the technical feasibility of timber frame structures at early design phases for highly seismic zones but did not perform or analyze cyclic loads in any timber element.

In other cases, the addition of the KeyWord Plus was very accurate and gave more information about the article. For example, the investigation of Lazzarini et al. [22] analyzed and discussed the structural dynamics of high-rise timber structures, specifically the constructed Mjøstårnet in Brumunddal (Norway). Their keywords mentioned high-rise timber buildings and tall timber structures, but there was no information about the specific construction method. In this case, KeyWord Plus integrated the word CLT, giving a more detailed word that can be related later in the analyses.

After the review, it was observed that words like *performance*, *behaviour*, *system*, *design*, and *model* were added multiple times by KeyWord Plus. Due to the general nature of these words, their relationship with mid- and high-rise timber building articles was mostly correct. Despite the general characteristics of these particular keywords, the consistency of the extra information addition was not precise, leaving articles that presented topics of *performance*, *design*, and *model* without these additional keywords.

2.2. Setting Up the Science Mapping Analysis

2.2.1. Definition of Stages and Inputs

Bibliometric analysis can be performed through a performance analysis and a science mapping analysis. Science mapping analyses the relations among these indicators through different methods, delimiting research areas and monitoring a science field to determine the cognitive structure and its evolutions [23]. A co-word analysis over a science mapping tool was made to investigate the evolution and future research themes of mid-rise and tall timber buildings. The software SciMAT [24] was selected to perform the bibliometric analysis. As mentioned in [25], SciMAT was widely applied in psychology, marketing, computer science, and management. After the database was prepared, a four-stage process began in SciMAT: (i) the building of the data set; (ii) the network's normalization; (iii) the cluster algorithm selection to map the results; and (iv) analysis selection to be performed on the generated map. The sub-process of each stage is presented in Table 1, with their respective inputs used in the article.

Table 1. Stage, sub-process, and inputs for SciMAT.

Stage	Sub-Process	Input
1	Select periods	1988–2010 2011–2016 2017–2020 2021–2022
2	Select the unit of analysis	Words
	Data reduction per period	2 3 3 3
	Network base	Co-occurrence
	Network reduction	0 2 2 2
	Similarity measure for network normalization	Equivalence index
3	Cluster algorithm	Clustering algorithm: Simple centre algorithm Max: 9 Min: 5
4	Document mapper	Core mapper Union mapper
	Quality measures	H-index Sum citations
	Longitudinal and temporal analysis	Evolution map: Inclusion index Overlapping map: Jaccard index

The selection of the study periods (1988–2010, 2011–2016, 2017–2020, and 2021–2022) was based on significant milestones in the evolution of mid- and high-rise timber building research. The first period (1988–2010) represents the foundational stage, where research primarily focused on timber-frame structures and life-cycle analysis. The second period (2011–2016) marks the emergence of mass timber products, such as CLT, and the initial structural feasibility studies of mid-rise timber buildings. The third period (2017–2020) reflects a rapid increase in scientific publications, a shift toward robustness and collapse prevention studies, and growing interest in the environmental and societal perceptions of timber buildings. Finally, the fourth period (2021–2022) covers the most recent research trends, including hybrid structural systems, refined life-cycle assessment (LCA) methodologies, and advanced structural modelling for taller timber buildings. These divisions were established to capture major research transitions and allow a more detailed analysis of evolving themes within the field.

2.2.2. Normalization Measures

In science mapping, normalization measures are required to detect the interaction between the functional unit and the significant elements, such as keywords, clusters, or periods. SciMAT has three instances to select normalization measures: the cluster network, overlapping map, and evolution map. According to [26] in the co-occurrence analysis, the most appropriated normalization measure was the equivalence index from Callon, shown in Equation (2). In this example, c_i and c_j represent a value associated with the elements *i* and *j*, respectively. On the other hand, c_{ij} represents a value shared by the elements *i* and *j*. For example, if we assume that *i* and *j* are keywords, c_{ij} can be the number of documents that present them, with c_i and c_j being the number of documents presented with the keywords *i* and *j*, respectively [23,27]

$$E_{i\,j} = \frac{(C_{i\,j})^2}{C_i * C_j} \tag{2}$$

For the overlapping and evolution maps, the Jaccard indexes and the inclusion indexes were used, respectively, as shown in Equations (3) and (4). Their selection was based on the increase in the edge thickness of the connection to identify more clearly the relationship between clusters [28,29].

$$E_{i\,j} = \frac{C_{i\,j}}{C_i + C_j - C_{i\,j}} \tag{3}$$

$$E_{i\,j} = \frac{C_{i\,j}}{\min(C_i; C_j)} \tag{4}$$

SciMAT applies these normalization measures in conjunction with advanced statistical principles to enhance the accuracy and interpretability of science mapping analyses. The software employs co-occurrence techniques (e.g., co-word analysis, co-citation, bibliographic coupling) to construct relational networks, which are further processed using clustering algorithms and centrality-density metrics to identify key thematic structures. Additionally, SciMAT integrates evolutionary analysis methods, such as overlapping and transition indices, to track thematic persistence and transformation across time periods. To refine data interpretation, descriptive statistics are used to quantify term frequencies and distribution patterns, while impact metrics like the h-index and citation counts provide insight into the influence of research topics. Furthermore, SciMAT leverages dimensionality reduction techniques, such as correspondence analysis, alongside linguistic normalization, ensuring a structured and coherent representation of bibliometric data.

2.3. Science Mapping Outputs

2.3.1. Overlapping Map

The overlapping map is a graphic representation of the unit of analysis evolution trough periods [23], see Figure 3. In this investigation, the horizontal arrow between periods (X1) represents the keywords that are maintained from one period to the next. The element in the parenthesis (Y1) on the horizontal arrow represents the value of the normalization measure used in the analysis, in this case, the Jaccard index. The periods were represented by A1 and A2.

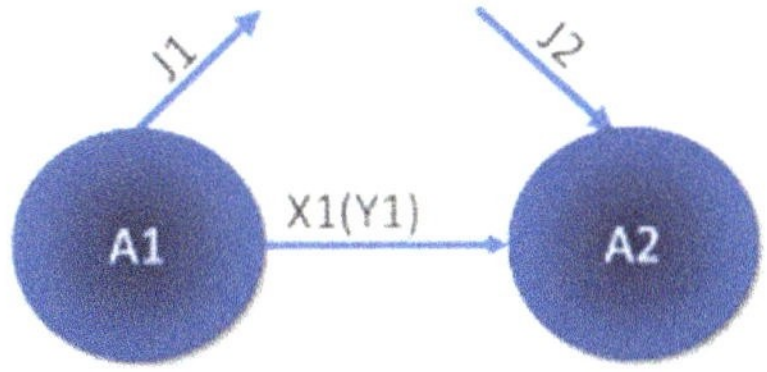

Figure 3. Theoretical overlapping map.

The arrow leaving the period in a cartesian Y positive direction (J1) represents the new keywords from the old period but are not included in the new period. The arrow entering the new period with a Y negative direction (J2) represents the new keywords integrated into the period.

2.3.2. Evolution Map

The evolution map shown in Figure 4 represents the theme development through the selected periods [30] in which the diameter represents the number of documents associated with the principal cluster.

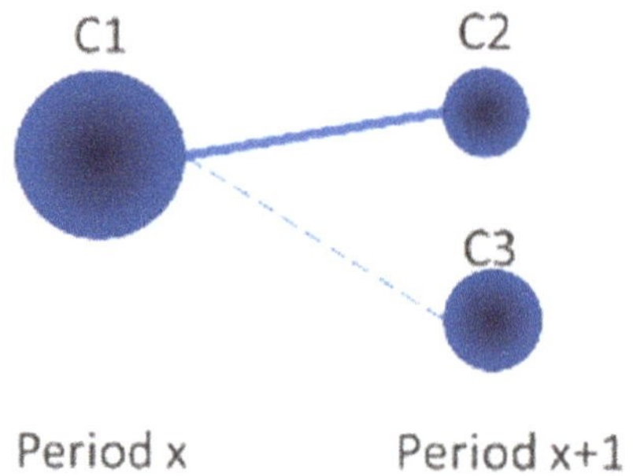

Figure 4. Theoretical evolution map.

In this research, the solid lines represent the fact that the cluster shares the main keyword with one of the clusters between periods. Conversely, the dotted line represents a connection between clusters (C1, C2, C3) with a keyword that is not the main one. The thickness of the line is proportional to the inclusion index, selected as a measure in the SciMAT interface.

2.3.3. Strategic Diagram

The strategic diagram is a graphic representation in a cartesian plane of the period's main clusters distribution. The placement is based on two variables, centrality (x-axis) and density (y-axis). Centrality represents the interaction degree of a single network with other networks and illustrates the external ties strength to other themes [23]. Density measures the strength of the internal ties among the unit of analysis, and it can be understood as a measure of topic development [23]. With the values of centrality and density, the cluster network can be classified in one of the fourth quadrants of the strategic diagram. The first quadrant is the motor theme, which presents a positive density and centrality and illustrates the cluster with the important topics in the research fields and their high development [31]. The second quadrant in a cartesian plane represents the highly developed but unconnected topics, mostly specific research topics. The third quadrant is the emerging or disappearing topics, with a low centrality and density. Finally, the fourth quadrant is the transversal and basic topics, which are topics not highly developed but important for the research fields [31]. The diameter of the cluster circle represents an indicator in the analysis. In this article, the diameter represents the number of documents per cluster network or cluster. Figure 5 shows the cartesian plane of the strategic diagram.

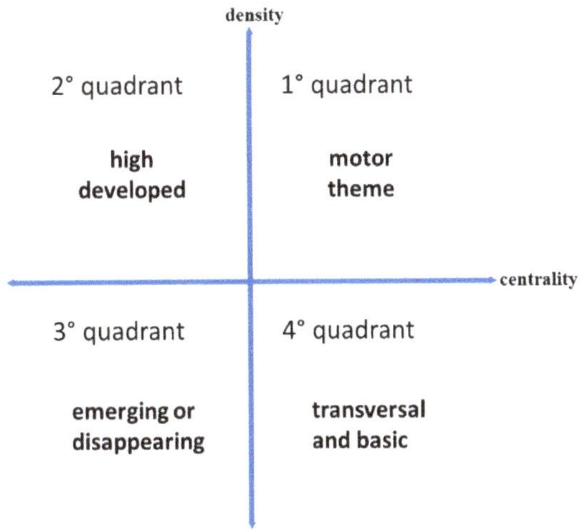

Figure 5. Theoretical strategic diagram.

2.3.4. Cluster Network

The cluster network represents the interaction between clusters that share a common unit of analysis. The name of the cluster network is represented by the most central unit of analysis [23]. The thickness of the edge represents the connection between two clusters and in this investigation, the edge thickness value corresponds to the equivalence index. As mentioned, the diameter of the circle represents the number of documents per cluster.

3. Results

3.1. Overlapping Map

In this investigation, the overlapping map reveals an increment in the number of keywords that are kept from one period to the following, from 55 to 282. The final period contains a reduced number of years for two main reasons: due to the increment of articles in the field and to identify current trends. Similarities in the cluster and periods are computed based on indexes or measures. In the case of the Jaccard index, shown between parenthesis in Figure 6, an increase through the periods was noticed. This increment indicates a potential common research topic in the past and future periods. Y positive and Y negative direction arrows show an increment of non-included and new keywords, showing high renovations on the investigation topics or keywords used.

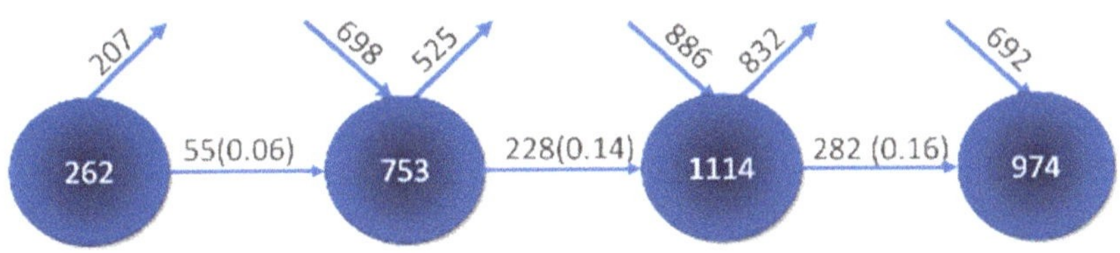

Figure 6. Overlapping map between periods.

3.2. Evolution Map

The evolution map evidences that the study of mid-rise and tall timber buildings progresses from general to more specific themes (see Figure 7). The initial main cluster keywords of period 1 were *timber-buildings* and *life-cycle*, which are more general than subsequent terms. As aforementioned, the words *building* and *timber-buildings* were merged into one group. *Timber-buildings* are connected to *timber-frame-building* in the second period through a solid line, sharing the core theme and indicating the initial relevance of frame structures in the early years of investigation. In the case of *timber-buildings*, the word was the core theme in periods one and two. The topics did not connect with the posterior clusters, suggesting more specific research.

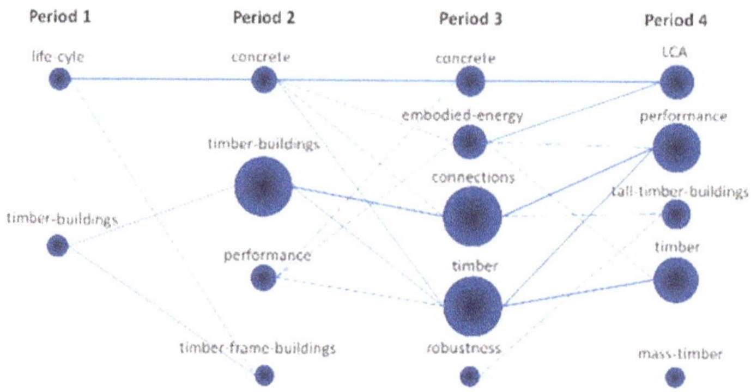

Figure 7. Evolution Map from period 1 to period 4.

In the second period, new clusters such as *concrete*, *performance*, *timber-frame-building*, and *timber-buildings* emerge. *Timber-buildings* are linked to *connections* through *cross-laminated-timber*, which indicates the research community's potential interest and necessity to study the structural elements and how the connections are made between them. *Timber-frame-building* did not connect in the next period, indicating a possible reduction in the investigation of timber-framed structures because those products do not allow for tall buildings. *Concrete* presents two solid-edge connections to the next period, *concrete* and *timber*. The connection with *timber* was through the main keyword *timber*. The clusters from the third period present more edge convergence than the second period, indicating that the next period cluster presents similar and not isolated themes. Also, *concrete* presented a dotted connection with *embodied-energy* and *connections*, through *life-cycle-analysis* and *design*, respectively. In the case of *performance*, the network presented three links, one solid and two segmented. The solid line is linked to *timber* under *performance*. Also, a connection was made by sharing a cluster of *impact* with the main cluster *concrete* and sharing the *residential-buildings* with the cluster *embodied-energy* in the third period.

The third period presented a new network which included the word *robustness*, and it was not connected with the previous period. *Robustness* appears as a structural network, with themes such as *structural-integrity*, *disproportionate-collapse*, *progressive-collapse*, and *failure*. A connection to the last period from *robustness* was via segmented lines to *tall-timber-buildings* and *mass-timber* through the cluster of *disproportionated-collapse* and *progressive-collapse*, respectively. *Concrete* and *embodied-energy* were both connected to the fourth period in a solid line to the cluster *LCA*. They indicate the constant research until the end of the analysis periods of environmental topics from timber in mid-rise and tall buildings. *Connections* were linked to *performance* in the last period, sharing the cluster *connection* and *tall-timber-buildings'* through *cross-laminated-timber*, the cluster with the most documents in both networks.

In summary, the evolution map (Figure 7) illustrates the progressive refinement of research themes in mid-rise and tall-timber buildings across four distinct periods, highlighting the shift from general foundational topics to more specialized and interconnected themes. During Period 1 (1988–2010), the research focused on general sustainability aspects like LCA and timber-frame buildings, which established a baseline for timber's environmental advantages. In Period 2 (2011–2016), the introduction of mass timber and studies on structural performance marked a significant transition towards understanding its feasibility for taller buildings. Period 3 (2017–2020) witnessed an emphasis on robustness, disproportionate collapse, and perceptions of timber buildings, reflecting efforts to address structural and societal challenges associated with tall structures. Finally, Period 4 (2021–2022) presented advanced investigations into hybrid construction methods, refined LCA methodologies, and high-performance evaluations for tall timber buildings, signalling maturity in both technical and environmental research. The connections between clusters, such as those linking concrete and timber with embodied energy or tall timber buildings with cross-laminated timber, underscore a growing interdisciplinary approach that integrates structural, environmental, and societal dimensions. This trajectory not only demonstrates the field's evolution but also aligns with global priorities for sustainable urban development.

3.3. *Period Analysis*

The evolution of the research in mid-rise and tall timber buildings is conducted by considering three aspects for each period: (1) a description of the three most cited articles, (2) an explanation of the strategic diagram with an explanation of the most iconic constructed timber buildings of the period, and (3) a description of the motor themes with their respective cluster network. Furthermore, only for the last period, (4) all the cluster networks are explained to analyze the current research topics and trends.

3.3.1. Period 1 (1988–2010)

Most Cited Articles

The first period (1998–2010) is the most extensive in number of years. The most cited research of the period was the investigation of Börjesson and Gustavsson [32] related to timber vs. concrete greenhouse gas balance. Different analyzed scenarios included the end of the life of the timber frame, the effect on forest emissions, and the carbonization of the concrete. The authors concluded that the efficiency of timber products in terms of greenhouse gas (GHG) depends not only on the production process but also on the timber waste handle after demolition. Also, the article concludes that concrete frame-building GHG would be higher than timber frame building, but the difference will be shorter if the carbonization process is considered from a long-term perspective. The second most cited article in the first period was written by Werner and Richter [33]. They presented a compilation and literature review of life-cycle analysis studies of timber-based building products compared to similar or equivalent products. The article sustains that there are increments in material emissions due to using impregnated products, particle boards, or fiberboards compared with standard solid timber products. Additionally, the authors mentioned the importance of maintaining representative and reliable life-cycle inventories (LCI) and implementing a new LCA data process to their database. The third most cited article of the period was the investigation of Puettmann et al. [34], which performed a LCI from cradle to the gate of different timber and engineering wood products (EWP), such as softwood lumber, soft plywood, laminated veneer lumber, and oriented strand board (OSB) based on CORRIM 2004 reports from the United States. In an LCI analysis, a separation of processes is made to identify and quantify the inputs and outputs involved in the product or method. The authors' results show that biomass (timber waste) was the primary fuel source to lower environmental emissions, releasing CO_2 from biomass instead of CO_2 from fossil fuels. Another conclusion was that based on European databases with different electricity production sources, resin production, and feedstock presented a high energy consumption [34].

Strategic Diagram

Regarding the construction of mid-rise and tall timber buildings in practice, it is interesting to note that by the end of this first period, the nine-storey 29 m-high Stadthaus building in London was constructed in 2009, by Waugh Thistleton Architects. This iconic building was a very important milestone for the discipline because it was the first timber building in the range of 30 m constructed with CLT, which greatly contributed to the widespread use of this mass timber product, demonstrating that it could be used to construct tall buildings. The development of standards and research at that time was poor, but the interest in life cycle and the first iconic CLT tall building served as a starting point for this building discipline.

As expected, for this incipient period between 1988 and 2010, the discipline was actually building itself and thus there was no clear main cluster location. The few articles in this period generate only two clusters. As shown in Figure 8, the first one was the *timber-buildings* cluster, which constituted the central topic of the discipline but was not developed enough nor had a significant number of related articles. The cluster network was located between the motor themes, which are the relevant and developed topics for the field and are located in the first quadrant of the strategic diagram, and the fourth quadrant, which are the basic and transversal topics. The second cluster network in the period was *life-cycle*. The cluster was located between the first quadrant and the second quadrant, which is the highly developed quadrant. Due to the science mapping analysis configurations and first-period characteristics, no main clusters were detected in the third quadrant, which is the emerging or disappearing topics.

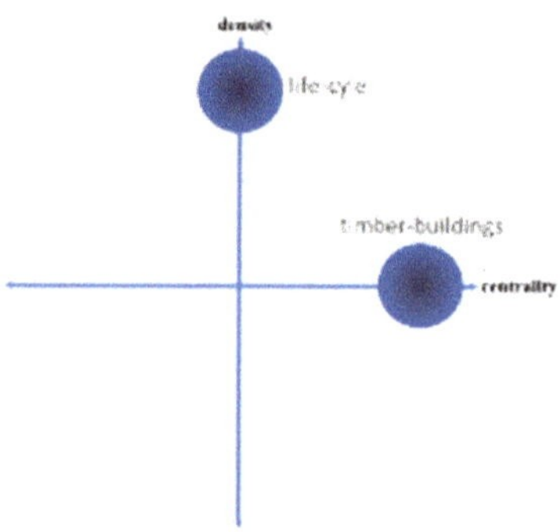

Figure 8. Strategic diagram for the first period (1988–2010).

Motor Themes

At this first incipient period, there were no clear motor themes, but topics such as *life-cycle, LCA, timber*, and *construction* may be regarded as the precursor of the first motor theme even when its transversality was not yet widespread enough in the discipline. As shown in Figure 9, the *life-cycle* constitutes a very connected cluster, which is related to environmental topics such as *LCA*, *energy*, and *emissions*.

Timber was the term with the most documents in this network. The appearance of *concrete* in the first period and in this topic investigation was due to the analysis of timber structures as an alternative to reduce environmental emissions. Despite the visual lack of connection in the network from *concrete* and *substitution* topics, the articles in the period of both topics focus either on timber utilization as an alternative for traditional constructions or compare its environmental properties [19,33,35]. In the case of *energy*, the topics were focused on calculating the environmental emissions and life-cycle primary energy of timber [19,32,34,36].

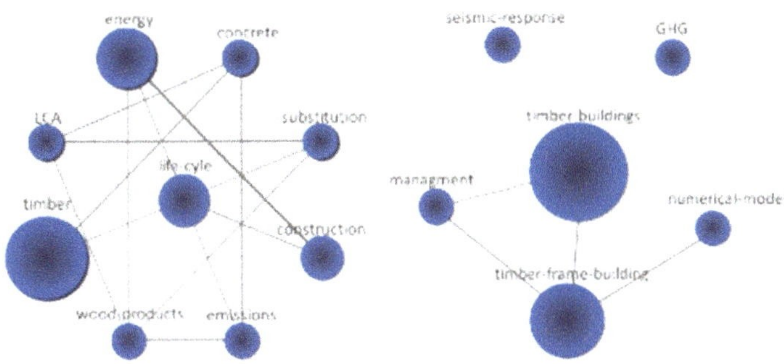

Figure 9. Cluster networks of the first period: life-cycle (**left**) and timber buildings (**right**).

As seen in Figure 9, the *timber-building* network was less connected and mostly included terms related to research on the structural safety of this type of building. Nevertheless, the term *GHG* was still a relevant topic as a significant amount of research focused on investigating different emission levels, from the impacts of timber building products [30] to analyses of the impacts of timber building, accounting for forest land perspectives [32] and carbon pools analysis [35,36].

Regarding the importance of structural aspects in this period, two aspects are important. The first one is that the most studied structural typology in this period covered timber-frame building articles [19,37–41]. This was expected since, only by the end of this period did mass timber products, and more particularly, CLT, start becoming popular. Mosalam and Mahin [40] performed large-scale seismic testing on a three-storey asymmetric timber frame building. The article concludes that hold-downs were essential to prevent the overturning collapse, finish material has to be considered in a seismic assessment of the existing building with a retrofitting focus, and an improvement in seismic performance of timber frame buildings can be achieved by adding a retrofit steel frame in open front [40].

3.3.2. Period 2 (2011–2016)

Most Cited Articles

A significant increment of publications in the field is evidenced during the second period, since the number of publications doubled, even when it only lasted about a quarter of the first period. The most cited article of this period was the work of Ceccotti et al. [42], which conducted an ambitious project to test the seismic performance of a full-scale seven-storey CLT building on a shake table. The article concludes that this newly proposed structural typology, strongly influenced by the Stadthaus during the first period, was structurally safe to construct even in high seismic regions. Yet, the authors registered very high accelerations at the top of the structure, which evidenced that even when these buildings can withstand strong earthquakes, more research is needed to enhance their seismic performance. The second most cited article also investigated CLT's structural behaviour in a multi-storey timber building. Fragiacomo et al. [43] evaluated different mass timber connections between CLT panels and performed an elastic seismic non-linear static analysis of a four-storey CLT building model. Based on the results of the investigation, the authors started proposing seismic performance factors for this new type of structure that allowed them to conduct practical seismic designs of buildings, such as overstrength. Also, the article focused on analyzing the local and overall ductility to demonstrate the performance of these type of buildings during extreme loading events. The third most cited article was the research of Dodoo et al. [20], which analyzed the carbon life-cycle implications of three types of timber building systems on a four-storey building with two approaches: conventional and low-energy. The building system types were CLT, beam and column using laminated veneer lumber (LVL), and prefabricated modules using light-frame elements. The low-energy design aimed to meet the Swedish passive house criteria. The results show that in terms of energy efficiency, LVL with the conventional systems results in higher carbon emissions, and low-energy CLT alternatives achieve the lowest life-cycle carbon emissions [20].

Strategic Diagram

During this period, more than 50 tall buildings were constructed worldwide, with two of them becoming iconic milestones for the discipline [44]. The first one was the 10-storey 30 m-high Forté building, finished in 2012 in Melbourne, Australia. Despite not being much taller than previous projects, this building rapidly became a worldwide icon because of its rapid and highly prefab CLT construction system. The second was the 15-storey 52 m-high Treet building, finished in 2015 in Bergen, Norway. Apart from the great increase in height in comparison with previous buildings, the Treet used a post and beam system based on LVL and glulam, which demonstrated that other structural timber typologies are also feasible for tall buildings. In addition, the Treet was also a great milestone in terms of policymaking because Scandinavian authorities relaxed the height limits for timber structures, allowing the construction of tall projects. This great achievement was, again, coincidental with the end of a relevant period in research since the end of the second period is concurrent with the Treet finishing.

Concrete and *timber-buildings* were the motor themes of the period. *Timber-frame-buildings was* located in the emerging or declining themes as seen in the strategic diagram in Figure 10. The location of *the timber-frame-buildings* network in the third quadrant could be associated with the structural typology limitation to construct buildings taller than six–seven storeys [45]. This evidence in the field of timber construction led researchers to switch their main focus to tall buildings rather than low-rise construction. Performance was located in the centre of the strategic diagram.

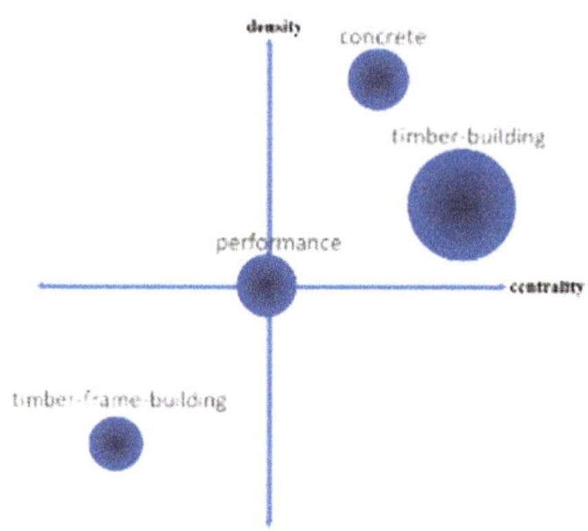

Figure 10. Strategic diagram for the second period (2011–2016).

Motor Themes

Although timber is the main topic of this investigation, *concrete* appears as a principal cluster in Figure 11. As seen in Figure 11, a major interest in *concrete* relates to comparing environmental impacts with timber [20,46,47], as evidenced by the strong links with terms such as *balance*, *LCA*, or CO_2 *emissions*, which at the same time are strongly interrelated with each other. The keyword *balance* was connected to almost all clusters, and research associated with the second period focused on the LCA, carbon emissions, and primary energy building [20,47–50]. *Concrete* also starts playing a significant role as a material necessary for casting hybrid timber–concrete buildings. At this time, researchers realized that the combination of timber with concrete could be very beneficial for enabling tall timber construction, mainly because timber could benefit from the large strength, stiffness, cost-effectivity, and fire resistance of concrete [51]. *Mass-timber* emerged in the *concrete* network during this period with strong connections to *structural-systems* and *balances*. Regarding the relationship with *balances*, the articles in the period were related to environmental emissions quantification in articles [20,47]. Both articles allude to mass-timber CLT in their content. The *structural-system* cluster comprised three articles [20,47,52] and all of the primary structural elements to be analyzed were CLT or mass-timber, confirming the clusters' strong connections.

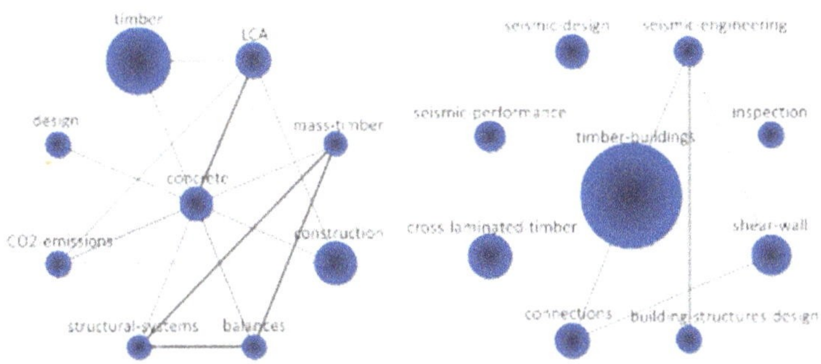

Figure 11. Cluster networks of the second period: concrete (**left**) and timber buildings (**right**).

Regarding the *timber-buildings* motor theme, as expected by the most cited publications, a great body of research focused on structural topics, especially on the seismic performance of timber buildings, with the expectation that this can be one important advantage for these buildings.

3.3.3. Period 3 (2017–2020)

Most Cited Articles

The most cited article of the period was the investigation of Churkina et al. [53], which explained the different emission rates of traditional building materials and illustrated the difference between light timber and massive timber elements in terms of emissions. The article analyses four future transition scenarios from concrete and steel urban building construction to mass timber urban buildings from 2020 to 2050. The cases were first business as usual, which considers 99.5% of buildings constructed by concrete and steel, and other cases with an increment from 10%, 50%, and 90% in timber urban buildings. All the scenarios considered the manufacturing capacity of countries and their potential industry development for mass timber factories. The article demonstrated the capacity of mass timber urban cities to achieve carbon pools but also remarked on the need to ensure sustainable forest management due to increased timber demand. The second most cited work of the period was by Asdrubali et al. [54], which reviewed and analyzed different timber structural topologies for residential buildings, structural properties, thermo-physical properties, acoustic properties, and environmental aspects. General information on the timber properties is explained in the different sub-topics. For example, from a seismic point of view, the authors mentioned the non-ductile behaviour of the timber stress–strain, which is a disadvantage for a seismic aspect. This particularity brings the necessity of concentrating the ductility of structures on the steel joints Asdrubali et al. [54]. The article also describes tall timber building projects, such as the Treet, Forté, and Brock Commons, and mentions that improving scientific research would generate a new era for timber buildings. The research concluded that with the technological development of timber construction, a possible increment of timber buildings would occur under low energy and passive goals. Finally, the third most cited article of the period was the research of Bartlett et al. [55]. The article reviews the different processes and subprocesses of the burning behaviour in timber and explains the analytical expression in the literature that represents these phenomena. The review also explains how different properties, such as species, moisture, system orientation, encapsulation, delamination, and others, can affect the charring rate described in the literature. For example, the authors mentioned a few investigations under the delamination process of laminated timber, such as CLT, in which they concluded that with a higher lamellae thickness, a lower overall charring rate could be seen due to a lower occurrence of delamination phenomena. The article concluded that incident heat flux, which in the investigation case represents and quantifies the fire scenario, significantly affects the charring rate [55]. These three articles suggest a shift in research interest in the field from the predominantly seismic-structural most cited articles of the previous period to an emphasis on the environmental benefits and scenarios of building the cities with timber, rather than concrete or steel. The period also suggests more emphasis on performance topics different to structural, such as fire performance.

Strategic Diagram

This third period included the construction of three very iconic timber buildings. The first was the 18-storey 53 m-high UBC Brock Commons building, finished in 2017 in British Columbia, Canada. This building was one of the most iconic timber projects ever constructed not only because of its height, but also because it used a very effective structural and construction system. The building, which consists of a student residence of the UBC, was realized by first setting two concrete cores and a podium, to subsequently erecting a very rapid construction system based on glulam columns and CLT slabs. The second was the iconic 18-storey 86 m-high Mjøstårnet building, finished in 2019 in Brumunddal, Norway. This building, which can be seen as an evolution of the Treet building as it was constructed with a similar structural system, meant not only a great increase in height, but also demonstrated that such height was possible by only using timber. The third was the 24-storey 84 m-high HoHo building in Wien, Austria. In contrast to Mjøstårnet, this was a very hybridized building comprising multiple timber–concrete composite components. These three buildings served to prove that most buildings in cities can be feasibly constructed totally or partially with timber, which may help to expand the scope of sustainability investigations covering the assessment of the environmental impact of constructing with timber at community and city levels, not only on specific components or buildings. The strategic diagram of the third period is shown in Figure 12.

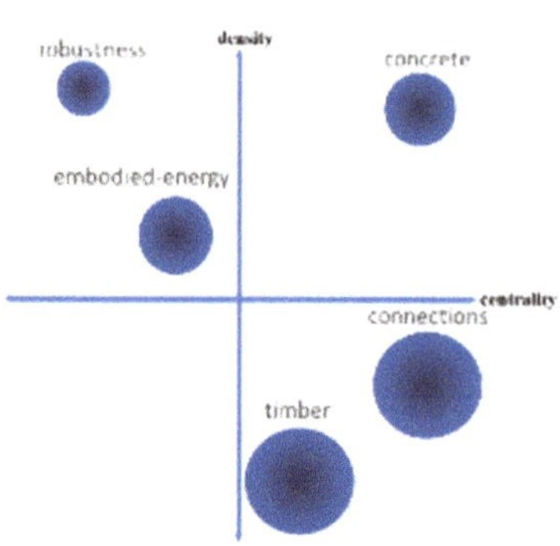

Figure 12. Strategic diagram for the third period (2017–2020).

A new topic in the highly developed quadrant was the main cluster of *robustness*, which focused on the structural subjects related to mid- and high-rise timber building [56–63]. The clusters in the network were *structural integrity, proportioned-collapse, failure*, and *progressive collapse*. Also, the cluster *embodied-energy*, can be found in the highly developed quadrant with a lower density and centrality [64–70]. This particular cluster network contains mostly environmental topics in the period related to *LCA, sustainable-assessment, operational-energy, GHG, waste, residential-buildings, passive-house and life-cycle.*

Timber was classified as a transversal-basic, containing two topics with an increase in frequency: *innovations* and *perceptions*. *Perceptions* are a more recurrent topic in this period, addressing research themes such as how the Swedish actors in the multi-storey residential building world see timber products [71] and other topics such as barriers identified at the time like cost, building codes, lack of expertise, durability, technical aspects, culture in the industry and material availability [72] of the timber industry and products. The topic *Innovation* was analyzed for new products such as EWP and their perception of Swedish architects [73] or elements such as CLT, but in the context of North America focused on the behaviour under water and moisture control of different envelopes innovations [74]. The cluster network of *connections* falls in the transversal-basic with *cross-laminated-timber* as the cluster with the most articles in the network.

Motor Themes

Concrete was the only motor network in the third period. As illustrated in Figure 12, the explanation, almost without exceptions, is that the research interest strongly focused on the potential environmental benefits of constructing with timber since all the terms in the cluster network directly or indirectly relate to this topic. Moreover, the topics show multiple connections with each other. Especially, for this period, environmental research starts to cover and gain significance at macro levels, studying the environmental impact on economies, communities, and cities [71,75,76]. The term *concrete* in the network was related mainly to environmental topics such as carbon footprint, life-cycle cost, energy analysis, Co2-emission, and emissions, as seen in Figure 13.

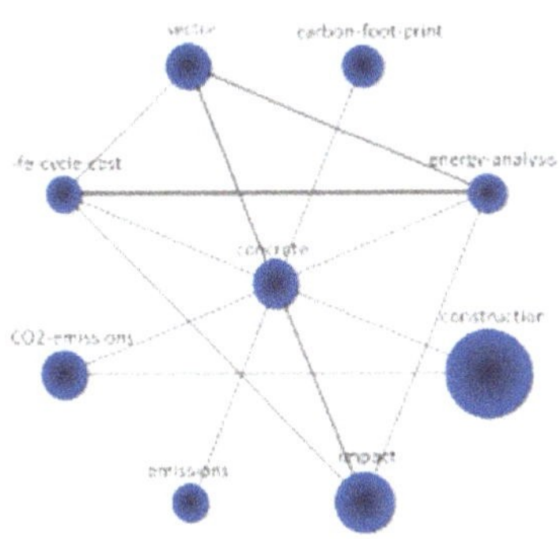

Figure 13. Cluster network of concrete for the third period.

The topics *impact* and *sector* were mainly associated with environmental topics [53,67,69,75,77–79]. In all period articles, the keyword was added via Keyword Plus, which does not add specific information. Other topics related to *impact* including *environmental-impact, sound-impact*, and *climate-impact*, due to the low frequency of the keywords, were not added to the cluster analysis.

3.3.4. Current Trends (2021–2022)

Most Cited Articles

Analyzing the current trends of the last period (2021–2022), the most cited article was the research of Gustavsson et al. [80]. The article analyzed the climate implications of different forest management, bioenergy, and timber construction emissions under various scenarios regarding forestry, type of building construction, and energy resource alternatives. The article shows the net CO_2 emissions in different periods on a 201-year timespan. The authors mentioned that the key factors guiding the results were forest management intensity, harvested biomass, time horizon, and non-timber product replacement. Gustavsson et al. [80] conclude that in the long run, instead of setting aside forest land to reduce the amount of biomass harvest and stored carbon, an active forestry with high harvest levels and efficient use of harvest biomass to replace non-timber products and fuels can provides significant climate mitigation. The second most cited article from the period was the seismic-structural investigation of Aloisio et al. [81], in which a CLT element was analyzed, and fragility functions with behaviour factors were addressed. Based on an extended energy-dependent generalized Bouc-Wen (EEGBW) model, the authors generate an analytical and numerical model of multi-storey CLT building as a shear plane structure. The main results estimated the different storey buildings' behaviour factors and fragility curves. The authors concluded that based on the fragility curves, the vulnerability to earthquakes heavily depended on the seismic scenario and storey number rather than the CLT wall panel typology [81]. The third most cited article of the fourth period was the study of Xia et al. [82]. The article analyzed the application of epoxy resin (EP) and phase-change material (PCM) compound into delignified timber to create transparent timber that additionally stores heat during the day to release it during night. The authors conclude that with the increment of PCM, the transmittance energy storage wood (TESW), with an 80% weight ratio of EP and PCM, reaches 134.1 and 122.9 j/g on the melting and crystallizing temperature. In addition, the TESW archives a light transmittance of up to 80.89%. These three articles are diverse but demonstrate an evolution from fundamental to more advanced topics. From the environmental perspective, analyses also start more explicitly covering the implications of forest management and biomass utilization on the sustainability of using timber for construction. On the structural seismic field, fragility functions of the new structural systems and new hysteresis models allow for more precise and in-depth analysis of timber buildings under different load scenarios. Finally, investigations like the one from Xia et al. [82] demonstrate the possibility of improving timber performance and extending construction applications by modification of its material properties.

Strategic Diagram

In terms of iconic buildings, this period included the completion of two iconic buildings. The first was the 21-storey 73 m-high Haut building completed in 2022 in Amsterdam, the Netherlands. Like some previous buildings, this project also used a highly hybridized timber–concrete structural system, but the particularity was that it was the Netherlands' first building to achieve the BREEAM Outstanding classification. The second was the hybrid timber–concrete 25-storey 87 m-high Ascent building, completed in 2022 in Milwaukee, WI, USA, which is currently considered the tallest timber building in the world. Although these two iconic buildings undoubtedly represent outstanding technological and engineering achievements, neither the increase in height nor the shift in construction technologies suffered notable changes concerning the previous period. Therefore, it could be argued that from the point of view of the construction of iconic projects, the construction technologies of tall timber buildings are achieving a certain degree of maturity. The strategic diagram from the last period is shown in Figure 14.

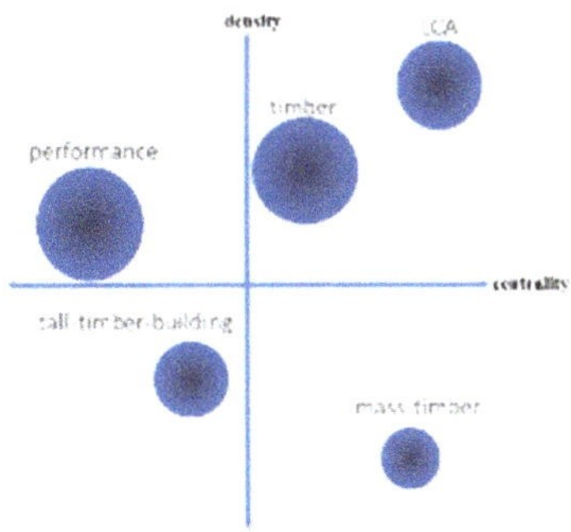

Figure 14. Strategic diagram 2021–2022.

Motor Themes

LCA was the main cluster in this network, connected to environmental topics, except for *Concrete* and *Construction*, as seen in Figure 15. The presence of *Concrete* can be explained due to timber comparison and *Construction* to the transversal origin of the topic.

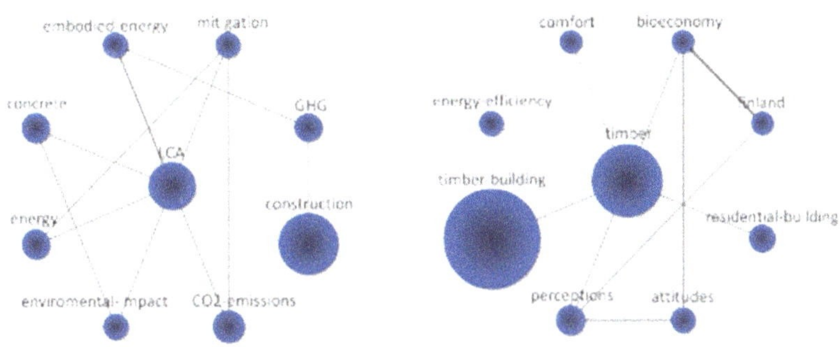

Figure 15. Motor themes from fourth period (2021–2022).

It is interesting that a change was also observed in the articles related to the term *GHG* in which, currently, the researchers are emphasizing the modelling of the final stages of building, such as demolition or recycling [4,83–85]. The estimation of the emissions through the LCA methodology has been presented throughout the periods and appears as a motor theme due to the necessity to quantify the benefits and improve the mitigation of emissions through timber as a building material of mid- and high-rise buildings. For example, in the last period, Younis and Dodoo [86] investigated the background of LCA in CLT buildings, exploring past CLT projects and analyzing how to improve the LCA method to obtain more accurate results and optimize the carbon footprint in these mass timber buildings.

The second cluster network in the motor theme was *timber*, one of the main topics in the research formula gathering common topics such as *timber-buildings* and *residential-building*, illustrated in Figure 14. Surprisingly, *Finland* appears as a relevant term for current research, which is strongly linked with *bioeconomy* and *perceptions*. This reflects the relevance of this country for its development in fostering timber construction from a bioeconomic and societal point of view, which serves as a worldwide reference.

The topics in this duo, *perceptions* and *attitudes*, were associated with professionals [87,88], government [89], and society's perspectives [90–93] on timber buildings. The keyword *bioeconomy* was linked to this duo through a few articles in which the main relation was the perception of the residents [93] and professionals [87] on timber housing to archive a forest-based bioeconomy. Another study in the period related to *bioeconomy* and the duo of *perceptions* and *attitudes* is the investigation of Toivonen et al. [89], which analyzed the policy narrative on timber multi-storey construction (TMC) and contrasted to the governance of the technological innovation systems around TMC. The investigation identifies that one of the policy narratives related to TMC was the bioeconomy narrative, which focused on TMC as a method to achieve a more sustainable bioeconomy [89].

It should be noted that even though the environmental assessment of timber construction is currently a strongly developed and transversal topic whose investigations are strongly interrelated, which is reflected by the leading position of the term *LCA* in the strategic diagram at the first quadrant, Figure 14, and the multiple links of its cluster network, Figure 15, the societal economic aspects are still not so developed. This could indicate that both investigations, including the multidisciplinary assessment of environmental and economic/societal aspects, and investigations linking different societal/economic aspects such as comfort with perceptions, energy-efficiency, and attitudes, may be of great impact for the discipline.

Current Trends in Other Relevant Cluster Networks

The term *mass-timber* emerges for the first time as the main topic in cluster networks. This term covers a great variety of topics that are poorly interconnected, from very specific structural topics such as *progressive-collapse*, to very general topics such as *systems*, *steel sustainability*, and *embodied carbon* see Figure 16.

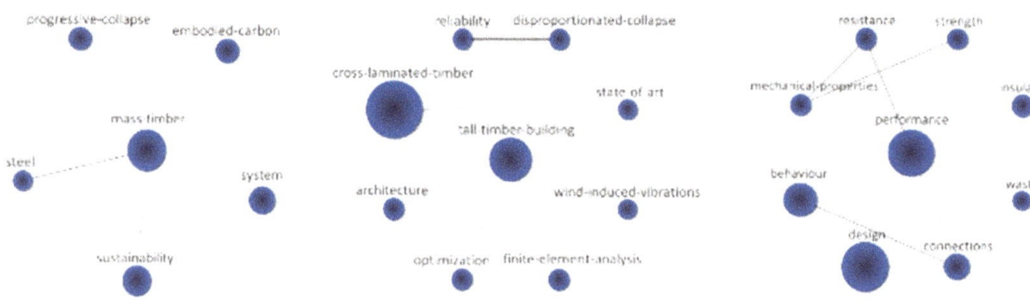

Figure 16. Current trends in research of other topics different from motor themes including *mass timber* (**left**), *tall timber buildings* (**centre**), and *performance* (**right**).

The new emerging connection between *mass-timber* and *steel* is also interesting, represented by the investigation by Allan and Phillips [94]. The authors perform an LCA of different buildings to understand the difference in using mass timber instead of North American conventional frame steel structures. The article concludes that the utilization of mass timber reduces emissions in several categories, especially in the global warming potential. Finally, the authors mentioned the potential to include mass timber as a floor solution in steel buildings, thus, also starting investigations on hybrid steel timber construction [94].

The topic *embodied carbon* is the main component in the LCA, which can be defined as the GHG emissions of a building over its life during the stage of extraction, transportation, manufacture of construction materials, building construction, and inclusive operations, demolitions, disposal, or recycling [4]. Another interesting new connection relates the term *mass-timber* with *embodied carbon*. The fact that more timber is being used for the new proposed structural mass-timber typologies also implies a more profound impact of timber construction in embodied carbon, which is currently an important research trend for timber structures. For instance, the article by Chaggaris et al. [95] investigated the cost, emissions, and carbon sequestration potential for different beam-column configurations. The article concluded that the cost of gypsum board for fire protection and the CLT for floor material tends to dominate the beam-column independent of the building configurations. Also, the article mentioned that carbon sequestration and cost objectives can be in disagreement because both increase with the increase in timber volume [95].

An alternative interesting topic in mass timber research is represented by the concept of biogenic carbon. The results in this bibliometric analysis show the use of the terminology *biogenic carbon* in the second period [96], but the notion of carbon balances, carbon capture, or carbon pools has been presented since the first period [32]. In the last period, a few articles mentioned biogenic carbon in their respective keywords [83,95,97,98].

A second relevant cluster that appears for the first time as a main topic in the network and that also could be seen as an evolution of the term *timber-buildings*, is the cluster *tall-timber-buildings*. The cluster is classified into declining and emerging topics, which makes sense because in this period more specific topics related to tall buildings are emerging.

In general, structural investigations in tall timber buildings are shifting from a more seismic-focused origin, into specific topics such as *wind-induced-vibrations* and *disproportionate collapse loads*. The partial shift from seismic to wind interest was to be expected as wind loading tends to be more critical for tall buildings rather than seismic, which tends to govern the design of lower-rise buildings. The investigation of the wind effect of induced wind load has been presented in the second and third periods in additional keywords such as *wind-loads* and *wind-pressure-coefficient* [39,99–103]. Multiple articles related to wind topics have been detected in the period analysis [22,104–106]; for example, Binck et al. [104], investigated and developed an adaptable and flexible lateral stiffening system for tall timber buildings between 50 and 147 m. The system was subjected to multiple finite element simulations in different configurations to analyze the lateral stiffens subjected to wind loading. The article concludes that the proposed system was feasible for 147 m and the lateral stiffening system was sensitive to geometrical parameters such as bay number and length. On the other hand, the rising interest in *disproportionate collapse* and its strong link with *reliability* could also be expected [107–109]. One of the most utilized structural typologies that are prevailing for tall timber buildings is based on using isolated timber columns for resisting gravitational loads. Timber being a brittle material, reliability studies, therefore, become relevant for preventing the disproportionate collapse of tall timber buildings. For instance, the investigation of Voulpiotis et al. [107] analyses a timber post and beam tall building to quantify the robustness based on a published method. The authors concluded that in terms of preventing disproportioned collapse, design improvements at a global building scale, such as trussed strong floors, could be more effective than increasing the strength and ductility of the selected connections. The most cited topic in the *tall timber buildings* network was, however, *cross-laminated-timber*, with 34 documents in the period, further reinforcing the link of tall buildings with this specific timber product. *Finite element analysis* is also a new main term that relates mostly to the period with *cross-laminated-timber*, indicating the emphasis of research on proposing new structural calculation models for tall timber buildings constructed with CLT [22,110–113].

Not all articles related to *finite element analysis* in the period were focused only on CLT. For example, Labrecque et al. [114] modelled a structural solution of light frame timber modules in two hyper structures: concrete and glulam. The article concludes that both solutions were viable under different structural restrictions depending on the hyper structure; still, more investigation is needed for subjects like connections between the hyper structure and the modular structure [114].

The last cluster network of the current strategic diagram is *performance*, which currently shows the largest number of contributions. This term covers very diverse topics including structural [22,108,115–119], envelope design [120], acoustic performance [121], environmental and energy [112,116,122,123], and social aspect [92].

4. Discussion

4.1. General Discussion

The results of the bibliometric analysis show multiple insights into how mid- and high-rise timber building topics have evolved during the study periods.

During the first periods, most topics were based on light-frame timber building studies and developing new EWP such as glulam or laminated veneer lumber. At the end of the first period, *CLT* and *mass-timber* were more recurrent and positioned as main topics in the subsequent periods, showing their potential increase as investigation topics in the mid-rise and tall building area. An increment in specific topics such as *connections* and *robustness* were identified in the final periods, indicating the investigation focuses on the structural feasibility of tall timber buildings.

The environmental topics have been present in all periods, constantly quantifying the emissions of timber products from EWP to a building scale. The keywords representing the sustainable topic were *LCA*, *emissions*, and *GHG*. The environmental impact quantification of timber as a solution for mid-rise and tall buildings requires constant development and study due to the constant change in LCA stages, assumptions [124], and efficiency of inputs and outputs. The last period shows an increase in studies related to end-of-stage analysis and carbon storage, showing a potential research interest in harvest timber products as a temporal/long storage pool.

Building on this sustained focus on environmental concerns, one of the most significant transformations in mid- and high-rise timber building research has been the shift from a predominant focus on structural safety towards greater emphasis on sustainability and energy efficiency. This change has been driven by the development of new materials such as CLT and mass timber products, increasingly stringent environmental regulations, and the growing interest in reducing the carbon footprint of the construction sector. Timber has emerged as a viable alternative to concrete and steel due to its carbon storage capacity, leading to an increased research focus on its environmental impact.

The seismic topic was a recurrent research line that has always been present from light frame to mass timber. The systematic characterization of the seismic and mechanical behaviour of structural elements was intended to ensure the structural feasibility of mid- and high-rise timber buildings. Recurrent topics in this area were *seismic-performance, seismic-design, seismic-engineering,* among others. Regarding tall timber buildings, particular topics show up in the last periods: *collapse, robustness*, and *wind induced vibrations*, which increased in frequency since the third period, noting the importance of settling a solid base on tall building topics to achieve higher heights with timber as a structural material.

The concept of hybrid building appears slightly in the periods of analysis. It was noticed that there was different terminology to address the hybrid structure or components; for example, *hybrid-structures, hybrid-buildings, timber-concrete,* and *hybrid-CLT-building*. A common terminology of *hybrid-structures* or *hybrid-buildings* is needed to join the research of materials combinations on mid- and high-rise timber buildings. In recent periods, a collaborative investigation with the topic of *Concrete* has been performed, leading to potential hybrid building from the element scale such as composite to the interaction of concrete core with mass timber walls and floors to achieve higher and cost-efficient buildings.

The *perception* and *attitude* towards mid- and high-rise timber buildings from the different actors could define how these structures have been integrated into standard alternatives of society's needs. The analysis results show an increment from the third period to the actual one on articles related to the *perception* and *attitude* of timber buildings as a main topic, showing the necessity of studies on how timber structures are perceived in the world and market. The insights of *perception* and *attitude* research on timber as a building material help architects, engineers, and constructors to prioritize which aspects should improve to increase the interest of stakeholders and citizens.

Fire investigations obtained from the research formula do not appear as relevant and common keywords or main clusters. Fire investigations were dated from this analysis since 1995 under the investigation of light-frame buildings. Topics of fire-related timber buildings were historical timber buildings' [125,126] fire performance of materials [127–133]. The study of fire behaviour in tall timber buildings is crucial to improving and promoting the safe construction of high-rise timber structures aligned with other aspects such as structural safety and mechanical electrical plumbing codes.

One of the most significant transformations in mid- and high-rise timber building research has been the shift from a predominant focus on structural safety towards greater emphasis on sustainability and energy efficiency. This change has been driven by the development of new materials such as CLT (Figure 17) and mass timber products, increasingly stringent environmental regulations, and the growing interest in reducing the carbon footprint of the construction sector. Timber has emerged as a viable alternative to concrete and steel due to its carbon storage capacity, leading to an increased research focus on its environmental impact.

Figure 17. CLT construction. Health centre in Abegondo, La Coruña, Spain. January 2025.

4.2. Limitations and Future Work

This study has several limitations related to the time span, database selection, search formula, and bibliometric analysis settings. First, the selection of WOS as the sole database for this bibliometric analysis was based on its well-structured metadata, standardized citation tracking, and compatibility with science mapping tools such as SciMAT. WOS provides high-impact, peer-reviewed publications that ensure data consistency and reduce redundancy. However, this choice also presents certain limitations. Other databases, such as Scopus and Google Scholar, contain additional sources that could contribute to a broader data set. Scopus includes a wider range of journals, particularly in engineering and sustainability fields, while Google Scholar provides access to grey literature, conference proceedings, and institutional reports that might not be indexed in WOS. Additionally, the keyword selection methodology may have influenced the results by prioritizing certain terms over others. Finally, the evolution of terminology in mid- and high-rise timber building research may have affected topic clustering, as certain terms have changed over time. To address these limitations, future research could expand the search to additional databases and employ machine learning techniques to enhance topic classification and detect emerging research themes.

A key next step in this research is to analyze publications from the most recent period (2023–2024) in a yearly step approach. Examining this close period will help identify the latest trends and challenges in mid- and high-rise timber building research, particularly regarding sustainability, structural performance, and regulatory developments. Moreover, there is a growing need for research on the perception and attitude towards mid- and high-rise timber buildings in Chile, particularly among public, private, and key stakeholders involved in timber construction projects. This research could provide valuable insights into the factors influencing the adoption and funding of timber-based urban developments.

Recent studies have advanced the field by addressing sustainability, automation, structural adaptability, and damage detection in timber structures. Bucklin et al. [134], in 2023, explored the thermal transmittance and air permeability of solid timber envelopes, showing how design modifications like deep air chambers enhance energy efficiency while maintaining airtightness. Lauer et al. [135], in 2023, examined automated on-site assembly of timber structures using biomimetic shells, demonstrating increased precision, efficiency, and sustainability in construction. Öberg et al. [136], in 2024, introduced the design for the structural adaptation (DfSA) concept, which promotes flexibility in timber buildings to extend service life, improve carbon storage, and enhance resource efficiency while identifying key regulatory and economic barriers. Liu et al. [137], in 2024, developed a precision detection method using portable light detection and ranging (LiDAR) technology for assessing damage in historic timber buildings, contributing to better non-destructive maintenance and restoration strategies.

These recent contributions highlight the necessity of continuous updates in bibliometric studies to capture emerging themes and shifts in research focus. Expanding the data set with new publications will refine the understanding of timber's role in future urban construction, bridging the gap between theoretical advancements and real-world implementation.

5. Conclusions

1. Population growth and urban needs: The increase in population has motivated the construction and design of taller buildings that can respond to society's necessities. Timber buildings emerge as a sustainable solution that can address habitability and workspace demands while reducing carbon emissions. Technological advancements have driven the evolution of timber construction from light-frame elements to various engineered wood products (EWPs) such as glulam and CLT.
2. Research contributions and evolution: This study conducted a bibliometric analysis using a science mapping technique to categorize and analyze the evolution of research topics on mid- and high-rise timber buildings, identifying key trends and challenges. A total of 1233 articles were initially selected, and after filtering, 701 articles were analyzed across four distinct periods.
3. Challenges in timber high-rise construction: Despite the advancements in research and the increasing adoption of timber in high-rise construction, its implementation in urban contexts presents significant challenges. These include regulatory barriers that still limit the permissible height of timber structures in many jurisdictions, initial costs associated with the manufacturing and assembly of engineered wood products, and industry and public perceptions regarding the durability and fire resistance of timber. Overcoming these challenges requires a combination of technological advancements, adapted regulations, and effective communication strategies that highlight the environmental and structural benefits of timber.
4. Future research directions: Addressing these challenges will be crucial for aligning the environmental potential of mid- and high-rise timber buildings with structural feasibility and societal acceptance. Future studies should focus on updating bibliomet-

ric analyses, integrating multidisciplinary approaches, exploring hybrid construction innovations, and enhancing fire safety research to align with industry standards and regulatory frameworks.

5. Final remarks: This study contributes to a deeper understanding of how mid- and high-rise timber buildings have evolved as a research field. Future research should aim to bridge the gap between sustainability, structural performance, and societal perceptions to maximize the benefits of timber construction in urban contexts.

Author Contributions: Conceptualization, A.W., P.G. and M.C.; methodology, A.W., P.G. and M.C.; software, A.W., P.G. and M.C.; validation, P.G. and M.C.; formal analysis, A.W., P.G. and M.C.; investigation, A.W., P.G. and M.C.; resources, A.W., P.G. and M.C.; writing—original draft preparation, A.W., P.G. and M.C.; supervision, P.G. and M.C.; funding acquisition, P.G. and M.C. All authors have read and agreed to the published version of the manuscript.

Funding: This work was funded by the projects ANID BASAL FB210015 CENAMAD; BG23/00134; and the Pontificia Universidad Católica de Chile through the 2024 International Sabbatical Support Competition of the Academic Vice-Rectorship.

Institutional Review Board Statement: Not applicable.

Informed Consent Statement: Not applicable.

Data Availability Statement: The original contributions presented in this study are included in the article. Further inquiries can be directed to the corresponding author.

Conflicts of Interest: The authors declare no conflicts of interest.

References

1. Al-Kodmany, K. Sustainable High-Rise Buildings: Toward Resilient Built Environment. *Front. Sustain. Cities* **2022**, *4*, 782007. [CrossRef]
2. Ali, M.M.; Al-Kodmany, K. Tall Buildings and Urban Habitat of the 21st Century: A Global Perspective. *Buildings* **2012**, *2*, 384–423. [CrossRef]
3. Mostafavi, F.; Tahsildoost, M.; Zomorodian, Z.S. Energy Efficiency and Carbon Emission in High-Rise Buildings: A Review (2005–2020). *Build. Environ.* **2021**, *206*, 108329. [CrossRef]
4. Robati, M.; Oldfield, P. The Embodied Carbon of Mass Timber and Concrete Buildings in Australia: An Uncertainty Analysis. *Build. Environ.* **2022**, *214*, 108944. [CrossRef]
5. Mishra, A.; Humpenöder, F.; Churkina, G.; Reyer, C.P.O.; Beier, F.; Bodirsky, B.L.; Schellnhuber, H.J.; Lotze-Campen, H.; Popp, A. Land Use Change and Carbon Emissions of a Transformation to Timber Cities. *Nat. Commun.* **2022**, *13*, 4889. [CrossRef]
6. UNECE. *The Outlook for the Unece Forest Sector in a Changing Climate; A Contribution to the Forest Sector Outlook Study 2020–2040*; United Nations: New York, NY, USA, 2022; ISBN 9789210023733.
7. Svatoš-Ražnjević, H.; Orozco, L.; Menges, A. Advanced Timber Construction Industry: A Review of 350 Multi-Storey Timber Projects from 2000–2021. *Buildings* **2022**, *12*, 404. [CrossRef]
8. Fuglsang Nielsen, L. Lifetime, Residual Strength, and Quality of Wood and Other Viscoelastic Building Materials. *Holz Als Roh-Und Werkst.* **1991**, *49*, 451–455. [CrossRef]
9. Sheffield, J.W.; Barbera, J.J. The 1991 Uniform Building Code Wind Load Provisions. *J. Wind. Eng. Ind. Aerodyn.* **1992**, *44*, 2921–2932. [CrossRef]
10. Villaverde, R.; Koyama, L.A. Damped Resonant Appendages to Increase Inherent Damping in Buildings. *Eng. Struct. Dyn.* **1993**, *22*, 491–507. [CrossRef]
11. González-Retamal, M.; Forcael, E.; Saelzer-Fuica, G.; Vargas-Mosqueda, M. From Trees to Skyscrapers: Holistic Review of the Advances and Limitations of Multi-Storey Timber Buildings. *Buildings* **2022**, *12*, 1263. [CrossRef]
12. Donthu, N.; Kumar, S.; Mukherjee, D.; Pandey, N.; Lim, W.M. How to Conduct a Bibliometric Analysis: An Overview and Guidelines. *J. Bus. Res.* **2021**, *133*, 285–296. [CrossRef]
13. Han, J.; Kang, H.J.; Kim, M.; Kwon, G.H. Mapping the Intellectual Structure of Research on Surgery with Mixed Reality: Bibliometric Network Analysis (2000–2019). *J. Biomed. Inform.* **2020**, *109*, 103516. [CrossRef] [PubMed]
14. Silva, T.C.; Araujo, E.C.G.; da Silva Lins, T.R.; Reis, C.A.; Sanquetta, C.R.; da Rocha, M.P. Non-Timber Forest Products in Brazil: A Bibliometric and a State of the Art Review. *Sustainability* **2020**, *12*, 7151. [CrossRef]

15. Abejón, R.; Moya, L. Cross-Laminated Timber: Perspectives from a Bibliometric Analysis (2006–2018). *Wood Mater. Sci. Eng.* **2021**, *17*, 429–450. [CrossRef]
16. Pyzhev, A.I. Studies on the Russian Forest Industry: Bibliometric Analysis. *Terra Econ.* **2021**, *19*, 63–77. [CrossRef]
17. Rodríguez-Grau, G.; Marín-Uribe, C.; Cortés-Rodríguez, P.; Montero, C.; Rosales, V.; Galarce, C. Bibliometric Analysis of the Green Gluing Technique (2000–2020): Trends and Perspectives. *Forests* **2022**, *13*, 1714. [CrossRef]
18. De Iuliis, M.; Cardoni, A.; Paolo Cimellaro, G. Resilience and Safety of Civil Engineering Systems and Communities: A Bibliometric Analysis for Mapping the State-of-the-Art. *Saf. Sci.* **2024**, *174*, 106470. [CrossRef]
19. Mahapatra, K.; Gustavsson, L. Multi-Storey Timber Buildings: Breaking Industry Path Dependency. *Build. Res. Inf.* **2008**, *36*, 638–648. [CrossRef]
20. Dodoo, A.; Gustavsson, L.; Sathre, R. Lifecycle Carbon Implications of Conventional and Low-Energy Multi-Storey Timber Building Systems. *Energy Build.* **2014**, *82*, 194–210. [CrossRef]
21. Berwart, S.; Estrella, X.; Montaño, J.; Santa-María, H.; Almazán, J.L.; Guindos, P. A Simplified Approach to Assess the Technical Prefeasibility of Multistory Wood-Frame Buildings in High Seismic Zones. *Eng. Struct.* **2022**, *257*, 114035. [CrossRef]
22. Lazzarini, E.; Frison, G.; Trutalli, D.; Marchi, L.; Scotta, R. Comfort Assessment of High-Rise Timber Buildings Exposed to Wind-Induced Vibrations. *Struct. Des. Tall Spec. Build.* **2021**, *30*, e1882. [CrossRef]
23. Cobo, M.J.; López-Herrera, A.G.; Herrera-Viedma, E.; Herrera, F. An Approach for Detecting, Quantifying, and Visualizing the Evolution of a Research Field: A Practical Application to the Fuzzy Sets Theory Field. *J. Informetr.* **2011**, *5*, 146–166. [CrossRef]
24. Cobo, M.J.; Lõpez-Herrera, A.G.; Herrera-Viedma, E.; Herrera, F. SciMAT: A New Science Mapping Analysis Software Tool. *J. Am. Soc. Inf. Sci. Technol.* **2012**, *63*, 1609–1630. [CrossRef]
25. Díaz-López, C.; Carpio, M.; Martín-Morales, M.; Zamorano, M. Analysis of the Scientific Evolution of Sustainable Building Assessment Methods. *Sustain. Cities Soc.* **2019**, *49*, 101610. [CrossRef]
26. Heradio, R.; De La Torre, L.; Galan, D.; Cabrerizo, F.J.; Herrera-Viedma, E.; Dormido, S. Virtual and Remote Labs in Education: A Bibliometric Analysis. *Comput. Educ.* **2016**, *98*, 14–38. [CrossRef]
27. Callon, M.; Courtial, J.P.; Laville, F. Co-Word Analysis as a Tool for Describing the Network of Interactions between Basic and Technological Research: The Case of Polymer Chemsitry. *Scientometrics* **1991**, *22*, 155–205. [CrossRef]
28. Petković, M.; Škrlj, B.; Kocev, D.; Simidjievski, N. Fuzzy Jaccard Index: A Robust Comparison of Ordered Lists [Formula Presented]. *Appl. Soft Comput.* **2021**, *113*, 107849. [CrossRef]
29. Van Eck, N.J.; Waltman, L. How to Normalize Cooccurrence Data? An Analysis of Some Well-Known Similarity Measures. *J. Am. Soc. Inf. Sci. Technol.* **2009**, *60*, 1635–1651. [CrossRef]
30. Nalbandian, K.M.; Carpio, M.; González, Á. Analysis of the Scientific Evolution of Self-Healing Asphalt Pavements: Toward Sustainable Road Materials. *J. Clean. Prod.* **2021**, *293*, 126107. [CrossRef]
31. Santana, M.; Cobo, M.J. What Is the Future of Work? A Science Mapping Analysis. *Eur. Manag. J.* **2020**, *38*, 846–862. [CrossRef]
32. Börjesson, P.K.; Gustavsson, L. Greenhouse Gas Balances in Building Construction: Wood versus Concrete from Life-Cycle and Forest Land-Use Perspectives. *Energy Policy* **2000**, *28*, 575–588. [CrossRef]
33. Werner, F.; Richter, K. Wooden Building Products in Comparative LCA: A Literature Review. *Int. J. Life Cycle Assess.* **2007**, *12*, 470–479. [CrossRef]
34. Puettmann, M.E.; Associate, R.; Wilson, J.B. Life-Cycle Analysis of Wood Products: Cradle-to-Gate Lci of Residential Wood Building Materials. *Wood Fiber Sci.* **2006**, *37*, 18–29.
35. Werner, F.; Taverna, R.; Hofer, P.; Richter, K. Carbon Pool and Substitution Effects of an Increased Use of Wood in Buildings in Switzerland: First Estimates. *Ann. For. Sci.* **2005**, *62*, 889–902. [CrossRef]
36. Werner, F.; Taverna, R.; Hofer, P.; Richter, K. Greenhouse Gas Dynamics of an Increased Use of Wood in Buildings in Switzerland. *Clim. Change* **2006**, *74*, 319–347. [CrossRef]
37. Pei, S.; van de Lindt, J.W. Coupled Shear-Bending Formulation for Seismic Analysis of Stacked Wood Shear Wall Systems. *Earthq. Eng. Struct. Dyn.* **2009**, *38*, 1631–1647. [CrossRef]
38. Kilpeläinen, M.; Ukonmaanaho, A. Panelized Building System for Multi-Storey Timber Frame Houses. *IABSE Symp. Rep.* **2001**, *85*, 37–42.
39. van de Lindt, J.; Dao, T. Construction Quality Issues in Performance-Based Wind Engineering: Effect of Missing Fasteners. *Wind. Struct.* **2010**, *13*, 221–234. [CrossRef]
40. Mosalam, K.M.; Mahin, S.A. Seismic Evaluation and Retrofit of Asymmetric Multi-Story Wood-Frame Building. *J. Earthq. Eng.* **2007**, *11*, 968–986. [CrossRef]
41. Langlois, J.D.; Gupta, R.; Asce, M.; Miller, T.H. Effects of Reference Displacement and Damage Accumulation in Wood Shear Walls. *J. Struct. Eng.* **2004**, *130*, 470–479. [CrossRef]
42. Ceccotti, A.; Sandhaas, C.; Okabe, M.; Yasumura, M.; Minowa, C.; Kawai, N. SOFIE Project—3D Shaking Table Test on a Seven-Storey Full-Scale Cross-Laminated Timber Building. *Earthq. Eng. Struct. Dyn.* **2013**, *42*, 2003–2021. [CrossRef]

43. Fragiacomo, M.; Dujic, B.; Sustersic, I. Elastic and Ductile Design of Multi-Storey Crosslam Massive Wooden Buildings under Seismic Actions. *Eng. Struct.* **2011**, *33*, 3043–3053. [CrossRef]
44. Salvadori, V. *Multi-Storey Timber-Based Buildings: An International Survey of Case-Studies with Five or More Storeys Over the Last Twenty Years*; Technische Universität Wien: Vienna, Austria, 2021.
45. Guindos, P. *Fundamentos Del Diseño y La Construcción Con Madera*; Ediciones UC: Santiago, Chile, 2019; Volume I.
46. Liu, Y.; Guo, H.; Sun, C.; Chang, W.S. Assessing Cross Laminated Timber (CLT) as an Alternative Material for Mid-Rise Residential Buildings in Cold Regions in China-A Life-Cycle Assessment Approach. *Sustainability* **2016**, *8*, 1047. [CrossRef]
47. Dodoo, A.; Gustavsson, L.; Sathre, R. Lifecycle Primary Energy Analysis of Low-Energy Timber Building Systems for Multi-Storey Residential Buildings. *Energy Build.* **2014**, *81*, 84–97. [CrossRef]
48. Skullestad, J.L.; Bohne, R.A.; Lohne, J. High-Rise Timber Buildings as a Climate Change Mitigation Measure—A Comparative LCA of Structural System Alternatives. *Energy Procedia* **2016**, *96*, 112–123. [CrossRef]
49. Hossaini, N.; Reza, B.; Akhtar, S.; Sadiq, R.; Hewage, K. AHP Based Life Cycle Sustainability Assessment (LCSA) Framework: A Case Study of Six Storey Wood Frame and Concrete Frame Buildings in Vancouver. *J. Environ. Plan. Manag.* **2015**, *58*, 1217–1241. [CrossRef]
50. Tettey, U.Y.A.; Dodoo, A.; Gustavsson, L. Effects of Different Insulation Materials on Primary Energy and CO_2 Emission of a Multi-Storey Residential Building. *Energy Build.* **2014**, *82*, 369–377. [CrossRef]
51. Erdle, A.; Weckendorf, J.; Asiz, A.; Smith, I. Effectiveness of Distributed Mass Damper Systems for Lightweight Superstructures. *J. Perform. Constr. Facil.* **2014**, *28*, A4014005. [CrossRef]
52. Foster, R.M.; Reynolds, T.P.S.; Ramage, M.H. Proposal for Defining a Tall Timber Building. *J. Struct. Eng.* **2016**, *142*, 02516001. [CrossRef]
53. Churkina, G.; Organschi, A.; Reyer, C.P.O.; Ruff, A.; Vinke, K.; Liu, Z.; Reck, B.K.; Graedel, T.E.; Schellnhuber, H.J. Buildings as a Global Carbon Sink. *Nat. Sustain.* **2020**, *3*, 269–276. [CrossRef]
54. Asdrubali, F.; Ferracuti, B.; Lombardi, L.; Guattari, C.; Evangelisti, L.; Grazieschi, G. A Review of Structural, Thermo-Physical, Acoustical, and Environmental Properties of Wooden Materials for Building Applications. *Build. Environ.* **2017**, *114*, 307–332. [CrossRef]
55. Bartlett, A.I.; Hadden, R.M.; Bisby, L.A. A Review of Factors Affecting the Burning Behaviour of Wood for Application to Tall Timber Construction. *Fire Technol.* **2019**, *55*, 1–49. [CrossRef]
56. Mpidi Bita, H.; Tannert, T. Experimental Study of Disproportionate Collapse Prevention Mechanisms for Mass-Timber Floor Systems. *J. Struct. Eng.* **2020**, *146*, 04019199. [CrossRef]
57. Mpidi Bita, H.; Currie, N.; Tannert, T. Disproportionate Collapse Analysis of Mid-Rise Cross-Laminated Timber Buildings. *Struct. Infrastruct. Eng.* **2018**, *14*, 1547–1560. [CrossRef]
58. Mpidi Bita, H.; Tannert, T. Disproportionate Collapse Prevention Analysis for a Mid-Rise Flat-Plate Cross-Laminated Timber Building. *Eng. Struct.* **2019**, *178*, 460–471. [CrossRef]
59. Huber, J.A.J.; Ekevad, M.; Girhammar, U.A.; Berg, S. Structural Robustness and Timber Buildings—A Review. *Wood Mater. Sci. Eng.* **2019**, *14*, 107–128. [CrossRef]
60. Mpidi Bita, H.; Tannert, T. Tie-Force Procedure for Disproportionate Collapse Prevention of CLT Platform-Type Construction. *Eng. Struct.* **2019**, *189*, 195–205. [CrossRef]
61. Huber, J.A.J.; Mpidi Bita, H.; Tannert, T.; Berg, S. Finite Element Analysis of Alternative Load Paths to Prevent Disproportionate Collapse in Platform-Type CLT Floor Systems. *Eng. Struct.* **2021**, *240*, 112362. [CrossRef]
62. Alih, S.C.; Vafaei, M. Performance of Reinforced Concrete Buildings and Wooden Structures during the 2015 Mw 6.0 Sabah Earthquake in Malaysia. *Eng. Fail. Anal.* **2019**, *102*, 351–368. [CrossRef]
63. Pina, J.C.; Saavedra Flores, E.I.; Saavedra, K. Numerical Study on the Elastic Buckling of Cross-Laminated Timber Walls Subject to Compression. *Constr. Build. Mater.* **2019**, *199*, 82–91. [CrossRef]
64. Escamilla, E.Z.; Habert, G.; Daza, J.F.C.; Archilla, H.F.; Echeverry Fernández, J.S.; Trujillo, D. Industrial or Traditional Bamboo Construction? Comparative Life Cycle Assessment (LCA) of Bamboo-Based Buildings. *Sustainability* **2018**, *10*, 3096. [CrossRef]
65. Padilla-Rivera, A.; Amor, B.; Blanchet, P. Evaluating the Link between Low Carbon Reductions Strategies and Its Performance in the Context of Climate Change: A Carbon Footprint of Awood-Frame Residential Building in Quebec, Canada. *Sustainability* **2018**, *10*, 2715. [CrossRef]
66. Bahramian, M.; Yetilmezsoy, K. Life Cycle Assessment of the Building Industry: An Overview of Two Decades of Research (1995–2018). *Energy Build.* **2020**, *219*, 109917. [CrossRef]
67. Jayalath, A.; Navaratnam, S.; Ngo, T.; Mendis, P.; Hewson, N.; Aye, L. Life Cycle Performance of Cross Laminated Timber Mid-Rise Residential Buildings in Australia. *Energy Build.* **2020**, *223*, 110091. [CrossRef]
68. Li, J.; Rismanchi, B.; Ngo, T. Feasibility Study to Estimate the Environmental Benefits of Utilising Timber to Construct High-Rise Buildings in Australia. *Build. Environ.* **2019**, *147*, 108–120. [CrossRef]

69. Tettey, U.Y.A.; Dodoo, A.; Gustavsson, L. Effect of Different Frame Materials on the Primary Energy Use of a Multi Storey Residential Building in a Life Cycle Perspective. *Energy Build.* **2019**, *185*, 259–271. [CrossRef]
70. Bruno, R.; Bevilacqua, P.; Cuconati, T.; Arcuri, N. Energy Evaluations of an Innovative Multi-Storey Wooden near Zero Energy Building Designed for Mediterranean Areas. *Appl. Energy* **2019**, *238*, 929–941. [CrossRef]
71. Markström, E.; Kitek Kuzman, M.; Bystedt, A.; Sandberg, D. Use of Wood Products in Multi-Storey Residential Buildings: Views of Swedish Actors and Suggested Measures for an Increased Use. *Wood Mater. Sci. Eng.* **2019**, *14*, 404–419. [CrossRef]
72. Gosselin, A.; Blanchet, P.; Lehoux, N.; Cimon, Y. Main Motivations and Barriers for Using Wood in Multi-Story and Non-Residential Construction Projects. *BioResources* **2017**, *12*, 546–570. [CrossRef]
73. Markström, E.; Kuzman, M.K.; Bystedt, A.; Sandberg, D.; Fredriksson, M. Swedish Architects View of Engineered Wood Products in Buildings. *J. Clean. Prod.* **2018**, *181*, 33–41. [CrossRef]
74. Riggio, M.; Alhariri, N.; Hansen, E. Paths of Innovation and Knowledge Management in Timber Construction in North America: A Focus on Water Control Design Strategies in CLT Building Enclosures. *Archit. Eng. Des. Manag.* **2020**, *16*, 58–83. [CrossRef]
75. Lu, H.R.; El Hanandeh, A.; Gilbert, B.P. A Comparative Life Cycle Study of Alternative Materials for Australian Multi-Storey Apartment Building Frame Constructions: Environmental and Economic Perspective. *J. Clean. Prod.* **2017**, *166*, 458–473. [CrossRef]
76. Himes, A.; Busby, G. Wood Buildings as a Climate Solution. *Dev. Built Environ.* **2020**, *4*, 100030. [CrossRef]
77. Dodoo, A.; Gustavsson, L. Energy Use and Overheating Risk of Swedish Multi-Storey Residential Buildings under Different Climate Scenarios. *Energy* **2016**, *97*, 534–548. [CrossRef]
78. Vidal, R.; Sánchez-Pantoja, N.; Martínez, G. Life Cycle Assessment of a Residential Building with Cross-Laminated Timber Structure in Granada-Spain. *Inf. De La Constr.* **2019**, *71*, 289. [CrossRef]
79. Leskovar, V.Ž.; Žigart, M.; Premrov, M.; Lukman, R.K. Comparative Assessment of Shape Related Cross-Laminated Timber Building Typologies Focusing on Environmental Performance. *J. Clean. Prod.* **2019**, *216*, 482–494. [CrossRef]
80. Gustavsson, L.; Nguyen, T.; Sathre, R.; Tettey, U.Y.A. Climate Effects of Forestry and Substitution of Concrete Buildings and Fossil Energy. *Renew. Sustain. Energy Rev.* **2021**, *136*, 110435. [CrossRef]
81. Aloisio, A.; Alaggio, R.; Fragiacomo, M. Fragility Functions and Behavior Factors Estimation of Multi-Story Cross-Laminated Timber Structures Characterized by an Energy-Dependent Hysteretic Model. *Earthq. Spectra* **2021**, *37*, 134–159. [CrossRef]
82. Xia, R.; Zhang, W.; Yang, Y.; Zhao, J.; Liu, Y.; Guo, H. Transparent Wood with Phase Change Heat Storage as Novel Green Energy Storage Composites for Building Energy Conservation. *J. Clean. Prod.* **2021**, *296*, 126598. [CrossRef]
83. Piccardo, C.; Gustavsson, L. Implications of Different Modelling Choices in Primary Energy and Carbon Emission Analysis of Buildings. *Energy Build.* **2021**, *247*, 111145. [CrossRef]
84. Myllyviita, T.; Hurmekoski, E.; Kunttu, J. Substitution Impacts of Nordic Wood-Based Multi-Story Building Types: Influence of the Decarbonization of the Energy Sector and Increased Recycling of Construction Materials. *Carbon. Balance Manag.* **2022**, *17*, 4. [CrossRef] [PubMed]
85. Amoruso, F.M.; Schuetze, T. Life Cycle Assessment and Costing of Carbon Neutral Hybrid-Timber Building Renovation Systems: Three Applications in the Republic of Korea. *Build. Environ.* **2022**, *222*, 109395. [CrossRef]
86. Younis, A.; Dodoo, A. Cross-Laminated Timber for Building Construction: A Life-Cycle-Assessment Overview. *J. Build. Eng.* **2022**, *52*, 104482. [CrossRef]
87. Karjalainen, M.; Ilgın, H.E.; Tulonen, L. Main Design Considerations and Prospects of Contemporary Tall Timber Apartment Buildings: Views of Key Professionals from Finland. *Sustainability* **2021**, *13*, 6593. [CrossRef]
88. Ilgın, H.E.; Karjalainen, M.; Pelsmakers, S. Finnish Architects' Attitudes towards Multi-Storey Timber-Residential Buildings. *Int. J. Build. Pathol. Adapt.* **2021**, *42*, 352–368. [CrossRef]
89. Toivonen, R.; Vihemäki, H.; Toppinen, A. Policy Narratives on Wooden Multi-Storey Construction and Implications for Technology Innovation System Governance. *For. Policy Econ.* **2021**, *125*, 102409. [CrossRef]
90. Viholainen, N.; Franzini, F.; Lähtinen, K.; Nyrud, A.Q.; Widmark, C.; Hoen, H.F.; Toppinen, A. Citizen Views on Wood as a Construction Material: Results from Seven European Countries. *Can. J. For. Res.* **2021**, *51*, 647–659. [CrossRef]
91. Nagy, E.; Berg Rustas, C.; Mark-Herbert, C. Social Acceptance of Forest-Based Bioeconomy—Swedish Consumers' Perspectives on a Low Carbon Transition. *Sustainability* **2021**, *13*, 7628. [CrossRef]
92. Giorgio, B.; Blanchet, P.; Barlet, A. Social Representations of Mass Timber and Prefabricated Light-Frame Wood Construction for Multi-Story Housing: The Vision of Users in Quebec. *Buildings* **2022**, *12*, 2073. [CrossRef]
93. Karjalainen, M.; Ilgın, H.E. The Change over Time in Finnish Residents' Attitudes towards Multi-Story Timber Apartment Buildings. *Sustainability* **2021**, *13*, 5501. [CrossRef]
94. Allan, K.; Phillips, A.R. Comparative Cradle-to-Grave Life Cycle Assessment of Low and Mid-Rise Mass Timber Buildings with Equivalent Structural Steel Alternatives. *Sustainability* **2021**, *13*, 3401. [CrossRef]
95. Chaggaris, R.; Pei, S.; Kingsley, G.; Feitel, A. Carbon Impact and Cost of Mass Timber Beam–Column Gravity Systems. *Sustainability* **2021**, *13*, 12966. [CrossRef]

96. Peñaloza, D.; Erlandsson, M.; Falk, A. Exploring the Climate Impact Effects of Increased Use of Bio-Based Materials in Buildings. *Constr. Build. Mater.* **2016**, *125*, 219–226. [CrossRef]
97. Andersen, J.H.; Rasmussen, N.L.; Ryberg, M.W. Comparative Life Cycle Assessment of Cross Laminated Timber Building and Concrete Building with Special Focus on Biogenic Carbon. *Energy Build.* **2022**, *254*, 111604. [CrossRef]
98. Hawkins, W.; Cooper, S.; Allen, S.; Roynon, J.; Ibell, T. Embodied Carbon Assessment Using a Dynamic Climate Model: Case-Study Comparison of a Concrete, Steel and Timber Building Structure. *Structures* **2021**, *33*, 90–98. [CrossRef]
99. Bezabeh, M.A.; Bitsuamlak, G.T.; Popovski, M.; Tesfamariam, S. Probabilistic Serviceability-Performance Assessment of Tall Mass-Timber Buildings Subjected to Stochastic Wind Loads: Part I—Structural Design and Wind Tunnel Testing. *J. Wind Eng. Ind. Aerodyn.* **2018**, *181*, 85–103. [CrossRef]
100. Bezabeh, M.A.; Bitsuamlak, G.T.; Popovski, M.; Tesfamariam, S. Probabilistic Serviceability-Performance Assessment of Tall Mass-Timber Buildings Subjected to Stochastic Wind Loads: Part II—Structural Reliability Analysis. *J. Wind. Eng. Ind. Aerodyn.* **2018**, *181*, 112–125. [CrossRef]
101. Roueche, D.B.; Prevatt, D.O.; Haan, F.L.; Datin, P.L. An Estimate of Tornado Loads on a Wood-Frame Building Using Database-Assisted Design Methodology. *J. Wind. Eng. Ind. Aerodyn.* **2015**, *138*, 27–35. [CrossRef]
102. Edskär, I.; Lidelöw, H. Dynamic Properties of Cross-Laminated Timber and Timber Truss Building Systems. *Eng. Struct.* **2019**, *186*, 525–535. [CrossRef]
103. Tannert, T.; Connolly, T.J. Hybrides Tragwerk Des 18-Stöckigen Studentenwohnheims "Tall Wood Building" in Vancouver Und Alternative in Holz. *Bautechnik* **2020**, *97*, 56–63. [CrossRef]
104. Binck, C.; Cao, A.S.; Frangi, A. Lateral Stiffening Systems for Tall Timber Buildings–Tube-in-Tube Systems. *Wood Mater. Sci. Eng.* **2022**, *17*, 309–316. [CrossRef]
105. Cao, A.S.; Stamatopoulos, H. A Theoretical Study of the Dynamic Response of Planar Timber Frames with Semi-Rigid Moment-Resisting Connections Subjected to Wind Loads. *Eng. Struct.* **2021**, *240*, 112367. [CrossRef]
106. Zhao, X.; Zhang, B.; Kilpatrick, T.; Sanderson, I. Numerical Analysis on Global Serviceability Behaviours of Tall Clt Buildings to the Eurocodes and Uk National Annexes. *Buildings* **2021**, *11*, 124. [CrossRef]
107. Voulpiotis, K.; Schär, S.; Frangi, A. Quantifying Robustness in Tall Timber Buildings: A Case Study. *Eng. Struct.* **2022**, *265*, 114427. [CrossRef]
108. Lyu, C.H.; Gilbert, B.P.; Guan, H.; Underhill, I.D.; Gunalan, S.; Karampour, H. Experimental Study on the Quasi-Static Progressive Collapse Response of Post-and-Beam Mass Timber Buildings under an Edge Column Removal Scenario. *Eng. Struct.* **2021**, *228*, 111425. [CrossRef]
109. Voulpiotis, K.; Köhler, J.; Jockwer, R.; Frangi, A. A Holistic Framework for Designing for Structural Robustness in Tall Timber Buildings. *Eng. Struct.* **2021**, *227*, 111432. [CrossRef]
110. Xia, Z.; Van De Kuilen, J.W.G.; Polastri, A.; Ceccotti, A.; He, M. Influence of Core Stiffness on the Behavior of Tall Timber Buildings Subjected to Wind Loads. *Front. Struct. Civ. Eng.* **2021**, *15*, 213–226. [CrossRef]
111. Chen, J.; Xiong, H.; Ventura, C.E. Non-Linear Simplified Models for Seismic Response Estimation of a Novel Tall Timber-Concrete Hybrid Structural System. *Eng. Struct.* **2021**, *229*, 111635. [CrossRef]
112. Navaratnam, S.; Widdowfield Small, D.; Gatheeshgar, P.; Poologanathan, K.; Thamboo, J.; Higgins, C.; Mendis, P. Development of Cross Laminated Timber-Cold-Formed Steel Composite Beam for Floor System to Sustainable Modular Building Construction. *Structures* **2021**, *32*, 681–690. [CrossRef]
113. Rinaldi, V.; Casagrande, D.; Cimini, C.; Follesa, M.; Fragiacomo, M. An Upgrade of Existing Practice-Oriented FE Design Models for the Seismic Analysis of CLT Buildings. *Soil Dyn. Earthq. Eng.* **2021**, *149*, 106802. [CrossRef]
114. Labrecque, N.; Ménard, S.; Oudjene, M.; Blanchet, P. Finite Element Study of Hyperstructure Systems with Modular Light-Frame Construction in High-Rise Buildings. *Buildings* **2022**, *12*, 330. [CrossRef]
115. Shameli Derakhshan, S.; Ni, C.; Zhou, L.; Qiang, R.; Huang, D. Cyclic Test and Seismic Equivalency Evaluation of High-Capacity Light Wood-Frame Shear Walls for Midrise Buildings. *J. Struct. Eng.* **2022**, *148*, 04022168. [CrossRef]
116. Wenzel, A.; Vera, S.; Guindos, P. Integration of Energy and Seismic-Structural Design Variables through the Optimization of a Multi-Story Residential Light-Frame Timber Building with Different Seismic Lateral Connectors and Building Stories. *J. Build. Eng.* **2022**, *57*, 104831. [CrossRef]
117. Dimovska, S.; Saloustros, S.; Pelà, L.; Roca, P. Modelling of In-Plane Seismic Behaviour of One-Way Steel or Timber Jack Arch Floors in Existing Buildings. Application to the Eixample District of Barcelona. *Eng. Struct.* **2022**, *262*, 114343. [CrossRef]
118. Carrero, T.; Montaño, J.; Berwart, S.; Santa María, H.; Guindos, P. Seismic Behavior of Innovative Hybrid CLT-Steel Shear Wall for Mid-Rise Buildings. *Bull. Earthq. Eng.* **2021**, *19*, 5917–5951. [CrossRef]
119. Movaffaghi, H.; Yitmen, I. Multi-Criteria Decision Analysis of Timber–Concrete Composite Floor Systems in Multi-Storey Wooden Buildings. *Civ. Eng. Environ. Syst.* **2021**, *38*, 161–175. [CrossRef]
120. Julien, E.; Blanchet, P.; Gosselin, L. Case-Study: Fully Prefabricated Wood Wall Connection to Improve Building Envelope and On-Site Efficiency. *Buildings* **2022**, *12*, 2185. [CrossRef]

121. Gibson, B.; Nguyen, T.; Sinaie, S.; Heath, D.; Ngo, T. The Low Frequency Structure-Borne Sound Problem in Multi-Storey Timber Buildings and Potential of Acoustic Metamaterials: A Review. *Build. Environ.* **2022**, *224*, 109531. [CrossRef]
122. Amoruso, F.M.; Schuetze, T. Hybrid Timber-Based Systems for Low-Carbon, Deep Renovation of Aged Buildings: Three Exemplary Buildings in the Republic of Korea. *Build. Environ.* **2022**, *214*, 108889. [CrossRef]
123. Bajno, D.; Grzybowska, A.; Bednarz, Ł. Old and Modern Wooden Buildings in the Context of Sustainable Development. *Energies* **2021**, *14*, 5975. [CrossRef]
124. Ouellet-Plamondon, C.M.; Ramseier, L.; Balouktsi, M.; Delem, L.; Foliente, G.; Francart, N.; Garcia-Martinez, A.; Hoxha, E.; Lützkendorf, T.; Nygaard Rasmussen, F.; et al. Carbon Footprint Assessment of a Wood Multi-Residential Building Considering Biogenic Carbon. *J. Clean. Prod.* **2023**, *404*, 136834. [CrossRef]
125. Tung, S.-F.; Su, H.-C.; Tzeng, C.-T.; Lai, C.-M. Experimental and Numerical Investigation of a Room Fire in a Wooden-Frame Historical Building. *Int. J. Archit. Herit.* **2020**, *14*, 106–118. [CrossRef]
126. Huai, C.; Xie, J.; Liu, F.; Liu, J. Study on the Thermal Stability and Degradation Kinetics of the Historical Wood: Restoration Residues Collected from Chinese Heritage Buildings. *Int. J. Arch. Heritage* **2022**, *18*, 164–176. [CrossRef]
127. Létourneau-Gagnon, M.; Dagenais, C.; Blanchet, P. Fire Performance of Self-Tapping Screws in Tall Mass-Timber Buildings. *Appl. Sci.* **2021**, *11*, 3579. [CrossRef]
128. Zhang, L.; Chen, K.; Liu, W.; Liu, Y.; Wang, K.; Ge, W.; Guo, K. Fire Performance of Pultruded Wood-Cored GFRP Sandwich Components for Building Construction. *Case Stud. Constr. Mater.* **2022**, *17*, e01555. [CrossRef]
129. Engel, T.; Werther, N. Impact of Mass Timber Compartment Fires on Façade Fire Exposure. *Fire Technol.* **2023**, *59*, 517–558. [CrossRef]
130. Sudhoff, P.; Steeger, F.; Zehfuß, J.; Kampmeier, B. Fire Performance of Insulating Materials Made from Renewable Resources—Part 2: Investigations on the Use as Compartment Filling in Timber Panel Constructions. *Bauphysik* **2021**, *43*, 303–313. [CrossRef]
131. Steeger, F.; Kolb, T.; Tech, S.; Wichmann, H.; Kampmeier, B.; Zehfuß, J.; Kolb, M. Fire Performance of Insulating Materials Made from Renewable Resources—Part 1: Experimental Investigations of -Smouldering Processes and Emissions. *Bauphysik* **2021**, *43*, 222–230. [CrossRef]
132. Bahrami, A.; Edås, M.; Magnenat, K.; Norén, J. The Behavior of Cross-Laminated Timber and Reinforced Concrete Floors in a Multi-Story Building. *Int. J. Adv. Appl. Sci.* **2022**, *9*, 43–50. [CrossRef]
133. Engel, T.; Moosmüller, K.; Werther, N. Fire Hazard Due to Electrical Installations in Modern Multi-Storey Timber Buildings. *Bautechnik* **2021**, *98*, 353–364. [CrossRef]
134. Bucklin, O.; Müller, T.; Amtsberg, F.; Leistner, P.; Menges, A. Analysis of Thermal Transmittance, Air Permeability, and Hygrothermal Behavior of a Solid Timber Building Envelope. *Energy Build.* **2023**, *299*, 113629. [CrossRef]
135. Lauer, A.P.R.; Benner, E.; Stark, T.; Klassen, S.; Abolhasani, S.; Schroth, L.; Gienger, A.; Wagner, H.J.; Schwieger, V.; Menges, A.; et al. Automated On-Site Assembly of Timber Buildings on the Example of a Biomimetic Shell. *Autom. Constr.* **2023**, *156*, 105118. [CrossRef]
136. Öberg, V.; Jockwer, R.; Goto, Y. Design for Structural Adaptation in Timber Buildings: Industry Perspectives and Implementation Roadmap for Sweden and Australia. *J. Build. Eng.* **2024**, *98*, 111413. [CrossRef]
137. Liu, H.; Wu, Y.; Li, A.; Deng, Y. Precision Detection and Identification Method for Apparent Damage in Timber Components of Historic Buildings Based on Portable LiDAR Equipment. *J. Build. Eng.* **2024**, *98*, 111050. [CrossRef]

sustainability

MDPI

Article

Bibliometric Analysis of Urban Coastal Development: Strategies for Climate-Resilient Timber Housing

Andrés García-Ruiz [1], Manuel Díez-Minguito [2], Konstantin Verichev [1] and Manuel Carpio [3,4,*]

1 Instituto de Obras Civiles, Universidad Austral de Chile, General Lagos 2060, Valdivia 5090000, Chile; andres.garcia@uach.cl (A.G.-R.); konstantin.verichev@uach.cl (K.V.)
2 Andalusian Institute for Earth System Research, University of Granada, 18006 Granada, Spain; mdiezm@ugr.es
3 Department of Construction Engineering and Management, School of Engineering, Pontificia Universidad Católica de Chile, Avenida Vicuña Mackenna 4860, Santiago 7820436, Chile
4 Centro Nacional de Excelencia para la Industria de la Madera (CENAMAD), Pontificia Universidad Católica de Chile, Santiago 8320000, Chile
* Correspondence: manuel.carpio@uc.cl

Abstract: Urban development in coastal areas has become increasingly important due to the climate crisis and its effects on sea level rise and extreme events, which increased the vulnerability of coastal zones. Therefore, it is important to analyze possible sustainable development techniques in urban planning and residential housing construction based on low-carbon footprint materials such as timber. These techniques should be capable of mitigating the effects of flooding and uncontrolled rises in coastal areas, as well as identifying normative and economic differences in their application in the Chilean context. For this purpose, a bibliometric analysis of 3882 articles selected from the Web of Science database between 1987 and 2022 was conducted, allowing us to identify a range of possible solutions to be developed in the study area. This includes evaluating their potential for normative application and a cost analysis of these solutions. In this regard, housing solutions such as amphibious houses and houses on stilts are two types of flood-resistant homes that are gaining popularity worldwide. Following the technical–economic analysis, it was observed that the solution on stilts can be up to 50% more cost-effective to implement in Chile. However, both options offer a promising solution to minimize the risks of coastal flooding and should be taken into account in the urban planning of coastal areas.

Keywords: urban development; coastal areas; timber; flooding; resilient buildings

1. Introduction

Urbanization is a community process that favors socio-economic development but also generates environmental conflicts. The urban growth of cities has been the most important human activity in the development of irreversible environmental and ecosystem changes on a global scale in recent decades. This has influenced the social and ecological interaction of our societies with the environment [1,2]. In recent years, the literature has explored the motivations and mechanisms of urbanization [3–5]. Studies related to urban occupation in coastal regions are of great interest, as about 40% of the population lives in coastal areas [6,7]. Certain research identified that urban development is clearly influenced by environmental, geophysical, institutional and socioeconomic conditions throughout the world [8]. In addition, urban centers become focal points for the expansion of neighboring communities, which occupy a larger territory. This is especially significant in coastal areas, which have a terrestrial as well as an intertidal and marine area of influence [9], where a population increase of 20% is expected over the next three decades. In the case of areas of coastal influence (Figure 1) (e.g., coasts, estuaries and wetlands), they have historically been inhabited by humans due to the accessibility to various elements that provided basic needs for human communities [6,9]. At present, the areas of coastal influence favor industrial fishing and tourism development [10–12]. Human community development has been modifying coasts in multiple forms. Many of these changes are permanent, arising as a result of the many activities that take place in the coastal zone, which are carried out by coastal managers and engineers.

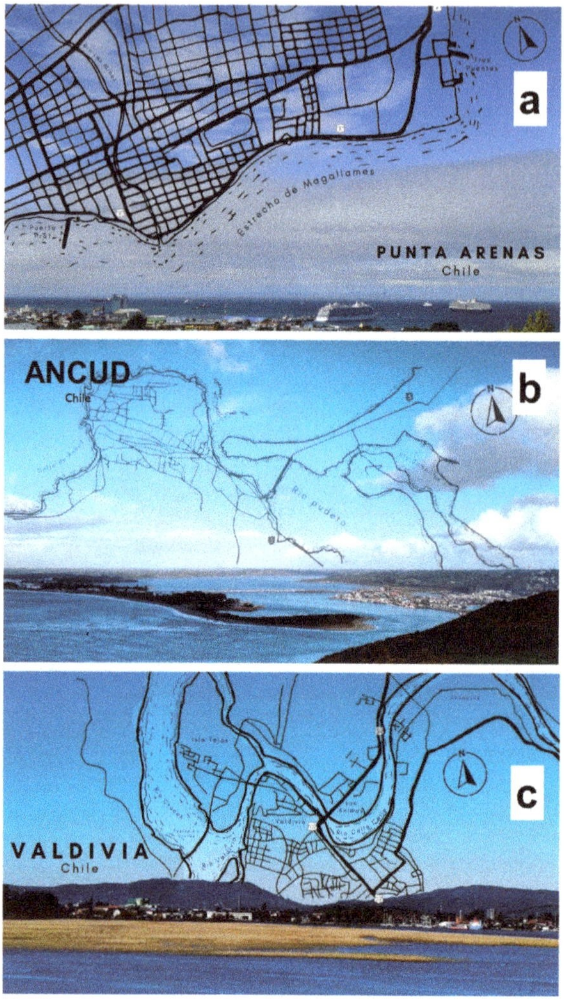

Figure 1. Example of coastal types considered in this work are (**a**) ocean–sea, (**b**) estuary, and (**c**) wetland.

The process of adapting to climate change, and in particular, to rising sea levels, requires strategic and adaptive planning and the forecasting of coastal urban development and the type of architecture for the projected building [13,14]. Therefore, it is essential to know the scientific evolution patterns associated with land use management in coastal cities (LUM-CC) (Figure 1), which will serve as a guide for the Architecture, Engineering and Construction (AEC) industry in the design of public policies and adaptation strategies, since the latter has a fundamental role in reducing the vulnerability of housing built on the coast.

Urban resilience can help address the challenges of climate change in cities by focusing on transforming the political and economic structures that perpetuate the vulnerability of urban communities. This involves putting people and their power at the center of decision-making and addressing the social and economic inequalities that make some communities more vulnerable than others [15]. Some aspects that improve urban resilience correspond to investment in infrastructure, including the construction of roads, bridges and other infrastructure that can withstand extreme events. On the other hand, developing early warning systems helps people to prepare for and respond to extreme weather phenomena. It is also essential to promote sustainable development, which helps to reduce the vulnerability of people and communities to climate change [13]. In this context, the type of residential architecture can help mitigate the risk to communities due to climate change [16]. In Chile, a study was developed to determine the risk of climate change impacts on its coasts, highlighting mainly the risk of flooding in residential areas near the coast [17]. The study concludes that for the entire national territory, a rise of (0.15–0.18 $\pm$ 0.1 m) of the Mean Sea Level (MSL) is expected in the 2026–2045 projection. In addition, an increase of 0.65 $\pm$ 0.3 m of the MSL is projected for the end of the century. Based on this, the study defines Valdivia as a critical commune due to the high number of inhabitants living below 10 m.a.s.l., in addition to the high amount of infrastructure, coastal infrastructure and equipment present below the mentioned elevation. The city of Valdivia presents 20.17% of its exposed area, with a coastline length of 437 km. In response to this need, the main objective of this work is to analyze the global state of knowledge in LUM-CC, identifying trends and key concepts to define sustainable adaptation measures and whether this is possible to realize with the use of timber. This, as a renewable material, offers significant environmental benefits by reducing the carbon footprint of buildings [18]. In addition, its versatility and durability make it an excellent choice for construction. By promoting the use of timber, we are contributing to a more sustainable and healthy future for generations to come [19].

The above, together with a technical–economic analysis for residential housing at risk of coastal flooding in southern Chile, and specifically for the city of Valdivia, will allow us to provide constructive and sustainable alternatives.

2. Materials and Methods

A bibliometric analysis will be carried out to establish the state of knowledge and the trends of scientific development in the last decades. We will then use the gaps in knowledge to develop technical applications in building solutions adaptive to the new climatic reality.

For the literature review, the Web of Science (WoS) databases were used. As shown in Figure 2, the methodology of this work considers two macro groups of analysis associated with coastal cities: "Management plan" and "Land use". The thematic search criteria were developed according to the following: (i) collect scientific articles from WoS databases in knowledge between 1987 and 2022 associated with LUM-CC; (ii) realize a quantitative review using bibliometric analysis and scientific mapping to find trends; and (iii) analyze the results obtained from the bibliometric analysis in order to identify topics of increasing interest.

Scientific mapping shows the structural and dynamic components of scientific work through graphical maps. These graphs represent a given field of knowledge and/or its specialties of individual documents and authors and how they relate to each other.

For the development of scientific mapping, an open-source software, SciMAT (Science Mapping Analysis Software Tool) Version v1.1.04, will be used to aggregate procedures, algorithms and measurements to evaluate each step of the overall scientific mapping workflow [20,21]. SciMAT applies bibliometric measures as important as the h-index [22], g-index [23] and hg-index [24], which describe the productive level of the development of the researcher and report on the number of articles defined. SciMAT was successfully applied in many areas, such as management [25,26], urbanism [27,28], construction [29,30] and health [31,32].

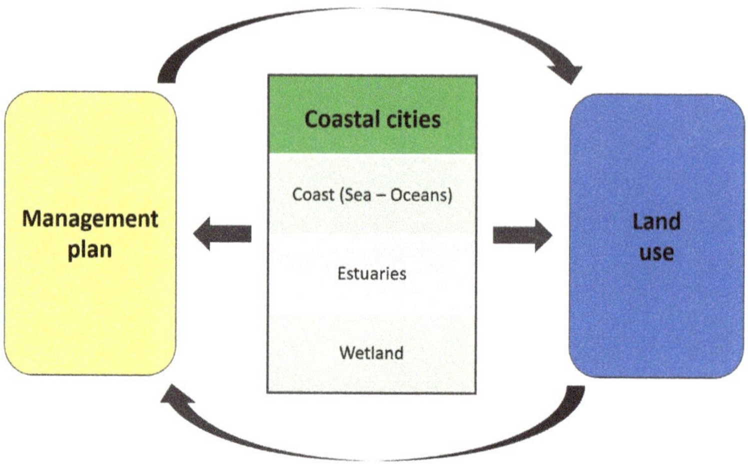

Figure 2. Relationship between the two search macro groups and coastline types.

2.1. Bibliometric Analysis

The bibliometric performance analysis corresponds to a statistical review of bibliometric indicators, such as the number of citations and their published papers (h-index) and the geographical distribution of the most cited papers and authors [33].

The literature search was carried out by identifying the relevant literature associated with the matrix concept LUM-CC (Figure 3), using specific criteria, which avoids any possibility of bias, indicated by Thomé et al. [34] and updated by Díaz-López et al. [29].

1.- Search planning : "Land-use management in coastal cities" (LUM-CC)

2.- Database selection: Web of Science (WOS)

3.- Keyword selection: (TS = ("estuar* management plan*" or "coastal* management plan*" or "shoreline* management plan*" or "coastal land use*" or "estuar* land use*" or "wetland* land use*"or "coastal cit*" or "coastal town*")) This formula as a result, found a total of 3882 items

4.- Final selection of literature: WoS Categories: Geosciences, Multidisciplinary; Engineering Environmental Studies; Water Resources; Geography, Physical; Oceanography; Environmental; Green & Sustainable Science & Technology; Geography; Urban Studies; Multidisciplinary Sciences; Engineering, Civil; Imaging Science & Photographic Technology; Regional & Urban Planning; Construction & Building Technology; Engineering, Ocean; Area Studies; Development Studies; Engineering, Multidisciplinary; Engineering, Marine; Transportation; Architecture; and Transportation Science & Technology. Number of articles to be studied is 1386.

5.- Determination of time period of analysis: the years proposed by WoS, between 1987 and 2023, were considered, but it has been decided to limit the range to 2022 in order to comply with full years. Four periods were then defined. These were distributed in order to ensure that no period exceeded 500 or was less than 300 documents, as follows (1987–2013) (437), (2014–2017) (319), (2018–2019) (306), 2020 (319), and finally, a period corresponding to (2021–2022) (475).

Figure 3. Diagram of selection of research papers. The "*" is used to indicate that the word is singular or plural.

The analysis of scientific maps corresponds to a graphical representation of the temporal evolution of a given field of knowledge and its authors [30], which in this case was developed using SciMAT software. This software performs bibliometric analysis of the content of papers based on scientific mapping using the methodology of Cobo et al., 2012 [20]; this software is a free and open-source scientific mapping tool. These features allow any user to analyze the conceptual, intellectual, and social evolution of a scientific field, as well as to define the importance of the works in the context of different time periods and the link between keywords. SciMAT after bibliometric analysis, generates flowcharts (Figure 4) [29].

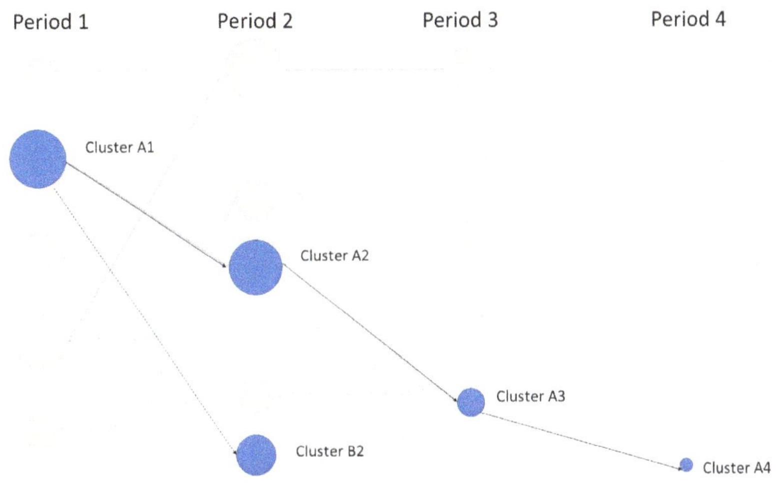

Figure 4. Evolution map. Solid lines indicate that related topics share a keyword; a dotted line shows that topics share words that are not the topic name. The size of the line is related to the inclusion rate.

The thematic evolution map (Figure 4) shows the development of the topics and the relationship between them from one period to another (i to i + 1). The circles represent the different groups, where the size of each group is directly proportional to the number of publications. Solid lines indicate that related topics share a keyword; a dotted line shows that topics share words that are not the topic name. Finally, the size of the line is related to the inclusion rate [35].

2.2. *Building Solutions for Urban–Residential Infrastructure against Flooding*

After the bibliometric analysis, the incorporation of solutions in buildings to increase resilience to floods in the short term, during and after the event, is estimated. A conscious design of buildings in flood zones includes the choice of building typologies or the integration of material technologies to improve their performance against flood vulnerability [36].

According to the U.S. Federal Emergency Management Agency and the Building Design Council of Canada, for a home to be less vulnerable to coastal flooding, it must have certain characteristics [37,38]. Some of the main features are as follows: (i) the dwelling should be constructed at an adequate height to prevent floodwater from entering the house. The height of the dwelling should be greater than the height of the expected flood level. (ii) The design of the dwelling should be adequate to resist the effects of water and wind. Windows and doors should be wind-resistant and have a watertight sealing system to prevent water from entering the house. (iii) Building materials should be water- and wind-resistant. Traditional materials, such as brick, stone, and concrete, are more resistant than more modern materials, such as plaster or plasterboard. (iv) The house should have an adequate drainage system that allows water to flow away from the house. This may include drainage channels, pumping systems, or elevated ground. (v) The location of the dwelling is important to avoid coastal flooding. The home should be built on an elevated area or on a hill to prevent water from pooling around the house.

Proposed Resilient Building Solutions in Chile

To determine a typical house, the average size of social housing in Chile is established, which varies according to the housing program and the region of the country. According to data from the Ministry of Housing and Urbanism (MINVU) of Chile, the average size of a social housing unit built under the Social and Territorial Integration Program (PIST) is approximately 50 m^2 (Figure 5).

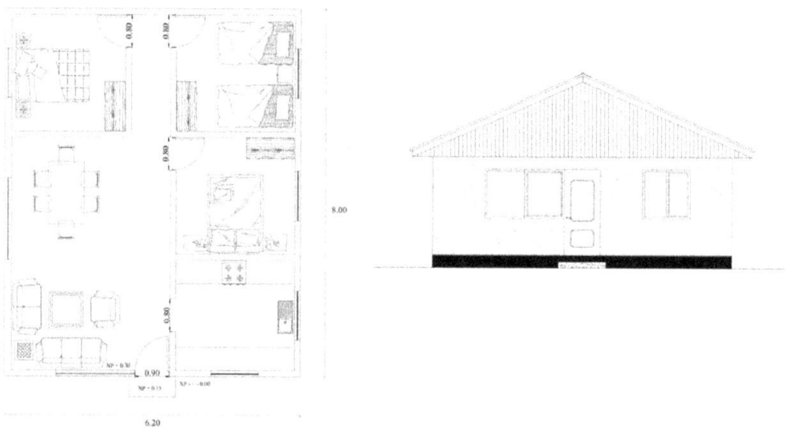

Figure 5. Two-Dimensional plans and elevation of the proposed house.

On the other hand, different construction solutions (Figure 6) were assigned for the walls based on the materials and constructions most commonly used for timber housing in the studied city. These materials and constructions were obtained from the title work "A data analysis of the Chilean housing stock and the development of modelling archetypes" [39]. Constructive solutions for walls in dwellings where the main structural material was timber were considered. The most representative solutions include a plasterboard, glass wool and Smart panel exterior cladding, as well as a plasterboard, expanded polystyrene and fiber cement siding exterior cladding.

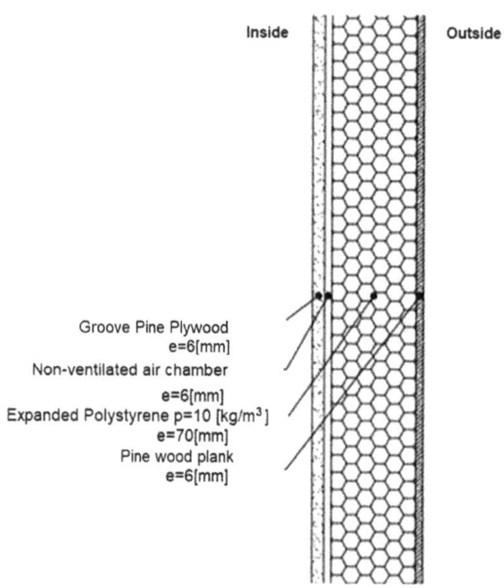

Figure 6. Construction solutions for perimeter walls.

In the case of the roofs, Figure 7 shows the scantlings of Conventional Solution 1 (SC1), which corresponds to a structural system of 2″ × 6″ pine timber for roof beams, with an OSB 11.1 [mm] bracing plate and asphalt shingles, with expanded polystyrene (EPS) insulation of variable thickness according to the thermal requirements of the area (Figure 6).

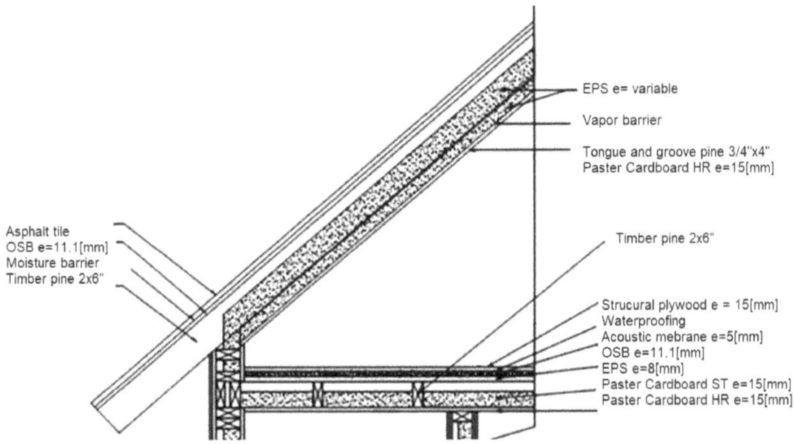

Figure 7. Construction solutions for roofing.

The structure of a floor of a house with timber supports consists of the following components:

- Beams: These are the horizontal supports that support the framing. They are usually made of timber or steel and are 60 cm to 120 cm apart.
- Truss: These are the horizontal beams that support the floor. They are made of impregnated timber and are 40 cm apart.
- Subfloor board: This is the layer of timber that is placed perpendicular to the joists and serves as a base for the finish floor. It will be made of 18 mm plywood.
- Finish Floor: This is the top layer of flooring that provides the walking surface; it will be 1 × 4″ boards. In addition, a 2 cm mortar veneer is recommended for leveling.

The framing, joists, and slab are nailed or screwed together. The subfloor is nailed or screwed to the framing, and the finish floor is nailed or glued to the subfloor (Figure 8).

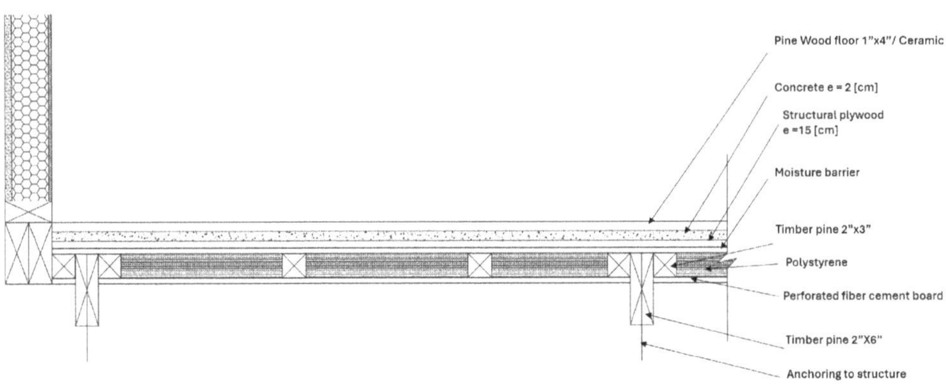

Figure 8. Construction solutions for floor.

2.3. Technical–Economic Analysis

Once the construction methods proposed by the bibliometric analysis and the respective structural design are established., an economic analysis of unit prices is performed, which is a tool used in the construction industry to break down and calculate the cost of each component or unit of a project. It consists of identifying and quantifying the materials, labor and equipment required to complete each activity or element of the project [40]. This type of analysis helps to determine the total cost of a project and is useful for decision-making in project management, so it will be used to establish which of the alternatives is the most economical in the Chilean market. Prices were obtained through an analysis of the construction materials market in the Región de los Ríos of Chile in October 2023, and prices will be presented in USD as of 20 October, according to the official peso/dollar exchange rate reported by the Central Bank of Chile, 1 USD = 941.82 CLP.

3. Results and Discussion

The method of systematic literature research and bibliometric analysis described above was applied to perform an exhaustive analysis of the research field LUM-CC, results that are reflected in the following sections, and how they influence the development of alternatives for the implementation of urban resilience to climate change.

3.1. Bibliometric Analysis

Figure 9 shows the annual distribution of the 1386 publications on the topic under study between 1987 and 2008. It is worth noting that since 2015, when the 2030—Agenda for Sustainable Development agreement was signed [41], there has been a substantial increase in scientific production (100). Between 2019 and 2022, the scientific production was the highest, with more than 5% of the articles analyzed, consolidating the LUM-CC theme due to the interest in climate change, sea level rise and the increasing number of people living in coastal cities [6,9,13,14].

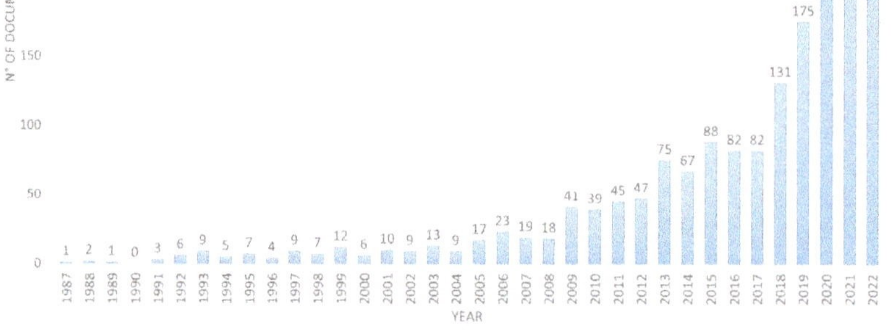

Figure 9. Time distribution of published studies by year.

3.2. Scientific Mapping Analysis

A SciMAT setup was carried out for bibliometric analysis and scientific mapping, in which the word was selected as the unit of analysis, co-occurrence analysis was selected as a tool to construct the networks, the equivalence index was selected as a measure of similarity to normalize the networks, and the k-means clustering algorithm was selected to detect topics. The documents were analyzed according to their year of publication, the journals used, the authors and the number of citations. The results obtained are presented below.

3.2.1. Overly Chart and Thematic Evolution Map for LUM-CC

For the thematic evolution map (Figure 10), the keywords are different in number and lexicography in each of the subperiods: Period 1 (P1): 1987–2013, Period 2 (P2): 2014–2017, Period 3 (P3): 2018–2019, Period 4 (P4): 2020 and Period 5 (P5): 2021–2022 (Figure 6). It can be observed how the clusters associated with LUM-CC have evolved using different keywords to explain the subject matter of the works in each period. For example, in P1, there is the "land-use" cluster, which remains unchanged in the second and third periods (continuous line) and even incorporates the "wetlands" cluster theme in P2, showing its intrinsic relationship with the land-use themes to finally evolve to the "flood" cluster in P4 and P5, which in turn incorporated the "sea-level-rise" cluster theme, generating a relationship of the matrix themes in this period.

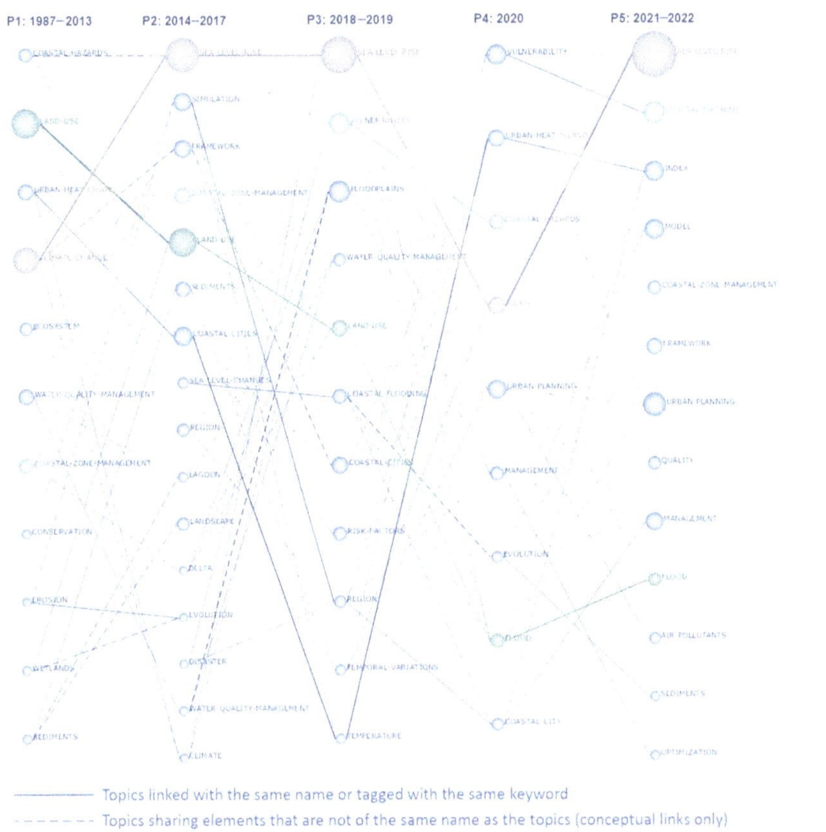

Figure 10. Map of the thematic evolution of research by period.

Considering the "climate-change" cluster, it evolves to "sea-level-rise", a theme that is maintained for P2 and P3, until it joins the "tides" cluster, which in turn returns to the central theme of "sea-level-rise" in P5. On the other hand, the evolution of the P1 cluster "coastal-zone-management" is maintained with this same keyword in P2, then this theme shares a conceptual link with "climate-change" and evolves towards "vulnerability", then to "coastal hazard" and, finally, it continues in "coastal-hazard" in P5.

3.2.2. Research Papers according to Type of Coastline

This section is devoted to analyzing the studies according to the shoreline types to which they refer, as represented in Figure 1: (a) ocean and sea, (b) estuaries, and (c) wetlands. The main studies in each section and the most representative clusters provided by SciMat for these keywords will be presented.

Focusing on the "sea and ocean" cluster analysis, it can be seen that there is a direct relationship between both keywords, which in turn are related to the topic of "prediction", which goes along with what was established in the previous section, presenting the importance of forecasting future coastal climate events, such as sea-level rise, and flood risks in coastal cities [42–45]. SciMat presents the relationship between the sea and land-use, which is reflected in how sea level change makes it necessary to predict the effects it will have on land-use planning in coastal cities [46].

The relationship between the driving theme and the concepts of "sediment" and "water", which in turn are interrelated, is relevant to account for the relationship between sediment transport and accumulation (both natural and artificial) in coastal waters [47].

Regarding the estuarine environment, SciMat shows an intrinsic relationship with wetlands, which is to be expected because of the ecosystem relationships that exist between them [48]. A connection between "management" and "water" was observed in terms of the impact that urban planning adjoining estuaries can have on water quality [49]. It is also linked to the keyword "impact", which points to the urban environment and how it causes changes in the estuarine food chain [50].

Research papers related to wetlands SciMat show the growing problem of wetland loss through the concept of "land reclamation" since land use "management" needs to be improved all over the world [51]. The relationship of wetlands to climate change largely coincides with the work discussed above in Section 3.2.2 [49,50,52]; in addition, a noteworthy work is described in [53], showing the effects of climate change and the state of vulnerability of these habitats to human interventions.

To better understand the temporal evolution of the work, scope and application on each type of coastal area described in this section, Table 1 is presented, showing the knowledge explored by researchers in each of the periods, the research methods used and the scientific debates and conclusions obtained over the years, such as that coastal development requires coordinated efforts from scientists, engineers and government officials, and that multifactorial assessment techniques should be used to consider social, economic and environmental factors in land planning decisions.

Table 1. Summary of major studies by periods and types of coastlines.

Type of Coast	Period	Major Studies	Analysis Tools	Main Conclusions
open ocean	P1	Concern about the economic, environmental and social consequences of climate change [42,54,55].	-Time series analysis. -Local urban database.	Vulnerability indexes against coastal hazard.
	P2 and P3	It is observed that the effects of climate change begin to become latent through the rise in sea level [14,44,45], and as the danger of flooding in coastal cities.	-Data analysis. -Projections new climate scenario. -Numerical models.	The need for joint scientific, technical, but mainly governmental efforts to assess coastal development [56]. Integrating ecological and social engineering.
	P4 and P5	There is mainly an idea of the need for adaptation, indicating that we must assume our new climatic reality [57].	-Analysis of effects of climate forcing on coastal cities. -Satellite mapping.	Deliver new protection tools, not just barriers and/or retaining walls, but updated building codes and nature-based measures that will be more effective in reducing flooding on a longer time scale.

Table 1. *Cont.*

Type of Coast	Period	Major Studies	Analysis Tools	Main Conclusions
Estuaries	P1	Human actions affecting the estuarine environment, construction materials and urban development [49,58,59].	-Satellite visualization. -Analysis of data recorded.	Develop estuarine management planning strategies to integrate urban development through integrated coastal zone management.
	P2	Focus on monitoring both urban expansion [60] and flood risks due to sea-level rise [61,62].	-Satellite visualization. -Multi-criteria analysis [63]. -Hydrodynamic Models [63].	Apply multifactorial assessment techniques for the integration of social, economic and environmental criteria governing land planning.
	P3, P4 and P5	Studies currently assess the flood risk of cities located on rivers close to the sea [64,65] and the use of effective tools to protect cities from flood hazards [66], in addition to resilient housing [36].	-Analysis of data recorded. -Analysis GIS. -Coastal bathymetric measurements.	Improving the integration of field parameters to obtain more realistic simulations. Generate proposals for active mitigation and adaptation, such as the use of mobile barriers and mechanical pumping equipment, which, combined and located closer to the sea, improve effectiveness. and propose different construction solutions for floating, amphibious or pile dwellings.
Wetlands	P1	The influence of climate change on the rise inMSL will affect coastal erosion, water salinization and wetland deterioration [43,46].	-Time series observation. -Local urban data.	The need for proactive adaptation plans, which evaluate (1) protecting, (2) accommodating or (3) removing infrastructure [53].
	P2 and P3	There is a concern about how the urban expansion of cities reduces the surface of wetlands [67,68].	-Analysis GIS. -Numerical models.	Implement intelligent management strategies to help solve the coastal erosion problem, considering a multi-criteria decision analysis to identify the most desirable management regimes [69].
	P4 and P5	Concerns about the risk of flooding of wetlands due to the rise in MSL are reopened [57]. In addition to the effects of pollution derived from human activities on these habitats [70]	-Evaluating satellite images.	Include coastal infrastructure adaptation measures, evaluating economic and social factors and considering soft strategies.

3.2.3. Evolution of Wetland Research

By evaluating the classification of published research articles by coastal type, it was found that urban planning in estuarine and wetland coastal zones accounts for 11.9% and 8.9% of all papers, respectively, while research in marine and oceanic coastal zones accounts for 79.2%. This gives us an idea that research on the effects and future conditions in coastal wetland zones is not yet consolidated and evolving over time. Furthermore, this shoreline type (wetlands) was chosen because these environments are often early indicators of ecosystem change because they are very sensitive to variations in MSL and the disturbances this can cause in their biogeochemical and hydrodynamic processes, as they represent biodiversity hotspots, supporting the presence of multiple habitat niches and, therefore, species of major environmental concern [71].

Therefore, wetland coastal urban (WCU) was included as a subsection of the CZM. Of the total number of articles analyzed in this study, 67 wetland-related articles are shown in Figure 11.

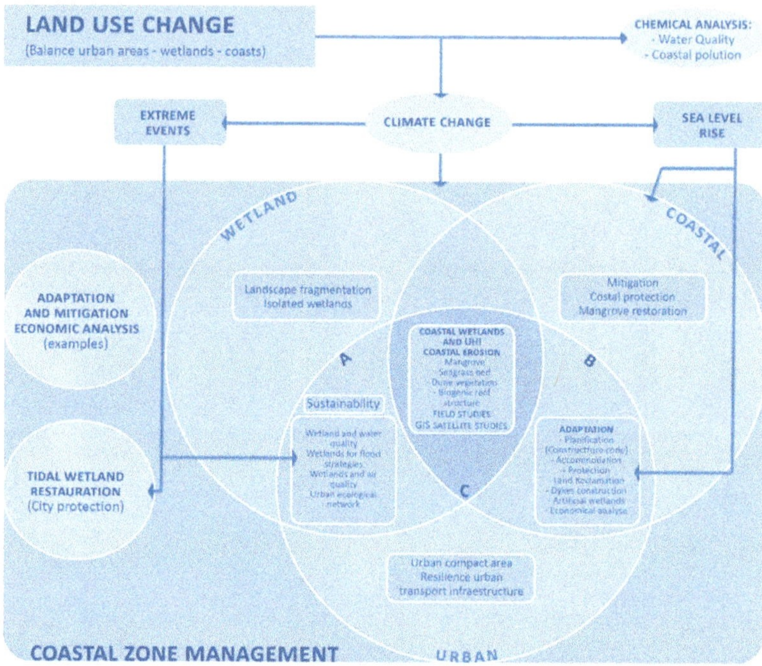

Figure 11. Summary of research trends in the entire time series of work on land use and management in coastal cities, focusing on wetland areas. A: Wetland–Urban Interaction; B: Urban–Coastal Interaction and C: Wetland–Coastal–Urban Interaction.

Figure 11 shows the connection between land use, urban wetland development, and their relationship with the riparian zone, all within the CZM. The figure shows how the works focused their analysis mainly on climate change and the various temporal dimensions of its effects and its impact on the riparian zone. On the one hand are extreme events (increasingly prevalent on the planet), namely the sustained mean sea-level rise, implications and problems for CZM, the first of which requires increased urban protection and economic problems generated by inadequate preparation for extreme events [53] and sustainability impacts (urban–wetland), in aspects such as water and air quality [72,73]. On the other hand, the rise in MSL, being a systematic condition of the planet, causes the need for adaptation in coastal cities and their housing infrastructure (urban-coastal), such as changes in building codes and regulations, creation of protection zones, artificial wetlands and amphibious housing [36,57,74,75].

Analyzing the problem of finding a balance between these three aspects, studies over all time periods showed a steady loss of wetland areas because of urban development. As early as the 1980s, it was shown that human settlement in the South Florida wetlands altered CH_4 sources, resulting in an area loss of about 33.6% between 1900 and 1973 [76]. Alternatively, using GIS methodology to estimate urban areas in Mazatlán, Mexico [49], an increase in urban areas between 1973 and 1997 (25.1 km^2 to 54.5 km^2 + 117.3%) and a decrease in mangroves (9.11 km^2 to 7.71 km^2 − 15.5%) was found. These were not the only works that showed this trend; other works, such as [71,77–80], highlight this problem worldwide. On the other hand, wetlands with coastal influence add to the anthropogenic threats of surface loss, the effect of climate change responsible for SLR. Early work in this area warned that a sea-level rise could reduce 2800 km^2 of wetlands, requiring a protection cost of USD 43.6 billion (1990) [53]. Therefore, recent decades have seen work showing numerous alternatives for protecting wetlands and coastal infrastructure [26,57,64,74,81], as shown in Figure 12.

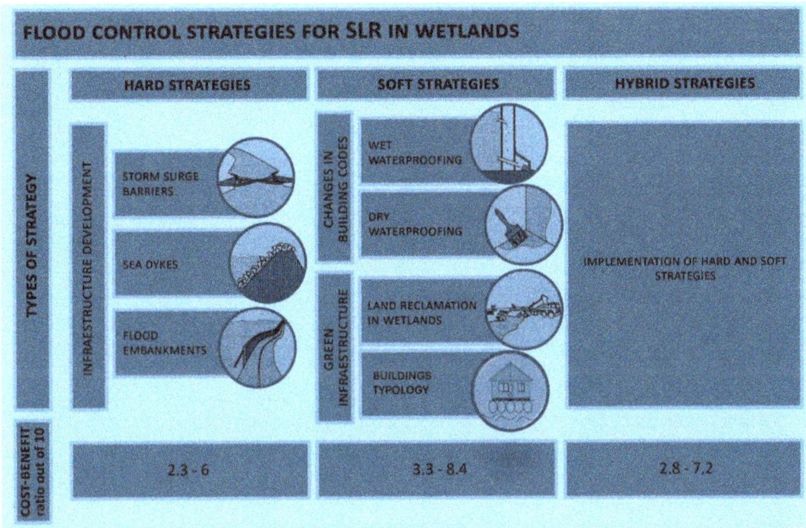

Figure 12. Flood control strategies for SLR in wetlands and its cost/benefit ratio of 10 according to [57].

All of these flood control strategies described above were subjected to a cost–benefit analysis to assess the economic feasibility of their implementation [82]. In the case of the cost–benefit scale in Figure 12, it is the element of economic evaluation proposed by Aerts et al., 2014 [83], that estimates both the initial investment and the cost of maintaining each solution in Figure 12. Benefits are expressed as a reduction in the expected annual damage of the evaluated site; if the cost/benefit ratio > 1, then the NPV (net present value) > 0, and the strategy is economically attractive [57]. In all strategies shown in Figure 12., the cumulative benefits can recover the total costs (i.e., initial investment and maintenance costs).

As for hard strategies, they mainly correspond to storm surge barriers, sea dikes and floodwalls. The main purpose of these elements is to prevent the area in question from flooding during a storm or high tide. In the case of a storm surge barrier, it is usually a mobile structure that usually closes before a storm and then opens again to facilitate the transport of goods and boats, while seawalls and flood control walls are rigid structures that are used to protect coastal communities, tidal inlets, rivers and estuaries from extreme weather events [84]. These solutions were used in projects such as the Venice Lagoon, the Port of Rotterdam, Shanghai and New York City [64,84–86].

On the other hand, soft strategies are related to indirect flood risk protection planning, such as the improvement of building codes, including, for example, wet flood protection materials, dry flood protection or the provision of flotation and/or rising of dwellings [16,82,83,87].

These measures are aimed at protecting individual property, while artificial wetlands, which correspond to land reclaimed from the sea (through sand reclamation) [81], and the different types of flood resilient housing, which, according to the U.S. Federal Emergency Management Agency and the Canadian Building Design Council, for a home to be less vulnerable to coastal flooding, it must have certain characteristics [37,38], like height; the dwelling should be built to an adequate height to prevent floodwater from entering the house.

The flood risk can be greatly reduced by hard protection strategies (e.g., storm surge barriers). In comparison, soft strategies (e.g., wetlands) have a higher cost/benefit ratio but result in higher residual risk. A hybrid strategy that combines the elements of hard and soft strategies is superior to both single-strategy approaches in terms of reduced future risk and a higher cost/benefit ratio [57]. As shown in Figure 12, it can be seen that the greatest cost–benefit can be obtained from soft solutions that originate from both material solutions and housing construction strategies [57]. Measures related to the reclamation of land redirected to anthropogenic construction sites led to a significant reduction in natural habitat. This, combined with economic activities associated with massive construction, can lead to contamination of the entire coastal ecosystem [74]; therefore, these types of soft solutions are not recommended for use by themselves, and solutions with different housing typologies are recommended.

This can start with houseboats, which range from simple structures to luxury homes with modern conveniences. Some houseboats are small and compact, whereas others are large and spacious [88]. Houseboats can also be permanent or mobile, meaning that they can move from one location to another. Houseboats are typically built on a floating platform made of timber, metal or fiberglass. Houseboats can also be connected to an electrical and water grid, or they can be self-sufficient with solar panels and rainwater-harvesting systems [75].

3.3. *Analysis of Building Typologies*

Four housing typologies were previously identified in the bibliometric analysis summarized in Figure 12: housing with an open first floor, floating housing, amphibious housing, and housing on stilts.

For open first-floor dwellings, physical vulnerability was observed, implying possible damage to the buildings. The intrinsic characteristics of the building design are relevant, such as the quality and quantity of the openings, as well as the orientation of the walls with respect to water flow [89]. In addition, their main structural construction material is reinforced concrete and ventilated roofs are made of aluminum or PVC plus glass, whose carbon footprint is very high [90]. Another aspect that affects the selection of this type of housing for the study area is the seismic component, which affects the perimeter protection of this type of housing, which is not necessarily flooded. Floating homes are typically built on floating platforms made of timber, metal or fiberglass [75]. The problem with this type of housing is that in Chile, there are no regulations that regulate it, in addition to being at the mercy of the effects of tsunamis that typically hit the heights of this country [91].

According to the aforementioned, the implementation of amphibious and resilient pile housing was selected for a deeper analysis. A cost–benefit analysis was performed to evaluate the economic feasibility of its application [82], and one of the amphibious types and the other built on piles were analyzed. Their construction is considered in Chile, and both share the same architecture and precepts of the Ministry of Housing and Urbanism. The area of application will be the city of Valdivia, as it is one of the main cities in the country and a center of tourism and economic development in the Los Ríos Region [92]. In addition, its geomorphological conditions, being immersed in a wetland area and having one of the rivers with the highest rainfall discharge during the year [93], as well as suffering from the effects of the earthquake and subsequent tsunami of 1960 [91], make it a promising study area for applying resilient construction solutions.

3.3.1. Structural Weights and Materials for Typical Housing

Table 2 shows the calculation results of the structural weight of each of the elements described in Section Proposed Resilient Building Solutions in Chile for the selected SERVIU house, as well as its overloading according to Chilean standard NCh1537 [94], establishes the requirements for the design and construction of earthquake-resistant buildings in Chile. It defines parameters to evaluate the strength and stability of structures, considering the seismic activity of the country. In addition, it addresses aspects such as soil classification, seismic loads and requirements for safe construction in seismic events.

Table 2. Weights of superstructure and secondary elements for studied house.

	Wall		
Component	Area (m^2)	Unit weight (kg/m^2)	Total weight (kg)
Groove Pine Plywood	120.72	9.8	1183.1
Non-ventilated air chamber	120.72	0	0
Expanded polystyrene	120.72	0.7	84.504
Pine wood plank	120.72	10.26	1238.5872
		Total:	2506.19
	Roof		
Component	Area (m^2)	Unit weight (kg/m^2)	Total weight (kg)
Asphalt shingles	27.36	7	191.52
OSB	27.36	7.65	209.304
Moisture barrier	27.36	0	0
EPS	27.36	3	82.08
Vapor Barrier	27.36	0	0
Cardboard plaster RH	27.36	35	957.6
Pine $3/4 \times 4'' \times 4''$ tongue-and-groove joint	294.12	0.94	276.47
Pine 2×6	47.88	2.76	132.15
		Total:	1849.1256
	Ceiling		
Component	Area (m^2)	Unit weight (kg/m^2)	Total weight (kg)
Waterproofing	49.6	0.005	0.248
Concrete overlay	49.6	48	2380.8
Structural paving	49.6	7.74	383.904
EPS	49.6	2	99.2
Cardboard plaster St	39.5	11.8	466.1
Cardboard plaster Rh	10.1	35	353.5
		Total:	4014.95
	Floor		
Component	Area (m^2)	Unit weight (kg/m^2)	Total weight (kg)
Moisture barrier	-	-	-
Truss 2×3	160	1.21	193.6
Polystyrene	-	-	-
Pine main beam $2'' \times 6''$	48	2.76	132.48
Mortar	49.6	48	2380.8
Structural plywood	49.6	7.74	383.904
Perforated fiber cement board	49.6	3	148.8
Ceramic	49.6	1.11	55.056
		Total:	3294.64
		Total weight	11,664.91

An additional weight should be considered due to the overload of use (NCh 1537, 2009), which is 200 kg/m^2. This value refers to the weight of people and all objects in a dwelling. For this 50 m^2 house, the total live load weight would be 10,000 kg. Therefore, the total weight is, with the wall, 21,664.91 kg.

3.3.2. Design Amphibious House

The house is mounted on four stainless steel guide posts that allow the house to be raised and lowered with water. Residents can enter through the front steps, and a stop is placed on the guide rod at the height of the maximum flood level. The waterline is the point where a building begins to float due to Archimedes' principle of buoyancy (Equation (1)) [16].

$$U = \frac{G}{(A * P)} \quad (1)$$

where U is the vertical drop below the waterline in m, G is the dead load of a built structure in KN, A is the area of the floating body in m^2 and P is the density of water in kN/m^3. If the weight of the vane is less than the thrust force it experiences, it will float. For this purpose, it is necessary to implement buoyancy systems, which are described below.

With the background of the weight of the structure, we proceed to determine the elements for buoyancy using the method developed by [16]. A total of 11 fiberglass tubes 8.2 m long and 0.6 m in diameter are recommended, as they can displace an amount of water equivalent to 24,860 m^3. The weight of the displaced water is 24,860 kg, which is greater than the total weight of the structure, including the floating base, which is 21,330 kg. The difference between the two most unfavorable weights is 3530 kg in favor of buoyancy. The arrangement of the piping system can be seen in Figure 13.

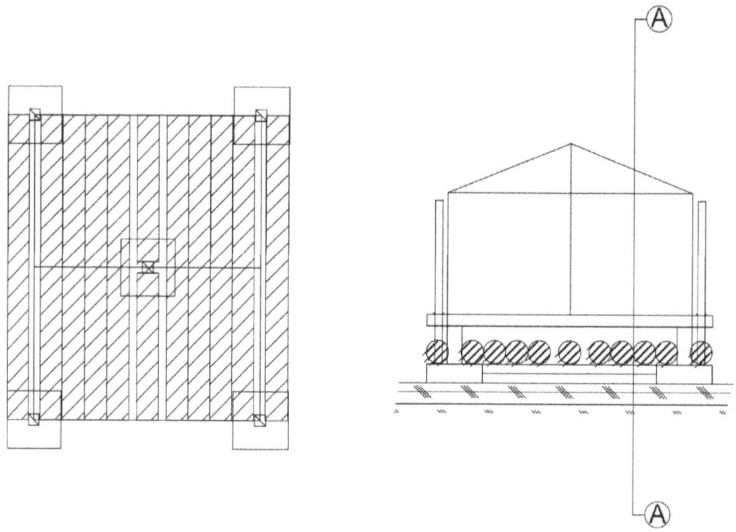

Figure 13. Arrangement of the fiberglass tubes, both in plan and in elevation view.

For the Archimedes equation for vertical drop (Equation (1)), the vertical drop below the waterline, U in meters, is equal to the dead load of the constructed structure, G in KN, divided by the area of the floating body, A in m^2, times the density of the water, P in kN/m^3. For this structure, G = 216.66 kN, A = 50 m^2 and P = 10 kN/m^3, so U = 216.66/(50 × 10) = 0.43 m.

According to Archimedes' principle, the upward force exerted by a fluid that opposes the weight of an object immersed in it is equal to the weight of the fluid that is displaced by the object. The depth of a building's foundation is calculated by considering the weight of the building and the weight of any object that will be placed in it.

Therefore, it is advisable to adopt a minimum drop of 1 m below ground level. It is not necessary to check the stability of the proposed house because it is attached to vertical guide posts.

Footings were used to calculate the foundations, as shown in Figure 14a. These footings had a dimension of 1.0 m by 1.0 m and a thickness of 0.5 m and were checked for shear and overturning. The footings were placed outside the floating structure, as shown in Figure 14b.

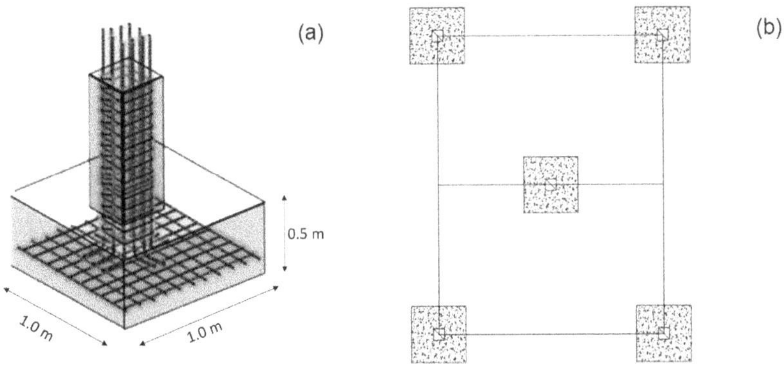

Figure 14. (**a**) Diagram of the type of isolated footing and its proposed dimensions. (**b**) Distribution of isolated footings under the surface of the amphibious house.

For the guide posts, there will be four stainless steel posts of section 20 × 20 cm, which will be interconnected to the floating structure of the house. The vertical mooring system will be anchored to the platform by means of the fender rollers [95]. Figure 15 illustrates this system, which shows how the pile rollers transmit the horizontal loads to the lateral system. These allow the house to move up and down through permanent connections to the water level. Residents can access the house via steps leading to the porch and front door. In addition, a stop is placed on the guide post at the height of the maximum flood level from the top of the platform [96]. The maintenance actions of the amphibious house elements were exposed to a humid–salty environment. For the fiberglass tubes, it is necessary to use epoxy resin paint, as it provides a coating and resistance to water and reagents such as chlorine, salt and PH variations. For the metal structure, an anticorrosive paint is applied, which forms a protective layer on the surface of the metal that prevents contact with oxygen and water, and finally, spray lubrication for the fender rollers [16].

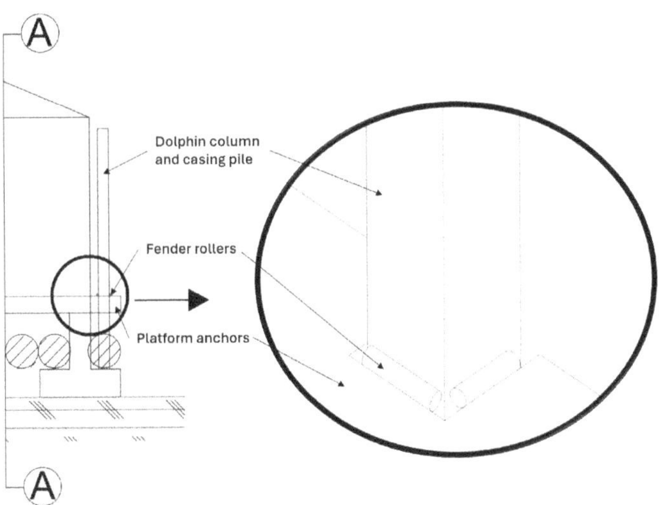

Figure 15. Example of anchoring and sliding guide posts to the floating house structure design house on stilts.

In Chile, construction on stilts is highly popular on the island of Chiloé, where they are called "palafitos", a series of houses built on timber pillars found on the coast. These houses originated in the 19th century when the fishermen of the area needed a place to store their tools and fishing nets. Over time, the "palafitos" became permanent homes for the fishermen and their families [87]. As indicated in the "palafitos" protection manual, the posts of the houses on piles must receive preservative paint that protects the supporting structure from the action of humidity and salinity of the environment. In this case, we selected the paint called carbolineum, also known as vegetable creosote, a liquid of natural origin that has antiseptic and preservative properties [97].

In the case of the design of the piles for this house, a distribution of elements will be followed, as shown in Figure 16. With a total of 13 piles, verification will be carried out for an unfavorable study area soil of the sandy silt type with σ_{adm} = 0.5 kg/cm^2.

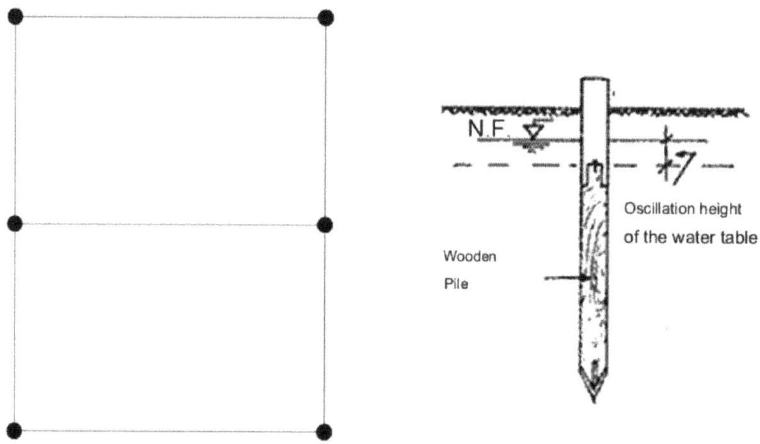

Figure 16. Pile layout and detail of timber pile to be installed.

The following values show that, initially, there is a permanent load and overload of use for small housing: $P_D = 113.76$ kN and $P_L = 98.1$ kN.

When opting for a solution consisting of timber piles, it should be taken into consideration that the maximum length of most timber piles is between 10 m and 20 m. The timber must be straight, strong, and free of defects. The American Society of Civil Engineers' "Manual of Practice", No. 17 (1959) stipulates that the type of function required would demand the use of Class B piles. Class B piles must have a minimum diameter of 305 to 330 mm and a service pressure of $P_{serv} = 34.81$ kN. This value should be less than the capacity of the pile tip.

Timber piles do not withstand high driving forces, so the capacity of the pile is limited to avoid crushing the fibers due to hammer blows, a phenomenon also known as splintering.

Assuming a low bearing capacity in the soil, for which it would be uneconomical to overextend the length of each pile until a more resistant bedrock or substrate is found, the design of each element can be considered for frictional resistance and lateral adhesion, in addition to the capacity at the tip, which is usually low in these cases.

Regarding the resistance or bearing capacity of a pile, considering Vesic's theory (Equation (2)) for cavity expansion, one would have (for a 12 m long timber pile):

$$Q_p = A_p\, \overline{\sigma}_0{}' N_\sigma^* \quad (2)$$

where Q_p is the pile tip capacity, A_p corresponds to pile tip area, $\overline{\sigma}_0{}'$ is the mean normal effective stress of the soil at the pile tip level. N_σ^*, corresponds to the carrying capacity factor. In the case of the study area

Loamy soil in Valdivia [98]: $\gamma \approx 8.04\ \frac{\text{kN}}{\text{m}^3}$

Internal friction angle: $\phi \approx 27°$
Cohesion: $c \approx 2.13$ kPa
Pile area: Ap = 0.07 m^2
Effective pressure (Equation (3)):

$$q' = \gamma \cdot h = 8.04 \frac{\text{kN}}{\text{m}^3} \cdot 12\ \text{m} = 96.48\ \text{kPa} \quad (3)$$

Normal effective stress, e.g., for sandy silt (Equation (4)):

$$K_0 = 1 - sin(27) = 0.5616 \quad (4)$$

$$\overline{\sigma}'_0 = \frac{1 + 2K_0}{3} q' == \frac{1 + 2 \cdot 0.5616}{3} \cdot 96.48\ \text{kPa} = 68.28\ \text{kPa}$$

According to Vesic's theory (Equation (5))

$$N_\sigma^* = f(I_{rr}) \quad (5)$$

where I_{rr} corresponds to the reduced stiffness index for the soil, which is obtained from Equation (5):

$$I_{rr} = \frac{I_r}{1 + I_r \Delta} \quad (6)$$

Δ corresponds to the average volumetric unit strain in the plastic zone below the pile tip. On the other hand, the intervals of Ir for a sandy silt soil are in the order of 70 [99]. With these data, it is possible to obtain Δ from Equation (6):

$$\Delta = 0.005(1 - \frac{\phi' - 25}{20}) \frac{q'}{p_a} \quad (7)$$

With p_a as atmospheric pressure ($\approx$100 kPa)

With the above, we have a value of Δ = 0.0044 m, which makes it possible to calculate the value of I_{rr}, = 58.46. Now, from Table 11.7, the "Carrying capacity factor" of the *Fundamentals of Foundation Engineering* book [99] can be obtained N_σ^* with $\phi \approx 27°$; this support factor will be 29.29.

Then, we replace, in Equation (2), to determine pile tip capacity, Q_p = 123.71 kN. Chilean regulations require a safety factor of 3 to 5 in these cases; we take 4.5 so that the Q_{adm} = Qp/(3.5) = 31.28 kN.

Finally, it is verified that $Q_{adm} > P_{serv}$; therefore, the section and length of the pile meet the load requirements.

3.4. Economic Analysis for Proposed Houses

The following is an analysis of construction prices of the supporting structures for the same type of 50 m^2 houses as described in the previous sections. As the geometry and materiality of the house are the same for both the pile structure and the amphibious support, this value will not be considered in the precise analysis. Tables 3 and 4 summarize the main construction elements of both options, compared to a traditional foundation structure, which considers a precast concrete support system, for a value of only USD 125, including installation.

When analyzing both budgets and their respective estimates, a lower construction complexity is observed in the case of houses on piles. This translates to a lower construction cost of more than 50%. This, together with the existence of construction regulations in Chile for this timber pile construction methodology, raises this alternative solution to flooding as feasible in this country.

Houses on stilts and amphibious dwellings were designed for use in flood-prone areas. It is worth mentioning that houses on stilts are usually more affordable and easier to build, while amphibious houses are usually more durable and able to withstand the forces of water and waves. A resumed comparison between the two construction solutions is shown in Table 5.

Table 3. Construction values of the supporting structure for an amphibious house.

Amphibious Housing Structure				
Description	**Unit**	**Amount**	**Unit Price**	**Total**
Excavations	m^3	2.5	USD 61	USD 153
Fiberglass Tubes 8.2 m (Diameter 0.6 m)	un	5	USD 320	USD 1600
Stainless steel guideposts	m	42	USD 70	USD 2940
Elevating roller systems	un	8	USD 280	USD 2240
Metal structure for pipes	kg	350	USD 5	USD 1750
Concrete foundations	m^3	2.5	USD 200	USD 500
Foundation armor	kg	200	USD 2	USD 400
Molded foundations	m^2	10	USD 15	USD 150
Manufacture and assembly of amphibious system	MHrs	1080	USD 8	USD 8640
Epoxy paint for fiberglass	gl	5	USD 30	USD 150
Anticorrosive paint	gl	2	USD 35	USD 70
			Total:	USD 18,593

Prices in USD.

Table 4. Construction values of the supporting structure for house on stilts.

House on Stilts Structure				
Description	**Unit.**	**Amount**	**Unit Price**	**Total**
Timber piles	m	54	USD 50	USD 2700
Pile-driving system	hr	32	USD 40	USD 1280
Support structure house	m^2	50	USD 85	USD 4250
Carbolineum timber preservative	gl	6	USD 25	USD 150
			Total:	USD 8380

Prices in USD.

Table 5. Comparative table between amphibious houses and houses on stilts.

	Stilt Houses	**Amphibious**
Desing	Pylon houses are usually built with lightweight materials, such as timber or bamboo, and are designed to rise above the ground to protect them from floods and other natural disasters.	Amphibious houses are usually built with more durable materials, such as steel or concrete, and are designed to withstand the force of water and waves.
Adventages	Stilt houses offer a number of advantages, such as protection against floods, pests and alluvium. They can also provide better ventilation and privacy than houses built on the ground.	Amphibious housing offers a number of advantages, such as the ability to move with the tides, the ability to withstand flooding and the possibility of being used for a variety of water activities.
Disadvantages	Houses on pillars can be more expensive to build and maintain than those built on the ground. They can also be more vulnerable to storms and other natural disasters.	Amphibious housing can be more expensive to build than houses on stilts. They can also be more difficult to access and maintain.

4. Conclusions

This study examined the evolution of knowledge related to LUM-CC, showing 3882 articles searched in the WoS platform from 1987 to 2022. The analysis of results identified the main subject areas, the most important authors and the countries in which most LUM-CC-related research was conducted. This led to the most significant findings of this article, which are presented below:

In general, the analysis of published LUM-CC papers shows that the risk of inundation due to sea level rise is a dominant theme across all time periods and coastline types.

This is reflected in the evolution of the predominant keywords in the scientific literature analyzed in five time periods, as it goes from "Land use" to "Floods", then the evolution from "Coastal cities" to "Risk factor" and ends in the last period with the concept of "Vulnerability". It is reflected in the early years of research, when it referred to the fight against climate change, then to the development of mitigation measures, and today to the creation of adaptability to new climate scenarios.

This will directly affect our way of conducting coastal urbanism because it highlights the need to explore proactive models of adaptation to this new climate reality that are more effective in reducing flooding in the long term, taking into account technical, administrative, economic and social factors. These adaptation plans must include the following: (1) protecting, (2) accommodating or (3) removing coastal infrastructure. Implementing intelligent management strategies to help solve the coastal erosion problem, considering a multi-criteria decision analysis to identify the most desirable management regimes. In addition to climate change, human action catalyzes the loss of these ecosystems, generating environmental, economic and social problems, which, coupled with local morphological conditions, makes the study of estuaries and wetlands an important area of research related to LUM-CC.

Flood risk reduction can be achieved through rigid protection measures such as storm surge barriers. On the other hand, soft strategies, such as wetlands, offer a higher cost/benefit ratio but carry a higher residual risk. A hybrid strategy that combines the elements of both strategies is superior to either an individual approach in terms of reduced future risk and a higher cost/benefit ratio. Amphibious houses are a promising solution to flooding because they are highly mobile, durable, energy-efficient, and can be manufactured from sustainable materials. The challenges associated with their construction are the cost of acquisition and maintenance and the availability of suitable sites. However, these challenges are likely to be overcome as the technology continues to develop. It should be noted that in the case of Chile and its high seismicity, as this house is not anchored to the ground but rather has an integrated structure, it has good seismic and flood-proof performance.

Houses on poles are more common in southern Chile, and it is important to consider several factors, such as the soil type, height of the water table, wind and seismic loads, and materials to be used. In Chile, a regulatory framework was established by NCh433.Of96, guaranteeing the safety and resistance of these structures during seismic events.

In addition to the design of houses, it is also important to have user protection strategies in the case of flooding. This includes having an immediate emergency response plan as well as providing temporary housing and other assistance to people displaced by floods.

Author Contributions: Conceptualization, M.C., M.D.-M. and A.G.-R.; methodology, M.C. and A.G.-R.; software, A.G.-R.; validation, M.C., M.D.-M. and K.V.; formal analysis, A.G.-R.; research, A.G.-R.; resources, M.C.; data curation, A.G.-R.; writing—preparation of the original draft, A.G.-R.; writing—revision and editing, M.C., M.D.-M. and K.V.; visualization, M.C. and A.G.-R.; supervision, M.C. and M.D.-M.; project management, M.C. and K.V.; obtaining funding, M.C. All authors have read and agreed to the published version of the manuscript.

Funding: This research was funded by the Agencia Nacional de Investigación y Desarrollo (ANID) of Chile, grant numbers ANID FONDECYT 1201052 and ANID BASAL FB210015 CENAMAD and Plan Andaluz de Investigación, Desarrollo e Innovación, PAIDI 2020. Ref. ProyExcel_00375 EPICOS.

Institutional Review Board Statement: Not applicable.

Informed Consent Statement: Not applicable.

Data Availability Statement: The data presented in this study are available upon request from the corresponding author.

Conflicts of Interest: The authors declare no conflicts of interest.

References

1. Grimm, N.B.; Faeth, S.H.; Golubiewski, N.E.; Redman, C.L.; Wu, J.; Bai, X.; Briggs, J.M. Global Change and the Ecology of Cities. *Science* **2008**, *319*, 756–760. [CrossRef]
2. Seto, K.; Sanchez-Rodriguez, R.; Fragkias, M. The New Geography of Contemporary Urbanization and the Environment. *Annu. Rev. Environ. Resour.* **2010**, *35*, 167–194. [CrossRef]
3. Xiao, J.; Shen, Y.; Ge, J.; Tateishi, R.; Tang, C.; Liang, Y.; Huang, Z. Evaluating urban expansion and land use change in Shijiazhuang, China, by using GIS and remote sensing. *Landsc. Urban. Plan.* **2006**, *75*, 69–80. [CrossRef]
4. Dewan, A.M.; Yamaguchi, Y. Land use and land cover change in Greater Dhaka, Bangladesh: Using remote sensing to promote sustainable urbanization. *Appl. Geogr.* **2009**, *29*, 390–401. [CrossRef]
5. Luo, J.; Wei, Y.H.D. Modeling spatial variations of urban growth patterns in Chinese cities: The case of Nanjing. *Landsc. Urban Plan.* **2009**, *91*, 51–64. [CrossRef]
6. Golberg, E. *Coastal Zone Space: Prelude to Conflict?* United Nations Educational; United NaParis: New York, NY, USA, 1994; p. 134.
7. ONU. Las Personas y Los Océanos. 2017. Available online: https://www.onu.org.mx/las-personas-y-los-oceanos/ (accessed on 26 December 2023).
8. Li, G.; Sun, S.; Fang, C. The varying driving forces of urban expansion in China: Insights from a spatial-temporal analysis. *Landsc. Urban Plan.* **2018**, *174*, 63–77. [CrossRef]
9. De Andres, M.; Barragán, J. Desarrollo Urbano en el Litoral a Escala Mundial. Método de Estudio para su Cuantificación./Urban-Coastal Development. Study Method for Quantifying in a Global Scal. *Rev. Estud. Andal.* **2016**, *33*, 64–83.
10. Lai, S.; Loke, L.H.L.; Hilton, M.J.; Bouma, T.J.; Todd, P.A. The effects of urbanisation on coastal habitats and the potential for ecological engineering: A Singapore case study. *Ocean Coast. Manag.* **2015**, *103*, 78–85. [CrossRef]
11. Barragán, J.M.; de Andrés, M. Analysis and trends of the world's coastal cities and agglomerations. *Ocean Coast. Manag.* **2015**, *114*, 11–20. [CrossRef]
12. Elsharouny, M.R.M.M. Planning Coastal Areas and Waterfronts for Adaptation to Climate Change in Developing Countries. *Procedia Environ. Sci.* **2016**, *34*, 348–359. [CrossRef]
13. Abadie, L.M.; Jackson, L.P.; Sainz de Murieta, E.; Jevrejeva, S.; Galarraga, I. Comparing urban coastal flood risk in 136 cities under two alternative sea-level projections: RCP 8.5 and an expert opinion-based high-end scenario. *Ocean Coast. Manag.* **2020**, *193*, 105249. [CrossRef]
14. Lorie, M.; Neumann, J.E.; Sarofim, M.C.; Jones, R.; Horton, R.M.; Kopp, R.E.; Fant, C.; Wobus, C.; Martinich, J.; O'Grady, M.; et al. Modeling coastal flood risk and adaptation response under future climate conditions. *Clim. Risk Manag.* **2020**, *29*, 100233. [CrossRef] [PubMed]
15. Bahadur, A.; Tanner, T. Transformational resilience thinking: Putting people, power and politics at the heart of urban climate resilience. *Environ. Urban* **2014**, *26*, 200–214. [CrossRef]
16. Varkey, M.V.; Philip, P.M. Flood risk mitigation through self-floating amphibious houses—Modelling, analysis, and design. *Mater. Today Proc.* **2022**, *65*, 442–447.
17. Ministerio del Medio Ambiente (Gobierno de Chile). Determinación del Riesgo de los Impactos del Cambio Climático en las costas de Chile 1a ed. Pontificia Universdidad Catolica de Chile, editoSantiago, Chile: Centro de Cambio climatico Global UC. 2019, p. 1133. Available online: https://cambioclimatico.mma.gob.cl/wp-content/uploads/2020/04/2019-10-22-Informe-V02-CCCostas-Exposici%C3%B3n-Rev1.pdf (accessed on 26 December 2023).
18. Morales-Beltran, M.; Engür, P.; Şişman, Ö.A.; Aykar, G.N. Redesigning for Disassembly and Carbon Footprint Reduction: Shifting from Reinforced Concrete to Hybrid Timber–Steel Multi-Story Building. *Sustainability* **2023**, *15*, 7273. [CrossRef]
19. Svatoš-Ražnjević, H.; Orozco, L.; Menges, A. Advanced Timber Construction Industry: A Review of 350 Multi-Storey Timber Projects from 2000–2021. *Buildings* **2022**, *12*, 404. [CrossRef]
20. Cobo, M.; López-Herrera, A.G.; Herrera-Viedma, E.; Herrera, F. SciMAT: A new science mapping analysis software tool. *J. Am. Soc. Inf. Sci. Technol.* **2012**, *63*, 1609–1630. [CrossRef]
21. Verichev, K.; Zamorano, M.; Salazar-Concha, C.; Carpio, M. Analysis of Climate-Oriénted Researches in Building. *Appl. Sci.* **2021**, *11*, 3251. [CrossRef]
22. Alonso, S.; Cabrerizo, F.J.; Herrera-Viedma, E.; Herrera, F. h-Index: A review focused in its variants, computation and standardization for different scientific fields. *J. Informetr.* **2009**, *3*, 273–289. [CrossRef]
23. Egghe, L. Theory and Practice of the g-Index. *Scientometrics* **2006**, *69*, 131–152. [CrossRef]
24. Cabrerizo, F.J.; Alonso, S.; Herrera-Viedma, E.; Herrera, F. q2-Index: Quantitative and qualitative evaluation based on the number and impact of papers in the Hirsch core. *J. Inf.* **2010**, *4*, 23–28. [CrossRef]
25. Santana, M.; Cobo, M.J. What is the future of work? A science mapping analysis. *Eur. Manag. J.* **2020**, *38*, 846–862. [CrossRef]
26. Lv, T.; Wang, L.; Xie, H.; Zhang, X.; Zhang, Y. Evolutionary overview of water resource management (1990–2019) based on a bibliometric analysis in Web of Science. *Ecol. Inform.* **2021**, *61*, 101218. [CrossRef]
27. Carpio, M.; González, Á.; González, M.; Verichev, K. Influence of pavements on the urban heat island phenomenon: A scientific evolution analysis. *Energy Build.* **2020**, *226*, 110379. [CrossRef]
28. Sharifi, A. Urban sustainability assessment: An overview and bibliometric analysis. *Ecol. Indic.* **2021**, *121*, 107102. [CrossRef]
29. Díaz-López, C.; Carpio, M.; Martín-Morales, M.; Zamorano, M. Analysis of the scientific evolution of sustainable building assessment methods. *Sustain. Cities Soc.* **2019**, *49*, 101610. [CrossRef]

30. Nalbandian, K.M.; Carpio, M.; González, Á. Analysis of the scientific evolution of self-healing asphalt pavements: Toward sustainable road materials. *J. Clean. Prod.* **2021**, *293*, 126107. [CrossRef]
31. Casado-Aranda, L.-A.; Sánchez-Fernández, J.; Viedma-del-Jesús, M. Analysis of the scientific production of the effect of COVID-19 on the environment: A bibliometric study. *Environ. Res.* **2021**, *193*, 110416. [CrossRef] [PubMed]
32. Salazar-Concha, C.; Ficapal-Cusí, P.; Boada-Grau, J.; Camacho, L.J. Analyzing the evolution of technostress: A science mapping approach. *Heliyo* **2021**, *7*, e06726. [CrossRef] [PubMed]
33. Cobo, M.J.; Martínez, M.A.; Gutiérrez-Salcedo, M.; Fujita, H.; Herrera-Viedma, E. 25 years at Knowledge-Based Systems: A bibliometric analysis. *Knowl.-Based Syst.* **2015**, *80*, 3–13. [CrossRef]
34. Thomé, A.M.; Scavarda, L.; Scavarda, A. Conducting systematic literature review in operations management. *Prod. Plan. Control* **2016**, *27*, 408–420. [CrossRef]
35. Cobo, M.; López-Herrera, A.G.; Herrera-Viedma, E.; Herrera, F. An approach for detecting, quantifying, and visualizing the evolution of a research field: A practical application to the Fuzzy Sets Theory field. *J. Informetr.* **2011**, *5*, 146–166. [CrossRef]
36. Nillesen, A.L. Chapter 24—Designing and building flood proof houses. In *Coastal Flood Risk Reduction*; Brody, S., Lee, Y., Kothuis, B., Eds.; Elsevier: Amsterdam, The Netherlands, 2022; pp. 329–339.
37. American Society of Civil Engineers. *Flood Resistant Design and Construction*; American Society of Civil Engineers: Reston, VA, USA, 2014; p. 92.
38. Coulbourne, W.L.; Kriebel, D.L.; Behm, R.L.; McKenna, K.K. *Guide for Design of Flood-Resistant Buildings*; National Research Council of Canada: Ottawa, ON, Canada, 2021.
39. Molina, C.; Kent, M.; Hall, I.; Jones, B. A data analysis of the Chilean housing stock and the development of modelling archetypes. *Energy Build.* **2020**, *206*, 109568. [CrossRef]
40. Trinidad, M.A. *Precios Unitarios. Primera*; de Tabasco, U.J.A., Ed.; Universidad Juarez Autonoma de Tabasco: Tabasco, Mexico, 2005; p. 80.
41. New York United Nations. *Transforming Our World: The 2030 Agenda for Sustainable Development*; New York United Nations, Department of Economic and Social Affairs: New York, NY, USA, 2015.
42. Hallegatte, S.; Green, C.; Nicholls, R.J.; Corfee-Morlot, J. Future flood losses in major coastal cities. *Nat. Clim. Chang.* **2013**, *3*, 802–806. [CrossRef]
43. Sallenger, A.; Doran, K.; Howd, P. Hotspot of accelerated sea-level rise on the Atlantic Coast of North America. *Nat. Clim. Chang.* **2012**, *2*, 884–888. [CrossRef]
44. Wahl, T.; Jain, S.; Bender, J.; Meyers, S.; Luther, M. Increasing risk of compound flooding from storm surge and rainfall for major US cities. *Nat. Clim. Chang.* **2015**, *5*, 1093–1097. [CrossRef]
45. Vitousek, S.; Barnard, P.; Fletcher, C.; Frazer, N.; Erikson, L.; Storlazzi, C. Doubling of coastal flooding frequency within decades due to sea-level rise. *Sci. Rep.* **2017**, *7*, 1–9. [CrossRef] [PubMed]
46. Nicholls, R. Planning for the Impacts of Sea Level Rise. *Oceanography* **2011**, *24*, 144–157. [CrossRef]
47. Su, L.; Sharp, S.M.; Pettigrove, V.J.; Craig, N.J.; Nan, B.; Du, F.; Shi, H. Superimposed microplastic pollution in a coastal metropolis. *Water Res.* **2019**, *168*, 115140. [CrossRef]
48. Miranda, L.; Andutta, F.; Kjerfve, B.; Castro, B. *Fundamentals of Estuarine and Physical Oceanography*; Springer: Singapore, 2017; Volume 8.
49. Ruiz-Luna, A.; Berlanga-Robles, C. Land use, land cover changes and coastal lagoon surface reduction associated with urban growth in northwest Mexico. *Landsc. Ecol.* **2003**, *18*, 159–171. [CrossRef]
50. Deegan, L.A.; Johnson, D.S.; Warren, R.S.; Peterson, B.J.; Fleeger, J.W.; Fagherazzi, S.; Wollheim, W.M. Coastal eutrophication as a driver of salt marsh loss. *Nature* **2012**, *490*, 388–392. [CrossRef]
51. Bilskie, M.V.; Hagen, S.C.; Medeiros, S.C.; Passeri, D.L. Dynamics of sea level rise and coastal flooding on a changing landscape. *Geophys. Res. Lett.* **2014**, *41*, 927–934. [CrossRef]
52. Glasby, T.; Connell, S. *Urban Structures as Marine Habitats*; UN, FAO, Eds.; ROYAL Swedish Academic Sciences: Stockholm, Sweden, 1999; Available online: http://hdl.handle.net/2440/12039 (accessed on 26 December 2023).
53. Nicholls, R.J.; Hoozemans, F.M.J. The Mediterranean: Vulnerability to coastal implications of climate change. *Ocean Coast. Manag.* **1996**, *31*, 105–132. [CrossRef]
54. Balica, S.; Wright, N.; Meulen, F. A flood vulnerability index for coastal cities and its use in assessing climate change impacts. *Nat. Hazards* **2012**, *64*, 73–105. [CrossRef]
55. Schiller, A.; de Sherbinin, A.; Hsieh, W.-H.; Pulsipher, A. *The Vulnerability of Global Cities to Climate Hazards; Environ. Urba*; Routledge: London, UK, 2007; Volume 19.
56. Torabi, E.; Dedekorkut-Howes, A.; Howes, M. Adapting or maladapting: Building resilience to climate-related disasters in coastal cities. *Cities* **2018**, *72*, 295–309. [CrossRef]
57. Du, S.; Scussolini, P.; Ward, P.J.; Zhang, M.; Wen, J.; Wang, L.; Koks, E.; Diaz-Loaiza, A.; Gao, J.; Ke, Q.; et al. Hard or soft flood adaptation? Advantages of a hybrid strategy for Shanghai. *Glob. Environ. Chang.* **2020**, *61*, 102037. [CrossRef]
58. Farrapeira, C.; Melo, A.V.O.M.; Barbosa, D.; Silva, K. Ship hull fouling in the Port of Recife, Pernambuco. *Braz. J. Ocean.* **2007**, *55*, 207–221. [CrossRef]
59. Yang, X.; Liu, Z. Using satellite imagery and GIS for land-use and land-cover change mapping in an estuarine watershed. *Int. J. Remote Sens.* **2005**, *26*, 5275–5296. [CrossRef]

60. Pourebrahim, S.; Hadipour, M.; Mokhtar, M.; Taghavi, S. Application of VIKOR and fuzzy AHP for conservation priority assessment in coastal areas: Case of Khuzestan district, Ira. *Ocean Coast. Manag.* **2014**, *98*, 20–26. [CrossRef]
61. Wang, G.; Liu, Y.; Wang, H.; Wang, X. A comprehensive risk analysis of coastal zones in China. *Estuar. Coast. Shelf Sci.* **2014**, *140*, 22–31. [CrossRef]
62. Dyckman, C.; SJohn, C.; London, J. Realizing managed retreat and innovation in state-level coastal management planning. *Ocean Coast. Manag.* **2014**, *102*, 212–223. [CrossRef]
63. García-Ruiz, A.; Carpio, M.; Giesecke, R.; Bermúdez, M.; Díez-Minguito, M. Circulation and distribution of suspended mesozooplankton carcasses in a mid-latitude estuary. *J. Mar. Syst.* **2021**, *225*, 103646. [CrossRef]
64. Cheng, H.; Chen, J.; Chen, Z.; Ruan, R.; Xu, G.; Zeng, G.; Zhu, J.; Dai, Z.; Chen, X.; Gu, S.; et al. Mapping Sea Level Rise Behavior in an Estuarine Delta System: A Case Study along the Shanghai Coast. *Engineering* **2018**, *4*, 156–163. [CrossRef]
65. Dykstra, S.; Dzwonkowski, B. The Propagation of Fluvial Flood Waves Through a Backwater-Estuarine Environment. *Water Resour. Res.* **2020**, *56*, e2019WR025743. [CrossRef]
66. Hall, J.W.; Harvey, H.; Manning, L.J. Adaptation thresholds and pathways for tidal flood risk management in London. *Clim. Risk Manag.* **2019**, *24*, 42–58. [CrossRef]
67. Li, Y.; Shi, Y.; Zhu, X.; Cao, H.; Yu, T. Coastal wetland loss and environmental change due to rapid urban expansion in Lianyungang, Jiangsu, China. *Reg. Environ. Chang.* **2014**, *14*, 1175–1188. [CrossRef]
68. Duan, H.; Zhang, H.; Huang, Q.; Zhang, Y.; Hu, M.; Niu, Y.; Zhu, J. Characterization and environmental impact analysis of sea land reclamation activities in China. *Ocean Coast. Manag.* **2016**, *130*, 128–137. [CrossRef]
69. Li, X.; Bellerby, R.; Craft, C.; Widney, S. Coastal wetland loss, consequences, and challenges for restoration. *Anthr. Coasts* **2018**, *3*, 1–15. [CrossRef]
70. Balogun, A.-L.; Yekeen, S.T.; Pradhan, B.; Althuwaynee, O.F. Spatio-Temporal Analysis of Oil Spill Impact and Recovery Pattern of Coastal Vegetation and Wetland Using Multispectral Satellite Landsat 8-OLI Imagery and Machine Learning Models. *Remote Sens.* **2020**, *12*, 1225. [CrossRef]
71. Fastelli, P.; Marcelli, M.; Guerranti, C.; Renzi, M. Recent Changes of Ecosystem Surfaces and their Services Value in a Mediterranean Costal Protected Area: The Role of Wetlands. *Thalass. An. Int. J. Mar. Sci.* **2018**, *34*, 233–245. [CrossRef]
72. Gracia, A.; Rangel-Buitrago, N.; Oakley, J.A.; Williams, A.T. Use of ecosystems in coastal erosion management. *Ocean Coast. Manag.* **2018**, *156*, 277–289. [CrossRef]
73. Zhu, C.; Zeng, Y. Effects of urban lake wetlands on the spatial and temporal distribution of air PM10 and PM2.5 in the spring in Wuhan. *Urban For. Urban Green* **2018**, *31*, 142–156. [CrossRef]
74. Zhang, Y.; Chen, R.; Wang, Y. Tendency of land reclamation in coastal areas of Shanghai from 1998 to 2015. *Land Use Policy* **2020**, *91*, 104370. [CrossRef]
75. Poulsen Rydborg, M.; Lauring, M.; Brunsgaard, C. Vulnerabilities and resilience in Danish housing stock: A comparative study of architectural answers to climate change in Danish housing in relation to other oceanic climates. In *E3S Web of Conferences*; EDP Sciences: Les Ulis, France, 2019; Volume 111.
76. Harriss, R.; Sebacher, D.; Bartlett, K.; Bartlett, D.; Crill, P. Sources of atmospheric methane in the South Florida environment. *Glob. Biogeochem. Cycles* **1988**, *2*, 231–243. [CrossRef]
77. Xie, Z.; Liu, J.; Zhu, G.; Shao, Q.; Xu, X. Evaluating Habitat Change and Boundary Adjustment of a Nature Reserve in Coastal Wetlands: Case Study of Beidagang Nature Reserve, China. *J. Coast. Res.* **2011**, *27*, 966–972. [CrossRef]
78. Ellis, J.T.; Spruce, J.P.; Swann, R.A.; Smoot, J.C.; Hilbert, K.W. An assessment of coastal land-use and land-cover change from 1974–2008 in the vicinity of Mobile Bay, Alabama. *J. Coast. Conserv.* **2011**, *15*, 139–149. [CrossRef]
79. Enaruvbe, G.O.; Ige-Olumide, O. Geospatial analysis of land-use change processes in a densely populated coastal city: The case of Port Harcourt, south-east Nigeria. *Geocarto Int.* **2015**, *30*, 441–456. [CrossRef]
80. Halls, J.N.; Magolan, J.L. A Methodology to Assess Land Use Development, Flooding, and Wetland Change as Indicators of Coastal Vulnerability. *Remote Sens.* **2019**, *11*, 2260. [CrossRef]
81. Kim, S.K. The Economic Effects of Climate Change Adaptation Measures: Evidence from Miami-Dade County and New York City. *Sustainability* **2020**, *12*, 1097. [CrossRef]
82. de Ruig, L.T.; Barnard, P.L.; Botzen, W.J.W.; Grifman, P.; Hart, J.F.; de Moel, H.; Sadrpour, N.; Aerts, J.C. An economic evaluation of adaptation pathways in coastal mega cities: An illustration for Los Angeles. *Sci. Total Environ.* **2019**, *678*, 647–659. [CrossRef]
83. Aerts, J.C.J.H.; Botzen, W.J.W.; Emanuel, K.; Lin, N.; de Moel, H.; Michel-Kerjan, E.O. Evaluating Flood Resilience Strategies for Coastal Megacities. *Science* **2014**, *344*, 473–475. [CrossRef]
84. Degrieck, J.; Van Paepegem, W.; Van Schepdael, L.; Samyn, P.; De Baets, P.; Suister, E.; Leendertz, J.S. Characterization of Composites for the Maeslant Storm Surge Barrie. In *Fracture of Nano and Engineering Materials and Structures Dordrecht*; Springer: Amsterdam, The Netherlands, 2006; pp. 315–316.
85. Pirazzoli, P.A. Possible Defenses against a Sea-Level Rise in the Venice Area, Italy. *J. Coast. Res.* **1991**, *7*, 231–248.
86. Rosenzweig, C.; Solecki, W.D.; Blake, R.; Bowman, M.; Faris, C.; Gornitz, V.; Horton, R.; Jacob, K.; LeBlanc, A.; Leichenko, R.; et al. Developing coastal adaptation to climate change in the New York City infrastructure-shed: Process, approach, tools, and strategies. *Clim. Chang.* **2011**, *106*, 93–127. [CrossRef]
87. Rivera Campos, P.; Tendero Caballero, R. Vernacular architecture in the palafitos from Chiloé = Arquitectura vernácula en palafitos de Chiloé. *Build. Manag.* **2021**, *5*, 7. [CrossRef]

88. Laturnus, T. *Floating Homes: A Houseboat*; Handbook Harbour Publishing Company: Madeira Park, BC, Canada, 1986.
89. Postacchini, M.; Zitti, G.; Giordano, E.; Clementi, F.; Darvini, G.; Lenci, S. Flood impact on masonry buildings: The effect of flow characteristics and incidence angle. *J. Fluids Struct.* **2019**, *88*, 48–70. [CrossRef]
90. Labaran, Y.H.; Mathur, V.S.; Muhammad, S.U.; Musa, A.A. Carbon footprint management: A review of construction industry. *Clean. Eng. Technol.* **2022**, *9*, 100531. [CrossRef]
91. Lomnitz, C. Major Earthquakes of Chile: A Historical Survey, 1535–1960. *Seism. Res. Lett.* **2004**, *75*, 368–378. [CrossRef]
92. Garcés-Vargas, J.; Ruiz, M.; Pardo, L.M.; Nuñez, S.; Pérez-Santos, I. Hydrographic features of Valdivia river estuary south-central Chile. *Lat. Am. J. Aquat. Res.* **2013**, *41*, 113–125.
93. Pino, M.; Perillo, G.M.; Santamarina, P. Residual Fluxes in a Cross-section of the Valdivia River Estuary, Chile. *Estuar. Coast. Shelf Sci.* **1994**, *38*, 491–505.
94. *NCh 1537 2009*; Diseño Estructural—Cargas Permanentes y Cargas de Uso. Instituto Nacional de Normalización: Santiago, Chile, 2009.
95. Koh, H.; Lim, Y.; Seow, T.; Stocks, D.; Thapar, A. The Floating Performance Stage @ Marina Bay, Singapore: New Possibilities for Space Creation. In Proceedings of the International Conference on Offshore Mechanics and Arctic Engineering—OMAE, Estoril, Portugal, 15–20 June 2008; Volume 3.
96. Nekooie, M.A.; Mohamad, M.I.; Ismail, Z. Drag coefficient for amphibious house. *Urban Water J.* **2017**, *14*, 1045–1057. [CrossRef]
97. Hurtado Saldias, M.; Sills Garrido, P.; Manríquez Cárdenas, C. Metodología para una rehabilitación arquitectónica sostenible: El caso de los palafitos de Chiloé. *Arquit. Sur.* **2018**, *36*, 36–57. [CrossRef]
98. Alvarado, D.; Valdebenito, G.; Burgos, M. Evaluacion De Caracteristicas Dinamicas De Los Suelos De Valdivia Empleando Metodos Sismicos De Prospeccion Geofisica. In *Congreso Chileno de Sinmologia e Ingenieria Sismica*; ASHISINA: Valdivia, Chile, 2019; p. 9.
99. Das, B. *Fundamentos de Ingeniería de Cimentaciones*, 7th ed. 2010, p. 819. Available online: https://www.academia.edu/42603156/Braja_Das_Fundamentos_de_ingenieria_de_cimentaciones_7ed (accessed on 26 December 2023).

sustainability

MDPI

Article

Life Cycle Assessment and Cost Analysis of Mid-Rise Mass Timber vs. Concrete Buildings in Australia

Riley Jolly [1], Holly Fairweather [1], Scott Rayburg [1] and John Rodwell [2,*]

[1] Department of Civil Engineering, Swinburne University of Technology, Hawthorn, VIC 3122, Australia; srayburg@swin.edu.au (S.R.)

[2] Department of Management & Marketing, Swinburne University of Technology, Hawthorn, VIC 3122, Australia

* Correspondence: jrodwell@swin.edu.au

Abstract: The building and construction industry is one of the largest greenhouse gas producers, accounting for 39% of global emissions, most of these coming from concrete and steel. Mass timber construction (MTC) potentially offers a sustainable alternative to these traditional building materials. However, more research is needed to establish the sustainability credentials of MTC relative to traditional concrete and steel structures, especially for mid-rise structures. The aim of this study is to evaluate the environmental and cost performance of mid-rise mass timber buildings by conducting a life cycle assessment (LCA). The LCA uses a cradle-to-cradle approach, considering the global warming potential (GWP), freshwater use (FW), and total use of non-renewable primary energy resources (PENRT). Results indicated that mid-rise mass timber buildings have significantly lower impacts than concrete buildings, with their GWP approximately 30 times lower, FW about 20 times lower, and PENRT reaching a negative value. Additionally, the cost analysis revealed that MTC buildings can be cheaper to build and thus possibly more profitable than concrete buildings. These findings establish mass timber as a viable and sustainable option for the future of Australia's construction industry.

Keywords: mass timber construction (MTC); mid-rise buildings; life cycle assessment (LCA); cross-laminated timber (CLT); glue-laminated timber (glulam); sustainability; concrete

1. Introduction

Urban areas such as cities are responsible for around 72% of human-induced global carbon dioxide emissions, with the building and construction sector responsible for around 39% of energy and process-related carbon dioxide emissions [1,2]. Concrete and steel have dominated the construction industry due to their performance in large-scale projects across criteria such as cost, fire resistance, and structural adequacy. However, the concrete industry alone accounts for 7% of global carbon dioxide emissions, while steel is responsible for 6–9% of global emissions [3,4]. Although significant improvements have been made to the environmental performance of common building materials such as concrete and steel, they still embody large amounts of carbon dioxide emissions during their production [5]. Australia accounts for only 0.33% of the world's population yet is one of the highest fossil and greenhouse gas emitters per capita in the world [6]. Buildings account for around 18% of direct carbon emissions in Australia, so the selection of suitable materials and construction methods is vital to reducing emissions [7].

Mass timber offers a promising solution to the limitations of traditional construction materials, serving as a viable alternative for core building elements. There has been a growing interest in the use of MTC due to its technical capabilities, cost-competitiveness, and positive environmental properties compared to concrete and steel [8]. Mass timber has been shown to meet all the standard structural requirements such as fire, wind, and seismic capabilities, even outperforming traditional building materials in many ways [9].

In addition to its technical performance, mass timber can sequester carbon over its life cycle while also having the potential to lead to fewer emissions during the construction and operational stages of buildings [8].

Since the completion in 2012 of Lendlease's Forté, a 10-storey mass timber building, the popularity of mass timber construction has been steadily rising in Australia. The Forté building showcased mass timber's excellence and innovation and received a 5 Star Green Star As-Built rating by the Green Building Council of Australia [10]. Other examples of MTC include the Kind Street Timber Tower, a 10-storey engineered timber building built 20% faster than conventional buildings, and Gillies Hall, Australia's first mid-rise mass timber student accommodation building [11]. Australia has updated building codes to accommodate MTC, providing an opportunity to rethink urban structures and systems [12].

Most of Australia's major cities are growing at rates faster than many developed cities internationally, putting pressure on these cities to densify and/or expand [13]. The growing population poses a threat to the liveability and sustainability of urban life in Australia. The population in Australia is largely concentrated in urban hubs with more than 76% of the population living in eight major cities [13]. Between 2010 and 2020 the major cities grew by 3.1 million people, accounting for 84% of the country's total population growth [13].

Despite the growing population in Australia, it has some of the lowest population densities in the world. For example, the city of Melbourne is roughly three times the area of London yet has a third of its population [14], mostly because apart from Melbourne's inner city, the rest of Melbourne continues to sprawl out with low-rise buildings. The current urban environment needs to support higher density development to continue to sustain the large population growth in Australia's major cities. While Australia's major cities continue to expand, so do their impacts, with increases in urban heat, congestion, pollution, and waste, and impacts on the natural environment [13]. Encouraging high-density development through mid-rise mass timber buildings in Australia is crucial for addressing population growth and climate change while enhancing urban wellbeing and livability.

To help accelerate the use of mass timber, the Australian government has launched the Cleaner Energy Finance Corporation (CEFC) Timber Building Program to encourage the transformation of medium and high-rise buildings through funding for eligible projects with an AUD 300 million finance fund [15]. The federal government program, along with advancements in MTC, have brought about the planned construction of multiple tall mass timber hybrid buildings in Australia. Some examples of these are the Atlassian HQ in Sydney, a 40-storey hybrid timber tower with a steel and glass façade, and the C6 tower in Perth, which will be the tallest commercial hybrid timber tower in the world [16]. Tall buildings, however, make up only a small portion of the building industry in Australia, with an estimated 80% of office buildings alone being mid-rise [17]. Australia's current focus on tall hybrid timber buildings, despite their challenges in gaining building approval, overlooks the potential of mid-rise mass timber structures, which could significantly transform various building types and contribute to sustainability goals.

Despite its numerous advantages, the widespread adoption of mid-rise mass timber faces significant challenges. Factors such as a lack of education within the construction industry, resistance to change from traditional materials, and insufficient media attention compared to high-rise projects impede its adoption [18]. Historically, sourcing materials for MTC has also been a deterrence, but the emergence of local manufacturers such as XLam and Wood Solutions suggests promising growth in domestic supply capabilities that should overcome these concerns [19,20]. Consequently, perhaps the main inhibitor to a rapidly expanding use of MTC for mid-rise buildings in Australia is uncertainty in the construction industry about the relative environmental benefits of switching to MTC and whether such a shift would be cost-effective.

The aim of this project is to evaluate the environmental and cost impacts of mid-rise MTC compared to concrete through a comprehensive cradle-to-cradle life cycle assessment (LCA) analysis. Unlike prior studies limited to cradle-to-site, this research fills a knowledge gap by examining mid-rise construction's holistic performance, particularly in Australia, aiming to provide updated insights into its environmental footprint beyond carbon emissions. The study has the following objectives:

1. To quantify and compare the environmental and costs impacts of MTC vs. traditional concrete and steel builds for mid-rise structures using a cradle-to-cradle life cycle approach;
2. To identify critical stages of the building's life cycle where environmental and cost impacts are most significant, and inform the competitiveness of mid-rise mass timber buildings in the construction industry and promote their adoption in Australia.

2. Materials and Methods

This section describes the methods utilized to conduct a life cycle analysis (LCA) and a cost analysis of mass timber and concrete mid-rise buildings located in Melbourne, Australia. The aim of the use of the LCA method was to gain a better understanding of the environmental impacts and differences between mid-rise mass timber and concrete buildings. A cradle-to-cradle LCA model was used to ensure all aspects of the building's life were taken into consideration beginning from the extraction of raw materials and concluding with the energy recovery and reuse possibilities for both building types. To determine what would be needed in the construction of an eight-storey building, a bill of materials was adapted from an LCA study conducted by Gu et al. [21], which included a bill of materials for an eight-storey building for both mass timber and concrete construction.

There were several alterations made to the original bill of materials. First, the amount of glulam required for the columns and beams was reduced because the required amount quoted by Gu et al. [21] was considered excessive in comparison to other studies. Conservative estimates indicate that to replace reinforced concrete columns and beams, the amount of glulam in m^3 required would be one and a half times that of the concrete required; thus, the amount of glulam required was reduced to that amount. Additionally, mineral wool and fiberglass mats (both used as insulation) were excluded from the bill of materials because they are required for both mass timber and concrete buildings in similar quantities and so would not represent a point of difference between the two. Similarly, the PE (polyethylene) vapor barrier was also removed from the bill of materials because the amount of PE vapor barrier required was close to equal for both buildings, and again would not represent a point of difference. All of the remaining materials and amounts utilized remained the same as those presented by Gu et al. [21]. The LCA was conducted with a life cycle of 75 years following the EN 15978 [22] and EN 15804 [23] standards to ensure all stages were assessed correctly. Table 1 provides the sources of all the data used through every stage of the life cycle analysis. A cost analysis was also conducted in conjunction with the LCA to further understand the costs associated with constructing both a timber and concrete building; the sections below further describe the methods utilized for each stage in detail.

Table 1. List of sources for the LCAs.

Stage	Purpose	Source
A1–A3	Bill of Materials	Gu et al. [21].
	Concrete	EPD Australasia—Average of Boral, Barro and Adbri [24–26].
	Glulam	EPD Australasia—Wood Solutions [27].
	Rebar	EPD Australasia—Australian Reinforcing company [28]
	Steel stud (1.15 mm, #681)	EPD Australasia—Rondo [29]
	CLT	EPD Australasia—Xlam [30]
	Gypsum Board (Fireshield 13 mm)	EPD Australasia—Siniat—plasterboard [31]
	Foam insulation	EPD Australasia—Metecno [32]
	Acoustic insulation (Absorb black 25 mm)	EPD Australasia—Martini [33]

Table 1. *Cont.*

Stage	Purpose	Source
A4	Truck carbon emissions calculator	Greenstar [34]
	Concrete transportation	EPD Australasia—Hanson [35]
	Imported CLT transportation	Durlinger et al. [36], carboncare.org [37]
	Glulam transportation	EPD Australasia—Woodsolutions [27]
A5	Total diesel equation	Greene et al. [38]
	Diesel GWP coefficient	Department of Climate Change, Energy the Environment and Water [39]
C1	MTC vs. concrete demolition	Duan et al. [40]
C2	Transport to waste disposal	AusLCI Carbon Emission Factors [41]
C3–D	Concrete	EPD Australasia—BGC Concrete [42]
	Glulam	EPD Australasia—Wood Solutions [27]
	Rebar	EPD Australasia—Australian Reinforcing Company [28]
	Steel stud	EPD Australasia—Rondo [29]
	CLT	EPD Australasia—Xlam [30]
	Gypsum board	EPD Australasia—Siniat [31]
	Acoustic insulation	EPD International—Interfloor crumb rubber underlay EPD [43]

2.1. Product Stage

According to standards EN 15978 and EN 15804, the production stage marks the initial phase in the building life cycle and is divided into three distinct modules. Module A1 concerns the raw material supply, Module A2 is the transport of raw materials, and Module A3 is the manufacturing of a product for construction use. Equation (1) was used to calculate the global warming potential (GWP), freshwater use (FW), and total use of non-renewable energy resources (PENRT) totals for each material used in the building designs:

$$E_1 = \Sigma f_{material} * M_{material} \quad (1)$$

where $f_{material}$ represents the environmental impact factors of GWP, FW, and PENRT per unit of the specified material, while $M_{material}$ represents the quantity of that material used within the buildings. The environmental impact factors for this stage were sourced from environmental product declarations (EPDs) corresponding to each material used in the building designs, covering Modules A1 through to A3. For the calculation of the environmental impact of concrete, the impact factors were taken as an average of three different concrete suppliers. The environmental impact factors for all other materials were taken from one EPD per material (Table 1).

2.2. Construction Stage

Transportation to site, also referred to as Module A4, covers the global warming potential emissions from transporting the manufactured product to the construction site. Module A4 used a mixture of environmental product declarations (EPDs) from Table 1 and Equation (2) to calculate the global warming potential contributions:

$$E_2 = \Sigma D \cdot M_{load} \cdot f_{transport} \quad (2)$$

where D is the total distance travelled, M_{load} is the mass of the hauled material in tonnes, and $f_{transport}$ is the emission factor of the transport mode (kg CO_2e/tkm). The transport stage considered environmental impacts for concrete, steel, and mass timber components. The construction site was in Melbourne, so supplier distances were based on previously used EPDs. For concrete, it was assumed a supplier was within 20 km of the build site. For all road transport, a truck with a 16–28 tonne gross weight was used, except for concrete, where specific truck details were unknown, and thus EPD emission factors were used. The emission factor for a 16–28 tonne truck was taken from AusLCIv1.42.

Two sourcing scenarios for cross-laminated timber (CLT) were considered: local and imported from Austria. The GWP for CLT imported from Austria was based on the work of Durlinger et al. [36]. In their study, they showed that CLT travels from Katsch an der Mur, Austria, to Koper by truck, then by ship to Melbourne via Singapore, totalling 19,433 km [36]. The ship's GWP was calculated using carboncare.org, based on shipment weight and route. The truck distance from the Port of Melbourne to the site was calculated using the same method as for locally sourced CLT.

The construction and installation phase (Module A5) details the GWP from activities during building and construction. Equation (3) was used to calculate the total diesel requirements, and Equation (4) was used to calculate the total GWP of each building:

$$V = 0.000037 * M_{material} * h + \frac{M_{material}}{500} + 0.83 \tag{3}$$

$$E_3 = V_{diesel} * f_{diesel} \tag{4}$$

where $M_{material}$ is the mass of materials in kg, h is half the building height in meters, and V_{diesel} is the total diesel required and is calculated in Equation (3), while f_{diesel} is the GWP emission factor for diesel. These formulae only considered the crane's diesel requirements. Previous LCA analyses showed no major difference in earthworks between MTC and concrete buildings, so that phase was excluded. Other construction elements contributing to GWP were also excluded due to limited information. The main difference was crane usage, influenced by construction speed and material mass. To calculate the diesel impact, the mass of materials hauled by the crane was calculated from the bill of materials and material densities.

2.3. *Use Stage*

The use stage covers environmental impacts from building operation until end of life, as defined by EN 15978 and EN 15804. The use stage includes seven modules: use (B1), maintenance (B2), repair (B3), replacement (B4), refurbishment (B5), operational energy use (B6), and operational water use (B7). The LCA excluded the use stage following the findings from past studies where Eslami et al. [24] and Forest and Durlinger [36] found Module B1 impacts were similar for both types of buildings. Maintenance, repair, replacement, and refurbishment were also assumed to have negligible differences by building type because the LCA focuses on the main structural elements of the buildings. Both buildings are considered to have the same energy class, making operational energy and water use identical, and thus these were excluded from this study [36,44].

2.4. *End-of-Life Stage*

The end-of-life stage involves four modules: deconstruction (C1), waste transportation (C2), waste processing (C3), and disposal (C4) per EN 15978 and EN 15804. A cradle-to-cradle approach extends evaluation to include reuse, recovery, and recycling (Module D). Total GWP for each building was calculated using Equation (5):

$$E_4 = (f_{building} * A_{building}) \tag{5}$$

where $f_{building}$ is the GWP emission factor, and $A_{building}$ is the total building area in m^2. The GWP for deconstruction was sourced from Duan et al. [45], which indicated emission factors per m^2 for both buildings.

The transport to waste processing, otherwise known as Module C2, covers the emissions from transporting the construction waste from the building site to a waste processing site. Module C2 used Equation (2) to calculate the global warming potential contributions. For the calculation of the global warming potential, it was assumed that a truck with a 16 to 28 tonne gross weight will be used to transport the materials. The emission factor for a 16 to 28 tonne truck was taken from AusLCIv1.42. The buildings have a design life of 75 years and it was assumed that there would be a suitable location within 30 km of the site in Melbourne.

The waste processing, disposal, and reuse, recovery, and recycling modules, also referred to as modules C3, C4, and D, respectively, are the final stage of the life cycle assessment. Modules C3, C4, and D cover the environmental impact contributions of waste processing and the potential reuse, recovery, and recycling of specific material. Equation (6) was used to calculate the total environmental impact for each end-of-life scenario:

$$E_5 = \Sigma(f_1 * M_{material}) + (f_2 * M_{material}) \quad (6)$$

where f_1 represents the environmental impact factors GWP, FW, and PENRT per unit of the specified material for Module C3, f_2 represents the environmental impact factors GWP, FW, and PENRT per unit of the specified material for Modules C4 and D, and $M_{material}$ represents the quantity of that material used within the buildings. The data collected to find the GWP, FW, and PENRT totals were based on multiple environmental product declarations (EPDs) and are summarized in Table 1 for each material type used in the building designs.

The concrete building was split into two scenarios, landfill and recycling, while the MTC building was split into four scenarios, landfill, energy recovery, recycling, and reuse. The landfill scenarios for both concrete and MTC take f_2 as the environmental factor for the C4 Module, whereas the concrete recycling scenario and MTC recycling, reuse, and recovery scenario take f_2 as the environmental factor for Module D.

For the concrete recycling scenario, each material in the building was assumed to be recycled except for the foam insulation board, and because no EPDs could be found to cover a recycling scenario, it was assumed to go to landfill. The energy recovery, recycling, and reuse scenarios of MTC buildings only apply to the timber components and assume all other materials were recycled. The energy recovery scenario involved energy recovery through shredding and combustion, offsetting energy from natural gas with options including coal or electrical replacement. The reuse scenario involved the direct reuse of the timber components from the building without any further processing. The final scenario, recycling, involves downcycling the timber components into wood chips to be utilised in other products. These scenarios were based on assumptions made by their equivalent EPDs.

2.5. Totals

Equation (7) is used to calculate the environmental impact totals for both concrete and mass timber construction (MTC):

$$E_{Total} = \Sigma E_1 + E_2 + E_3 + E_4 + E_5 \quad (7)$$

where E_1 is the product stage total environmental impact, E_2 is the total environmental impact of both transportation modules, E_3 is the total environmental impact for the construction/installation stage, and E_4 is the total environmental impact of the end-of-life disposal, reuse, recovery, and recycling. The calculated totals were split up into different best and worse scenarios for each of the different building-type cases, with the first case being the concrete building, the second being the MTC building, and lastly, the MTC building with imported CLT. Totals were calculated for GWP, FW, and PENRT.

2.6. Cost Analysis

The cost analysis was conducted based on a previous cost analysis completed by Wood Solutions on an eight-storey residential building in Australia [46]. However, there were some alterations made to the cost analysis given in that report. First, wall finishes were excluded from the results because they do not contribute to differential costs between the building types. Next, the cost of the foundation was added to the analysis because the mass timber building does not require a foundation as thick as the concrete building does. Third, the construction costs including labour, site management, site sheds, and any equipment required were considered to be AUD 52,000 per week for the whole duration of construction for both buildings. Finally, it was conservatively assumed that six weeks would be saved in construction time for the mass timber building, thus reducing the construction time from twelve months for the concrete building to ten and a half months for the mass timber building.

3. Results

The following section outlines the results of the life cycle and cost assessments of the mass timber and concrete mid-rise buildings in Melbourne, Australia. Environmental impact factors of both buildings were calculated and compiled in accordance with the methods described above. The findings are presented as graphs, illustrating the environmental impacts for both building types during each stage of the life cycle assessment.

3.1. Product Stage

The product stage analysed the global warming potential (GWP), freshwater use (FW) and the total use of non-renewable primary energy sources (PENRT) for the production of materials required for the reinforced concrete and mass timber construction (MTC) buildings. The bill of material quantities for each building is shown in Table 2. The materials' characteristics as well as their quantities were used in calculating each environmental impact.

Table 2. Bill of material quantities.

Assembly	Material	MTC Quantity	Concrete Quantity	Unit
Columns and Beams	Concrete 40 MPa	n/a	93	m^3
	Glulam	140	n/a	m^3
	Rebar	n/a	59	tonnes
Exterior Walls	Steel stud	n/a	738	kg
	CLT	172	n/a	m^3
	13 mm Gypsum board	1638	1638	m^2
	70 mm Foam insulation	n/a	1638	m^2
Floors	25 mm Acoustic insulation	4115	n/a	kg
	CLT	1437	n/a	m^3
	Concrete 40 MPa	790	2293	m^3
	Rebar	5	42	tonnes
Foundation	Concrete 40 MPa	367	674	m^3
	Rebar	12	20	tonnes
Interior Walls	CLT	1152	n/a	m^3
	Concrete 40 MPa	n/a	516	m^3
	13 mm Gypsum board	13,724	24,041	m^2
	Rebar	n/a	73	tonnes
	Steel stud	8112	16,302	kg

Note: n/a = not applicable, MTC = mass timber construction, CLT = cross-laminated timber.

Figure 1a illustrates the total GWP for concrete and MTC buildings, showing MTC's negative GWP compared to concrete's positive GWP. In the MTC building, the utilization of CLT and glulam combined reduced the GWP by a total of −1,415,328 kg CO_2, signifying the beneficial climate impact of including MTC components within the building.

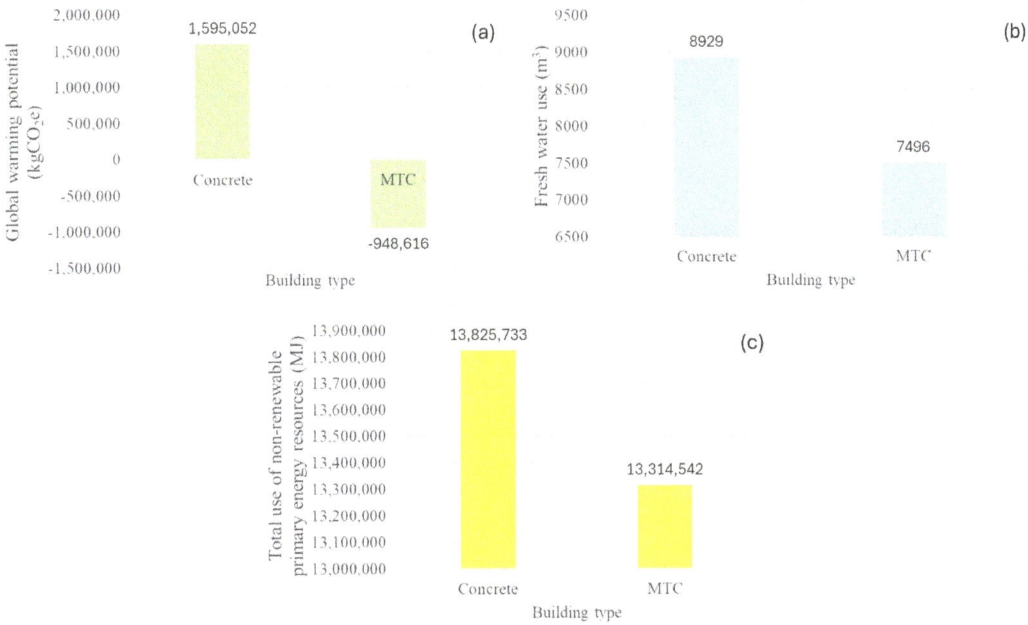

Figure 1. Totals for the product stage of the life cycle. (**a**) Global warming potential (kgCO_2e); (**b**) freshwater use (m^3); (**c**) non-renewable primary energy resources (MJ).

The total freshwater usage and total use of non-renewable primary energy resources are depicted in Figures 1b and 1c, respectively. In both cases, MTC outperformed the concrete structures, with 17% lower freshwater use by the MTC building and 500,000 fewer MJ of non-renewable energy resources consumed by the MTC building relative to that of the concrete structure.

3.2. Construction Stage

The GWP of the transport required to get materials to the site location is depicted in Figure 2a. Per the methods section, the concrete building components were considered to be transported from Metropolitan Melbourne, local CLT and glulam were considered to be transported from Regional Victoria, and imported CLT was considered to be transported from Austria. The concrete building exhibited the lowest global warming potential (GWP) for transportation. Locally sourced MTC materials demonstrated five times the GWP of the concrete building, while imported MTC materials showed 15 times the GWP of the concrete building.

The GWP of construction, specifically the diesel requirements associated with crane usage for the concrete and MTC buildings, is depicted in Figure 2b. The MTC building exhibited the lowest GWP for construction, releasing close to half the GWP emitted by the concrete building.

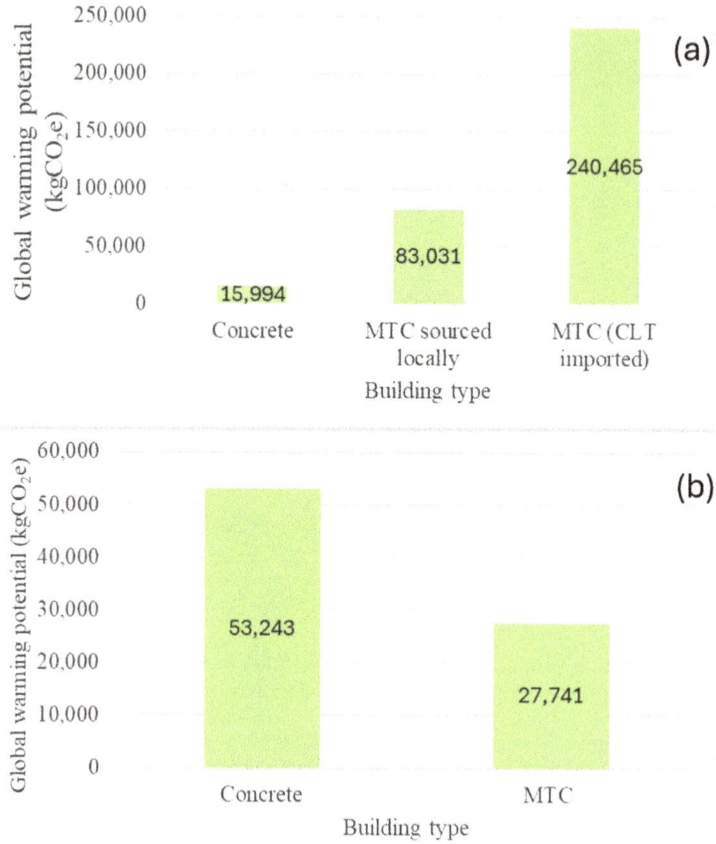

Figure 2. Totals for the construction stage of the life cycle. (**a**) Global warming potential ($kgCO_2e$) for transport to the construction site; (**b**) global warming potential ($kgCO_2e$) for construction.

3.3. End-of-Life Stage

Figure 3a depicts the GWP of the deconstruction of the concrete building in comparison to the MTC building. The concrete building produced the highest GWP, exhibiting more than double the GWP of the MTC building. Figure 3b depicts the GWP of transportation to processing or disposal sites. The concrete building had the highest GWP, approximately 40% higher than the MTC building. The results show that a total of 10,663 kg of CO_2e emissions were generated from the concrete components of the MTC building, accounting for 66% of the total emissions for MTC.

The data presented in Figure 4a–c highlight the GWP, FW, and PENRT totals associated with different end-of-life scenarios including landfill, energy recovery, recycling, and reuse. The MTC buildings consistently produced the highest GWP for each end-of-life scenario, where the GWP for the best-case scenario was approximately eight times higher than the concrete landfill scenario.

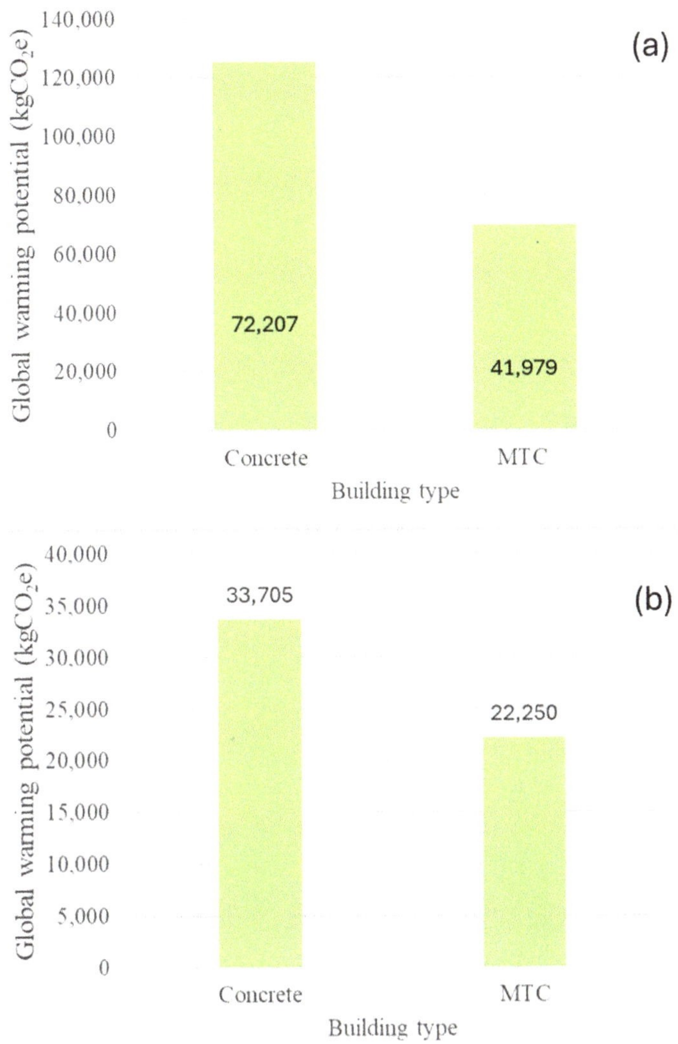

Figure 3. Totals for the deconstruction stage of the life cycle. (**a**) Global warming potential ($kgCO_2e$) for deconstruction; (**b**) global warming potential ($kgCO_2e$) for transport to waste processing.

The MTC reuse stage in Figure 4b produced the best freshwater use results, using approximately six times less water per m^3 than the best concrete building scenario. The MTC energy recovery scenario was the best performing scenario with respect to PENRT, using approximately 20 times fewer Megajoules than the best concrete building scenario (Figure 4c).

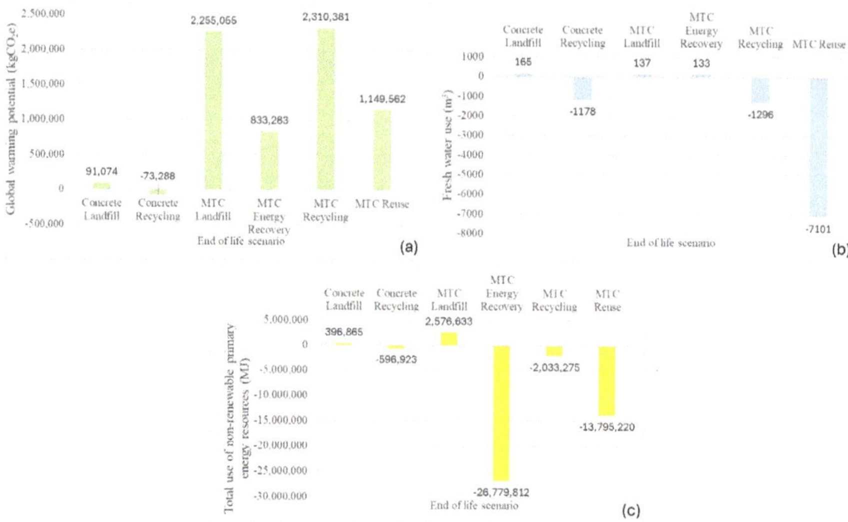

Figure 4. Totals for the end-of-life scenarios. (**a**) Global warming potential (kgCO_2e); (**b**) freshwater use (m^3); (**c**) non-renewable primary energy resources (MJ).

3.4. Totals

Figure 5a–c show the combined GWP, FW, and PENRT results for the life cycle assessment from cradle to cradle. Results show that the MTC building performs better in every scenario compared to the concrete building, with the exception of PENRT totals for both worst case scenarios. The concrete building's total GWP was approximately 30 times higher than the MTC building, its total FW was approximately 20 times more than the MTC building and the total PENRT for the MTC building was negative for its best-case scenario.

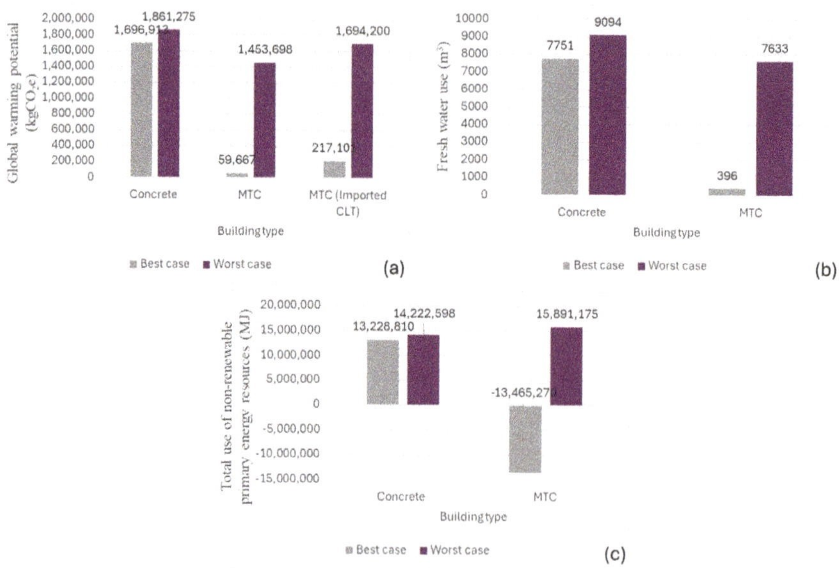

Figure 5. Totals for the entire life cycle. (**a**) Global warming potential (kgCO_2e); (**b**) freshwater use (m^3); (**c**) non-renewable primary energy resources (MJ).

3.5. Cost Analysis

The cost analysis of the elements that differentiated between the building types indicated that close to AUD 480,000 is saved when using mass timber in a building rather than concrete. Table 3 summarizes the prices broken down by material costs and construction costs for each element of the two buildings. The cost analysis indicates that the initial material price for both buildings is close to AUD 5,000,000 and that the majority of the savings for the MTC building comes from the ceiling finishes, columns and beams, and reduced construction time.

Table 3. Cost analysis of the differentiating elements between the two mid-rise building types.

Assembly	MTC	Concrete	Variance
Columns and beams	$34,935	$365,644	−$330,709
Exterior walls	$518,082	$416,165	$101,917
Floors	$2,539,961	$1,810,398	$729,563
Foundation	$364,350	$537,441	−$173,091
Interior walls	$1,286,436	$1,224,522	$61,914
Roof	$233,100	$356,617	−$123,517
Ceiling finishes	$0	$459,085	−$459,085
Termite protection	$25,000	$0	$25,000
Construction costs (labour, etc.)	$2,392,000	$2,704,000	−$312,000
Total	$7,393,864	$7,873,872	−$480,008

Six weeks was saved in construction time for the MTC building, reducing the construction time from twelve months for the concrete building to ten and a half months for the MTC building, ultimately reducing the total cost by around AUD 300,000 due to savings in labour, site management, site sheds, and any equipment required such as cranes and scaffolding. The cost analysis total indicated that the MTC building is cheaper to construct than the concrete building with a saving of 6%, or AUD 480,000, attained when constructing the building with MTC.

4. Discussion

This study aimed to evaluate the environmental and cost performance of mid-rise mass timber buildings and identify specific stages with the most significant impacts. The analysis seeks to inform stakeholders and enhance the competitiveness of mass timber buildings in Australia. The results showed that mid-rise mass timber buildings perform significantly better in several key areas compared to traditional concrete structures. The following section discusses the findings of the life cycle assessment and uses the current literature to support these findings, highlighting the strengths and limitations of each building type for each life cycle stage to contribute to informed decision making in the construction industry.

4.1. Product Stage

In the product stage of the life cycle assessment, three indicators of environmental impact were analysed: global warming potential (GWP), freshwater use, and total non-renewable energy (PENRT). MTC outperformed concrete in all three indicators. Concrete and rebar had negative impacts across all indicators, whereas glulam and CLT showed positive impacts. The MTC building included all four materials, while the concrete building did not include glulam or CLT. The concrete building required over three times the amount of concrete and more than ten times the amount of steel rebar compared to the MTC building, significantly impacting the outcome.

The GWP of the MTC building was negative, indicating CO_2 sequestration, while the concrete building had a large positive GWP, indicating CO_2 emissions. Trees absorb CO_2 through photosynthesis, binding carbon into sugars used for growth while releasing oxygen [47]. Glulam and CLT store more carbon than is emitted during the product stage, including harvesting, transporting, and manufacturing. MTC plantations promote carbon sequestration, as young trees capture carbon as they grow. Trees continue to store carbon until they decompose or are burned. About 50% of a tree is carbon, and decomposition takes many years, so MTC buildings store this carbon once built [40]. These findings confirm Harte's support regarding MTC's ability to reduce emissions [8].

Concrete and steel had significant impacts across all environmental indicators, with 90% of the GWP of the concrete building attributed to concrete and steel rebar, resulting in over 1 million additional kgCO_2 emissions compared to the MTC building. Freshwater use also favoured MTC, though the difference was smaller than for GWP. Concrete and steel accounted for 90% of freshwater use in the concrete building, while in the MTC building, 30% was attributed to concrete and steel, and 60% to glulam and CLT. Hardwood glulam used more freshwater than CLT, but using softwood glulam reduced water usage. Glulam and CLT water usage comes from plantations, pressing and curing, and post-lamination treatment [48]. These processes are efficient, and reducing freshwater use in MTC would require significant investment.

Non-renewable energy (PENRT) also favoured MTC, as CLT and glulam required less non-renewable energy per cubic meter than concrete and steel rebar. Less material is needed for MTC buildings, so overall, MTC buildings outperform concrete in PENRT. In addition, wood product production tends to involve biomass energy produced by waste products from the wood-harvesting process (a renewable energy source). This contrasts with the heavy reliance on fossil fuels for the production of concrete and steel. These factors combine to see MTC outperform concrete and steel in the consumption of non-renewable energy. The product stage concluded that MTC outperforms concrete in every stage, proving the superiority of materials like CLT and glulam over concrete and steel, supporting Abed et al.'s [9] argument for industry adoption.

4.2. Transport and Construction Stage

Transportation to the construction site was one area where the concrete building outperformed the MTC building. Concrete and steel required shorter transportation distances (about 30 km) to the construction site due to the availability of suppliers near Metropolitan Melbourne. This reduced fuel consumption, emissions, and logistical complexities, contributing to construction efficiency.

In contrast, MTC, particularly imported mass timber, required longer transportation distances. Mass timber travelled about 300 km to the site, with imported timber from Austria traveling approximately 20,000 km. This underscores the environmental impact of transportation and the importance of local sourcing. Transporting mass timber is challenging due to the difficulty of moving large timber beams and high transportation costs [18]. Imported timber needs specific processing, handling, and wrapping due to container size limitations. Local production capabilities and government incentives like the Cleaner Energy Finance Corporation (CEFC) Timber Building Program have facilitated the transition to local supply chains. This reduces the reliance on imported goods and promotes the use of regional suppliers.

The recent completion of Australia's first mass timber building under the CEFC program in 2023, using only local products, showcases the feasibility and benefits of local sourcing [49]. Local manufacturing reduces transportation distances and fosters economic growth, job creation, and regional development [50]. Local mass timber production supports both rural and urban economies, offering employment and mitigating climate change. XLam's establishment of the pioneering CLT manufacturing plant in 2017 served as a catalyst for local production, reducing the nation's dependence on imported products [19]. An increase in demand for mass timber will provide economic incentives to invest in forest plantations [51].

Australia's forestry supply chain supports domestic and export markets, vital for many rural communities [52]. While concrete is expected to outperform mass timber in transportation to construction sites, transportation is a small component of the overall environmental impact. However, the stagnant growth in new plantations threatens future local mass timber supply. Presently, Australia's total plantation area is at its smallest since 2003-04, primarily due to the ongoing conversion of hardwood plantations to alternative land uses [53]. This trend poses a looming threat to future local mass timber supply. The forest industry has the potential to address the increasing demand for wood products, while new plantations would support the environment, regional employment, and economic activity, and improve social outcomes [52].

The installation and construction stage showed the MTC building's superiority, with the concrete building emitting nearly double the CO_2 emissions. The main differentiation was the time of construction, impacting CO_2 emissions from diesel use for crane operations. The concrete building's material mass was nearly double that of the MTC building, leading to higher emissions. These results highlight the efficiency of constructing mass timber buildings, reducing GWP and construction time and cost.

4.3. End-of-Life Stage

At the end-of-life stage, deconstruction/demolition presented a significant difference in GWP, highlighting the advantages of MTC, which had about half the GWP compared to the concrete building. Due to MTC being a relatively new construction material, there is a knowledge gap regarding its environmental impacts at this stage. The first modern MTC building in Australia was constructed in 2012, so it has not reached end of life [54,55]. Duan et al. [45] used demolition energy data for CLT and concrete to estimate emissions, concluding that CLT has a demolition energy of 0.0359 MJ/kg and reinforced concrete 0.0612 MJ/kg with emissions equalling 7.62 and 3.63 $kgCO_2e/m^2$, respectively. These results show the MTC building's superiority in GWP.

Transportation to waste processing facilities highlighted the advantages of CLT and glulam due to their low density compared to concrete. Concrete, with a density of approximately 2400 kg/m^3, is significantly denser than CLT and glulam, which range from 480 kg/m^3 to 674 kg/m^3 [27,30,56]. Concrete's high density requires more trips, increasing GWP. Mass timber's lower transport costs and reduced GWP during transportation contribute to its environmental benefits [57]. However, urban environments like Melbourne can impact fuel consumption due to traffic conditions [58]. The impact of high traffic conditions can result in 1.1 to 1.5 times higher fuel consumption compared to free-flowing traffic [59].

In waste processing, concrete structures benefit from well-established end-of-life strategies, with steel and concrete being highly recyclable [38]. Most life cycle assessment studies of mass timber buildings focus on a cradle to site, or cradle to the construction system boundary, and do not include the end-of-life stage. Predicting the end of life and outcomes beyond the system boundary for mass timber buildings poses challenges because it necessitates forecasting decades into the future, resulting in some uncertainty regarding the accuracy of end-of-life assessments [60]. Mass timber end-of-life scenarios include reuse, recycling, combustion for energy recovery, or landfill [61]. Reusing mass timber extends its lifespan, maximizing sequestered carbon and delaying CO_2 release [62]. The extended lifespan of mass timber also reduces the demand for new timber extraction, mitigating the environmental impacts associated with the timber production process. The reuse of mass timber components also had a large impact on freshwater use, almost making the entire lifecycle, excluding the use stage, water neutral. However, it is important to note that not all mass timber can be recycled because the presence of nail holes or deconstruction-related damage can deem some material unsuitable for reuse [38]. Innovations in modular construction and prefabrication could facilitate easier reuse. While timber is a source of carbon sequestration, once it is incinerated or decays, the sequestered CO_2 stored in the wood is released [60]. Thus, where mass timber cannot be reused, it should be used to produce biomass energy through direct combustion so that it can offset the use of fossil fuels [9,30]. Although it is well-established that wood decomposes very slowly, resulting in minimal carbon loss over decades, the specific decomposition rate of mass timber remains largely unknown due to variations in landfill management practices and climates [63]. In landfills, mass timber decomposes slowly but releases biogenic methane, which has a much higher GWP than CO_2 and thus could have a greater effect on global warming [45]. However, one study in Australia on the decomposition of wood products found that after being buried for 46 years, softwoods and hardwoods had only lost 18% and 17% of their original carbon content, respectively [64]. Other research has shown that only up to 8% of the carbon in wood may be emitted as gas in landfill sites [65]. These studies suggest that the slow rate of timber decomposition minimises immediate carbon emissions.

4.4. Costs

The cost analysis indicated a 6% saving for MTC buildings, nearly AUD 500,000 for a mid-rise MTC building compared to a concrete building. Initial material costs are similar, but construction time savings make MTC more cost-effective. This supports Evison et al.'s [66] findings but contradicts studies suggesting MTC is more expensive [67]. Variability in results may explain builders' hesitation in adopting MTC, especially regarding construction phase time savings.

4.5. Limitations

This study's use of USA-based building design data introduces potential discrepancies, as it may not reflect Australian practices. Modifications were made for environmentally friendly choices, but data constraints limited options. Assumptions about waste processing facilities may not align with real-world scenarios, and uncertainties about end-of-life scenarios highlight the complexity of conducting comprehensive LCAs for mass timber construction. Conservative estimates in this study suggest MTC's superiority over concrete for mid-rise buildings.

5. Conclusions

This study addresses the urgent need to mitigate urban CO_2 emissions, particularly in Australia where buildings contribute significantly to the high per capita emissions [2,7]. Overall, the life cycle assessment indicated that mid-rise MTC buildings have significantly lower environmental impacts than concrete buildings, with GWP approximately 30 times lower, freshwater use about 20 times lower, and even lower total use of non-renewable energy resources, with MTC having a negative value. Further, the cost analysis revealed potential savings close to AUD 500,000 with MTC, mainly due to faster construction times, suggesting that MTC mid-rise buildings are more profitable than typical concrete and steel mid-rise buildings, with changes in the supply chain in Australia suggesting that MTC mid-rise buildings will be even more profitable in future.

The detailed findings indicated that mid-rise MTC outperformed concrete buildings in the product stage, demonstrating significantly lower global warming potential, freshwater use, and total use of non-renewable energy resources. During the construction stage, MTC exhibited lower GWP from construction activities, although for transportation, the GWP for MTC was higher, particularly for imported materials—although that situation will soon change and represented only a small difference between the building types. At end of life, the MTC building produced higher GWP but performed better in freshwater use and non-renewable energy resources.

These findings suggest that MTC construction should be a preferrable option for the construction industry in Australia, and globally. These structures are both more sustainable and cheaper, indicating that they have both an environmental and commercial advantage over traditional structures. As MTC products are already commercially available, a shift to increased use of MTC as a preferred building material will stimulate further growth in the production of MTC, helping to drive costs even lower, and limiting the potential adverse impacts that come with importing these materials from overseas. Future research should investigate potential barriers to MTC adoption in the Australian construction industry, to help expedite the transition to MTC for mid-rise structures.

Author Contributions: Conceptualization, R.J., H.F., S.R. and J.R.; methodology, R.J. and H.F.; data curation, R.J. and H.F.; formal analysis, R.J., H.F., S.R. and J.R.; writing—original draft preparation, R.J., H.F., S.R. and J.R.; writing—review and editing, R.J., H.F., S.R. and J.R.; supervision, S.R. and J.R. All authors have read and agreed to the published version of the manuscript.

Funding: This research received no external funding.

Institutional Review Board Statement: Not applicable.

Informed Consent Statement: Not applicable.

Data Availability Statement: The data are contained within the article above.

Conflicts of Interest: The authors declare no conflicts of interest.

References

1. World Green Building Council. Bringing Embodied Carbon Upfront. Available online: https://worldgbc.org/advancing-net-zero/embodied-carbon/ (accessed on 17 October 2023).
2. United Nations Human Settlements Programme (UN-Habitat). Urban Planning Law for Climate Smart Cities: Urban Law Module of the Law and Climate Change Toolkit. Available online: https://unhabitat.org/urban-planning-law-for-climate-smart-cities-the-urban-law-module-of-the-law-and-climate-change (accessed on 15 October 2023).
3. Pandit, J.K.; Watson, M.; Qader, A. *Reduction of Greenhouse Gas Emissions in Steel Production*; CO2CRC Ltd.: Melbourne, Australia, 2020.
4. Supriya; Chaudhury, R.; Sharma, U.; Thapliyal, P.C.; Singh, L.P. Low-CO_2 emission strategies to achieve net zero target in cement sector. *J. Clean. Prod.* **2023**, *417*, 137466. [CrossRef]
5. Crawford, R.H.; Cadorel, X. A Framework for assessing the Environmental Benefits of Mass Timber Construction. *Procedia Eng.* **2017**, *196*, 838–846. [CrossRef]
6. Yu, M.; Wiedmann, T.; Crawford, R.; Tait, C. The Carbon Footprint of Australia's Construction Sector. *Procedia Eng.* **2017**, *180*, 211–220. [CrossRef]
7. Department of Climate Change, Energy, the Environment and Water. Government Priorities: Buildings. Available online: https://www.energy.gov.au/government-priorities/buildings#:~:text=Data%20from%20the%20department%20shows, businesses%20and%20the%20wider%20economy (accessed on 17 September 2023).
8. Harte, M.A. Mass timber—The emergence of a modern construction material. *J. Struct. Integr. Maint.* **2017**, *2*, 121–132. [CrossRef]
9. Abed, J.; Rayburg, S.; Rodwell, J.; Neave, M. A review of the Performance and Benefits of Mass Timber as an Alternative to Concrete and Steel for Improving the Sustainability of Structures. *Sustainability* **2022**, *14*, 5570. [CrossRef]
10. Architecture and Design. Forte by Lend Lease. Available online: https://www.architectureanddesign.com.au/projects/multi-residential/forte-by-lend-lease (accessed on 8 October 2023).
11. Wood Solutions Mid-Rise. Available online: https://www.woodsolutions.com.au/case-studies/mid-rise?page=0 (accessed on 17 October 2023).
12. Australian Building Codes Board (ABCB). National Construction Code Volume One, NCC. 2019. Available online: https://ncc.abcb.gov.au/editions/2019/ncc-2019-volume-one (accessed on 22 July 2024).
13. Hill, S.; Cumpston, Z.; Quintana Vigiola, G. Australia State of the Environment 2021: Urban Independent Report to the Australian Government Minister for the Environment. Available online: https://soe.dcceew.gov.au/sites/default/files/2022-07/soe2021-urban.pdf (accessed on 17 October 2023).
14. Milman, O. Melbourne's Urban Sprawl: Just How Big Can the City Get? Available online: https://www.theguardian.com/australia-news/2015/sep/03/melbournes-urban-sprawl-just-how-big-can-the-city-get (accessed on 28 October 2023).
15. Clean Energy Finance Corporation/CEFC Timber Building Program. Available online: https://www.cefc.com.au/insights/investment-insights/cefc-timber-building-program/ (accessed on 17 October 2023).
16. Built Offsite Australia Leads the World in Tall Mass Timber Buildings. Available online: https://builtoffsite.com.au/news/mass-timber/ (accessed on 17 September 2023).
17. Green Building Council of Australia Mid-Tier Commercial Office Buildings in Australia, Green Building Council of Australia. Available online: https://www.gbca.org.au/uploads/97/36449/Mid-tier%20Commercial%20Office%20Buildings%20Sector%20Report_FINAL.pdf (accessed on 17 September 2023).
18. Zaman, A.; Chan, Y.Q.; Jonescu, E.; Stewart, I. Critical Challenges and Potential for Widespread Adoption of Mass Timber Construction in Australia—An Analysis of Industry Perceptions. *Buildings* **2022**, *12*, 1405. [CrossRef]
19. XLAM Our State of the Art Production Facility. Available online: https://xlam.co/our-facilities (accessed on 17 September 2023).
20. Wood Solutions Cross Laminated Timber. Available online: https://www.woodsolutions.com.au (accessed on 17 September 2023).
21. Gu, H.; Liang, S.; Pierobon, F.; Puettmann, M.; Ganguly, I.; Chen, C.; Pasternack, R.; Wishnie, S.; Maples, I. Mass Timber Building Life Cycle Assessment Methodology for the U.S. Regional Case Studies. *Sustainability* **2021**, *13*, 14034. [CrossRef]

22. *EN15978:2011*; CEN/TC 350. Sustainability of Construction Works—Assessment of Environmental Performance of Buildings—Calculation Method. BSI Standards: London, UK, 2011.
23. *EN 15804:2012 + A1: 2013/FprA2:2019*; CEN/TC 350. Sustainability of Construction Works—Environmental Product Declarations—Core Rules for the Product Cate-Gory of Construction Projects. 2019. Available online: https://www.usgbc.org/resources/en-15804%E2%80%942012-sustainability-construction-works-environmental-product-declarations-core-ru (accessed on 22 July 2024).
24. Environmental Product Declaration: Boral Victoria Region—Pre-Mix Concrete. Available online: https://epd-australasia.com/epd/boral-victoria-region-pre-mix-concrete-epd/ (accessed on 9 May 2024).
25. Environmental Product Declaration for Barro Group Pre-Mix Concrete. Available online: https://epd-australasia.com/epd/barro-group-pre-mix-concrete/ (accessed on 9 May 2024).
26. Adbri Concrete Products EPD—Victoria. Available online: https://epd-australasia.com/epd/adbri-concrete-products-victoria/ (accessed on 9 May 2024).
27. Forest and Wood Products Australia Environmental Product Declaration: Glued Laminated Timber (Glulam). Available online: https://epd-australasia.com/wp-content/uploads/2018/03/epd565-Glued-Laminated-Timber-v1.pdf (accessed on 8 May 2024).
28. Environmental Product Declaration—Reinforcing Bar and Mesh. Available online: https://epd-australasia.com/epd/steel-reinforcing-bar-and-mesh/ (accessed on 9 May 2024).
29. Environmental Product Declaration for Rondo Products: Ceiling, Wall Framing and Finishing. Available online: https://epd-australasia.com/company-epd/rondo/ (accessed on 9 May 2024).
30. XLam Australia Environmental Product Declaration: XLam CLT Panel. Available online: https://epd-australasia.com/wp-content/uploads/2021/03/SP02326-XLam-EPD_v1.1_Jun23.pdf (accessed on 8 May 2024).
31. Siniat Plasterboard Environmental Product Declaration. Available online: https://epd-australasia.com/epd/siniat-plasterboard/ (accessed on 9 May 2024).
32. Insulated Wall and Roof Panels with a PIR Core. Available online: https://epd-australasia.com/epd/metecno-insulated-wall-and-roof-panels-with-a-pir-core/ (accessed on 9 May 2024).
33. Decorative Acoustic, Acoustic and Thermal Insulation Products Manufactured by CSR Martini. Available online: https://epd-australasia.com/epd/decorative-acoustic-acoustic-and-thermal-insulation-products-manufactured-by-csr-martini/ (accessed on 9 May 2024).
34. Green Building Council Australia. Upfront Carbon Emissions Calculation Guide—Interim. Available online: https://new.gbca.org.au/resources/reports-and-publications/#guides (accessed on 10 May 2024).
35. Ecotera High Performance/Low Carbon Concrete (Victoria—Melbourne Metro). Available online: https://epd-australasia.com/epd/ecotera-high-performance-low-carbon-concrete-victoria-melbourne-metro/ (accessed on 9 May 2024).
36. Durlinger, B.; Crossin, E.; Wong, J. Life Cycle Assessment of a Cross Laminated Timber Building. Available online: https://fwpa.com.au/wp-content/uploads/2013/02/PRA282-1112_Life-Cycle_Assessment_of_a_cross_laminated_timber_building_0.pdf (accessed on 4 May 2024).
37. Carboncare. CO_2 Emissions Calculator. Available online: https://www.carboncare.org/en/co2-emissions-calculator (accessed on 23 April 2024).
38. Greene, J.; Hosanna, R.; Willson, B.; Quinn, J. Whole life embodied emissions and net-zero emissions potential for a mid-rise office building constructed with mass timber. *Sustain. Mater. Technol.* **2023**, *35*, e00528. [CrossRef]
39. Australian Government Department of Climate Change, Energy, the Environment and Water. Australian National Greenhouse Accounts Factor. Available online: https://www.dcceew.gov.au/climate-change/publications/national-greenhouse-accounts-factors-2023 (accessed on 10 May 2024).
40. Leedham, A.; Heppner, J.; Smith, M.; Cooke, T. Lever Architecture and Atelier Ten Mythbusting Mass Timber. Available online: https://www.architecturalrecord.com/ext/resources/Issues/2023/06-June/Mythbusting-Mass-Timber.pdf (accessed on 14 April 2024).
41. AusLCI Carbon Emission Factors. Available online: https://www.alcas.asn.au/auslci-emissions-factors (accessed on 9 May 2024).
42. Environmental Product Declaration—Ready-Mix Concrete. Available online: https://epd-australasia.com/epd/bgc-concrete-ready-mix-concrete/ (accessed on 9 May 2024).
43. Environmental Product Declaration—Crumb Rubber Underlay. Available online: https://www.environdec.com/library/epd7910 (accessed on 10 May 2024).
44. Eslami, H.; Yaghma, A.; Jayasinghe, L.; Waldmann, D. Comparative life cycle assessment of light frame timber and reinforced concrete masonry structural systems for single-family houses in Luxembourg. *Heliyon* **2024**, *10*, e26083. [CrossRef] [PubMed]
45. Duan, Z.; Huang, Q.; Sun, Q.; Zhang, Q. Comparative life cycle assessment of a reinforced concrete residential building with equivalent cross laminated timber alternatives in China. *J. Build. Eng.* **2022**, *62*, 105357. [CrossRef]
46. Wood Solutions Rethinking Apartment Building Construction—Consider Timber. Available online: https://www.woodsolutions.com.au/publications/rethinking-construction-consider-timber (accessed on 5 May 2024).
47. Norman, C.; Kreye, M. PennState Extension How Forests Store Carbon. Available online: https://extension.psu.edu/how-forests-store-carbon (accessed on 6 May 2024).
48. Ong, C.B. 7- Glue-laminated timber (glulam). *Wood Compos.* **2015**, 123–140. [CrossRef]
49. Australian Sustainable Hardwoods T3 Collingwood the Pinnacle of Mass Timber Architecture. Available online: https://ash.com.au/projects/t3-collingwood-the-pinnacle-of-mass-timber-architecture/ (accessed on 4 May 2024).

50. Scouse, A.; Kelly, S.S.; Liang, S.; Bergman, R. Regional and net economic impacts of high-rise mass timber construction in Oregon. *Sustain. Cities Soc.* **2020**, *61*, 102154. [CrossRef]
51. Nepal, P.; Johnston, C.; Ganguly, I. Effects on global forests and wood product markets of increased demand for mass timber. *Sustainability* **2021**, *13*, 13943. [CrossRef]
52. Australian Forests Products Association Inquiry into the Timber Supply Chain Constraints in the Australian Plantation Sector Submission. Available online: https://www.aph.gov.au/DocumentStore.ashx?id=8c3b3763-f9ac-49e1-8a77-1dc036adb8b3andsubId=6909899 (accessed on 7 May 2024).
53. Australian Bureau of Agriculture and Resource Economics and Sciences Australian Plantation Statistics 2023. Available online: https://www.agriculture.gov.au/abares/research-topics/forests/forest-economics/plantations-update#download-the-overview-report-and-datasets (accessed on 7 May 2024).
54. McGavin, R.L.; Dakin, T.; Shanks, J. Mass-timber construction in Australia: Is CLT the only answer? *BioResources* **2020**, *15*, 4642–4645. [CrossRef]
55. Duan, Z.; Huang, Q.; Zhang, Q. Life cycle assessment of mass timber construction: A review. *Build. Environ.* **2022**, *221*, 109320. [CrossRef]
56. *AS3600:2018*; Concrete Structures. Standards Australia: Sydney, NSW, Australia, 2018.
57. Zgonc, B.; Tekavcic, M.; Jaksic, M. The impact of distance on mode choice in freight transport. *Eur. Transp. Res. Rev.* **2019**, *11*, 10. [CrossRef]
58. Collier, S.; Ruehl, C.; Yoon, S.; Boriboonsomsin, K.; Durbin, T.; Scora, G.; Johnson, K.; Herner, J. Impact of heavy-duty diesel truck activity on fuel consumption and its implications for the reduction of greenhouse gas emissions. *J. Transp. Res. Board* **2019**, *2673*, 125–135. [CrossRef]
59. Sanchez-Balseca, J.; Pineiros, J.; Perez-Foguet, A. Influence of travel time on carbon dioxide emissions from urban traffic. *Transp. Res. Part D Transp. Environ.* **2023**, *118*, 103698. [CrossRef]
60. Allan, K.; Phillips, A. Comparative cradle-to-grave life cycle assessment of low and mid-rise mass timber buildings with equivalent structural steel alternatives. *Sustainability* **2021**, *13*, 3401. [CrossRef]
61. Liang, S.; Gu, H.; Bergman, R. Environmental life-cycle assessment and life-cycle cost analysis of a high-rise mass timber building: A case study in pacific northwestern united states. *Sustainability* **2021**, *13*, 7831. [CrossRef]
62. Hafner, A.; Rueter, S. Method for assessing the national implications of environmental impacts from timber buildings-an exemplary study for residential buildings in Germany. *Wood Fibre. Sci.* **2018**, *50*, 139–154. [CrossRef]
63. Hart, J.; Pomponi, F. More timber in construction: Unanswered questions and future challenges. *Sustainability* **2020**, *12*, 3473. [CrossRef]
64. Ximenes, F.; Gardner, W.; Cowie, A. The decomposition of wood products in landfills in Sydney, Australia. *Watse. Manag.* **2008**, *28*, 2344–2354. [CrossRef]
65. Robati, M.; Oldfield, P. The embodied carbon of mass timber and concrete buildings in Australia; An uncertainty analysis. *Build. Environ.* **2022**, *214*, 108944. [CrossRef]
66. Evison, C.; Kremer, D.; Guiver, J. Mass timber construction in Australia and New Zealand-status, and economic and environmental influences on adoption. *Spec. Issue—CLT/Mass Timber* **2018**, *50*, 128–138. [CrossRef]
67. Smith, T.; Fragiacomo, M.; Pampanin, S.; Buchanan, A. Construction time and cost for post-tensioned timber buildings. *Constr. Mater.* **2009**, *162*, 141–149. [CrossRef]

sustainability

MDPI

Article

Timber Biogenic Carbon Stock in the Urban Environment: Santiago City as a Second Forest

Felipe Victorero [1,2,*] and Waldo Bustamante [1,3]

1 School of Architecture, Pontificia Universidad Católica de Chile, El Comendador 1916, Santiago 7520245, RM, Chile; wbustamante@uc.cl
2 Centro Nacional de Excelencia para la Industria de la Madera (CENAMAD), Pontificia Universidad Católica de Chile, Santiago 7820436, RM, Chile
3 Centre for Sustainable Urban Development (CEDEUS), Pontificia Universidad Católica de Chile, El Comendador 1916, Santiago 7520245, RM, Chile
* Correspondence: favictor@uc.cl

Abstract: Urban environments significantly contribute to carbon emissions, both through operational processes and the embodied emissions of construction materials, thus exacerbating climate change. Nevertheless, urban timber structures offer a viable alternative by acting as carbon sinks, capable of sequestering carbon for decades or even centuries. This study develops and applies a methodology to quantify the biogenic carbon stored in Santiago City's timber-based buildings, conceptualized as a "Second Forest" that transfers and preserves the carbon capture capacity of trees in the built environment. The analysis estimates that Santiago's urban timber constructions have expanded their wood-built surface area by 192,831 m^2 over the past eight years, reflecting the growing adoption of timber in urban construction. During the same period, biogenic carbon storage increased from 199.78 kt to 202.73 kt, equivalent to 10.84 kt of CO_2 under average conditions. These findings highlight the potential of urban planning strategies, such as promoting taller timber buildings and adopting circular timber practices, to enhance carbon sequestration and reduce reliance on carbon-intensive materials. This research highlights the fundamental role that timber buildings play in urban climate change mitigation, positioning them as active contributors to global carbon management efforts.

Keywords: biogenic carbon stock; urban wood structures; sustainable urban planning; carbon sequestration; timber urban resources

1. Introduction

Harvest wood products, sourced from sustainably managed forests, have proven to be a renewable alternative with a lower environmental impact compared to materials that are intensive in fossil fuels and contribute to climate change with higher greenhouse gas (GHG) emissions [1–3]. Additionally, their potential to capture carbon dioxide from the atmosphere during photosynthesis and tree growth presents a unique opportunity as a material in which wood product manufacturing can store biogenic carbon over its lifespan. In simplified terms, for every two kilograms of anhydrous wood, approximately one kilogram of biogenic carbon can be estimated, around 50% of its mass, which is equivalent to storing approximately 3.67 kg of carbon dioxide from the atmosphere when considering the mass of carbon and oxygen involved [4,5]. This condition remains in time, unless the carbon is released through combustion and oxidation or degraded by biological agents, mainly being released as carbon dioxide and methane.

https://doi.org/10.3390/su17020529

This potential for atmospheric carbon dioxide storage over time depends directly on ensuring that any wood product remains in the environment for as long as possible and prevents the release of carbon, whether accidental or deliberate. This can be achieved through the manufacture of long-lasting products, reuse, and, to a lesser extent, recycling or modification for other uses, considering that any recycling or modification process tends to lose some material and release stored carbon to some extent.

Consequently, wood products used in construction, especially in structural applications, offer buildings the potential to store substantial amounts of biogenic carbon for decades, or even centuries [6]. Moreover, the lower GHG emissions and reduced environmental impact associated with wood construction offer a substantial potential for decreasing the environmental impact of this sector. Currently, the construction industry is responsible for approximately 37% of atmospheric GHG emissions [7]. Additionally, traditional construction materials such as concrete and steel currently account for approximately 11% of global GHG emissions due to their production processes, in which it is difficult to reduce emissions, and their global production volumes. In fact, approximately 70% of the planet's anthropogenic mass currently consists of materials such as concrete and aggregates, and it is expected that by 2040, concrete by itself would exceed the entire earth's biomass [8]. Moreover, these materials are increasingly scarce, requiring transportation over long distances for processing, resulting in emissions and environmental impacts associated with their extraction [9].

By contrast, forest resources are abundant in many regions worldwide and have the potential to grow even further through reforestation and afforestation strategies, offering substantial potential for combating climate change [10]. This is especially important in countries that consider forests and wood products as strategic resources for achieving GHG emission reduction targets and meeting raw material demands. Moreover, durable wood products harvested from sustainably managed forests can effectively help prevent CO_2 emissions, even when sourced from younger forests. For example, newly planted forests, when managed with a focus on denser planting schemes and rapid growth strategies, can capture CO_2 more effectively than older forests [11]. And although older forests capture less CO_2 overall, individual mature trees capture more CO_2 than several young trees; more importantly, they may have stored CO_2 for a longer time, potentially even predating the industrial revolution [12]. Therefore, cutting down old trees in mature forests can have a higher impact on CO_2 release compared to cutting young trees in young forests, where reforestation plans can help capture new CO_2.

Likewise, if the destination of wood products involves a life in construction elements, where the materials' duration in the built environment could span several replacement tree growth cycles in the environment, any release of carbon dioxide could eventually be compensated. Thus, it is desirable that the useful life of any wood product exceeds the time it took the tree to capture the stored biogenic carbon. Furthermore, forests with rapid growth and carbon dioxide capture, such as those with a regeneration cycle of just a couple of decades, are more appealing from this perspective than forests that are centuries old, where the slower regeneration of carbon capture may outweigh the benefits.

In this context, wooden constructions within the urban environment of a specific city act as a biogenic carbon stock that endures over time and varies depending on the longevity of wooden construction elements in the built environment. Moreover, the integration of new wooden buildings can potentially offset carbon released into the atmosphere or increase urban carbon storage. Furthermore, if wood surpasses the replacement rate of equivalent carbon harvested from the original tree through the planting of a new tree, it could provide a long-term benefit to the global carbon balance. This not only represents a transfer of biogenic carbon from the forest to the city but also creates a carbon reservoir independent of the original forest stock.

In this regard, previous research has attempted to quantify the amount of wood in urban environments using various methods but has encountered challenges that have constrained the results [13–18]. Most of these challenges are associated with limitations in the availability of public data on the materials used in buildings, making it difficult to represent material quantities on a large scale and track their evolution over time [14]; the restriction of results to a specific range of building types within particular timeframes and spatial constraints [15]; or the use of material quantity indicators that need to be extrapolated from other contexts [17].

These limitations highlight the need for methodologies that address gaps in the quantification of urban wood use and reduce uncertainty, particularly in developing countries with limited data on construction materials and their environmental impacts. This study focuses on Santiago City as a case study due to its unique urban characteristics, including a growing use of timber in construction and its role as the capital of a developing nation with ambitious carbon neutrality targets. By examining Santiago, this research provides insights applicable to other rapidly urbanizing regions with similar constraints, offering a replicable framework for understanding the environmental potential of urban timber.

The concept of Santiago as a "Second Forest" reflects the premise that urban timber structures can replicate the carbon storage capacity of natural ecosystems. This concept suggests that urban environments, when designed with an emphasis on wood-based construction, have the potential to become significant carbon sinks, simultaneously reducing emissions while enhancing carbon storage. By envisioning cities as anthropogenic extensions of natural carbon storage systems, this study highlights the need for a paradigm shift in urban planning where built environments actively contribute to climate change mitigation by retaining part of the original forest's carbon through harvested wood products while allowing the development of low-emission buildings in emerging nations.

Furthermore, this perspective emphasizes the broader potential of urban wood use in reducing reliance on carbon-intensive materials such as concrete and steel. Additionally, as a forest-rich country, Chile depends on the maintenance and expansion of its forests to offset at least 50% of its growing emissions and achieve carbon neutrality [19], underscoring the need for an industry that promotes long-lasting and renewable harvested wood products that foster sustainably managed forests. By integrating these principles into policy frameworks, nations like Chile could develop effective strategies to offset emissions while enhancing carbon storage. This approach positions timber construction not only as a sustainable alternative but also as a critical element in achieving carbon neutrality by 2050 and aligning with the nation's commitments under the Paris Agreement.

The primary purpose of this work is to present a methodology for quantifying the biogenic carbon stored in wooden buildings within the urban built environment of a city, particularly in contexts with limited or no available data on wood material use and carbon storage. Additionally, a secondary aim of this research is to provide an initial understanding of how biogenic carbon is distributed in a large developing city, such as Santiago, Chile. Considering its significant expansion in recent decades and its pressing need for new infrastructure, this study examines the temporal behavior of biogenic carbon and identifies future research directions to improve projections of wood usage and biogenic carbon storage in the coming years.

2. Materials and Methods

2.1. Case Study: Santiago of Chile

Chile has positioned itself as a regional leader in Latin America in terms of sustainability and development in the construction industry. Despite being a developing country, Chile has adopted policies that promote energy-efficient buildings through the implementation of sustainable certifications, alongside tools designed to encourage a more circular construction industry. These initiatives aim to reduce the environmental footprint of the rapid urban growth the country has experienced in recent decades—a characteristic common to emerging nations that face the need to quickly expand their infrastructure. Furthermore, Chile has gradually begun to adopt more complex models for assessing environmental impact, laying the groundwork for the development of life cycle assessments in construction and the quantification of embodied emissions.

Chile is a strong advocate of international climate commitments, being a signatory to the Paris Agreement, which reflects its determination to achieve carbon neutrality by 2050. This goal is supported by a comprehensive strategy that includes significant progress in emission reductions, accounting for 50% of the projected reduction, and the sustainable management of its extensive forests, which cover 17.9 million hectares [20]. These forests are expected to act as natural carbon sinks, offsetting the remaining 50% of national emissions. Within this strategy, the construction sector accounts for approximately 31% of national emissions [21] and is projected to contribute, through sustainable building strategies, at least 17% to the national emission reduction plan.

Over the past two decades, Chile has emerged as an international leader in the adoption of renewable energy, particularly solar and wind energy, with more than 50% of the installed capacity in its electrical system coming from renewable sources. These energy sources have transformed the country's energy landscape, significantly reducing greenhouse gas emissions associated with electricity generation and, consequently, the operational emissions of the built environment. While the country is expected to reach 100% renewable energy by 2050, and national industries like lithium and green hydrogen enable energy storage, the significant reduction in operational emissions from the built environment alone would not be sufficient. This is due to the construction sector's heavy reliance on fossil fuel-intensive materials such as concrete. Within this Chilean context, it is essential to understand the carbon-storing benefits of materials like wood in the construction sector, alongside their advantages in promoting a forestry sector that produces sustainable wood products capable of storing carbon and reducing embodied emissions.

The city of Santiago, represented in this study by the 32 municipalities of the Province of Santiago (Figure 1), with a total area of approximately 2110 km^2 and a population exceeding five million inhabitants, serves as a representative example of national building and public policy trends in Chile. Like many cities worldwide, Santiago faces significant challenges related to urban densification, transportation, and resource consumption. Nonetheless, the city has implemented pioneering initiatives in green infrastructure, urban planning, and efficient public transportation, positioning itself as a model for other cities in the Latin American region. However, the impact of these policies on wood-based buildings and the urban biogenic carbon stock remains unclear. More critically, it is uncertain whether these policies are contributing to biogenic carbon storage or facilitating its release into the atmosphere.

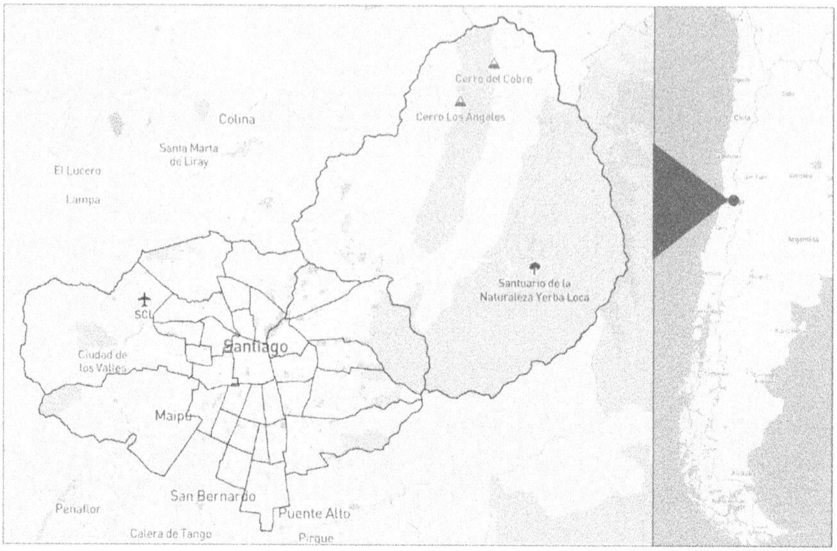

Figure 1. The study area consisting of 32 municipalities that make up the Santiago province (prepared using https://mapshaper.org/ (accessed on 8 August 2024) with ©Mapbox and ©OpenStreetMap).

2.2. *Quantifying Wood in the Built Environment*

Accurately quantifying the volume of wood present in buildings within the built environment is the key to understanding its extent and distribution patterns. This requires considering various factors, such as construction systems, building types, and wood species, among others. Moreover, the potential benefits of wood construction at the urban level can only be analyzed if its scope and variation in space and time are well understood.

However, determining how much wood exists in the built environment is not a straightforward task, and this information is not readily available in all countries. Various approaches can be used to obtain these data, including (i) data from national records, such as building permits or real estate property records; (ii) estimates made through sample surveys, such as censuses or sampling methodologies; (iii) approaches based on aerial imagery and model estimations; and (iv) commercial records of forest production and sales of wood products.

This study is based on the use of national real estate records to quantify wood in the built environment. The case study of Santiago in Chile benefits from tax records maintained by the country's Internal Revenue Service, which detail the primary materials used in building structures for the purpose of a property tax assessment. This database includes multiple records that allow for the identification of built square meters according to the materials used in structural walls, the intended use of the building (e.g., residential, commercial, educational, or other), the year of construction, records of modifications over time, the quality of construction, and others that are not relevant to this research.

Chilean records have been available since 2018, with updates being made every six months, and they include public access data on demolitions, modifications, and new constructions [22]. These records are gathered using detailed manuals and assessments carried out by professional property appraisers. The eight-year measurement period is due to the system's digitization in 2018; earlier records exist but lack the consistent structure found in the current dataset. Furthermore, while the Chilean National Statistics Institute (INE) keeps a record of building permit requests, which identifies the main materials used and conducts a national housing census that records the materials of surveyed buildings, these sources are less reliable than the real estate registry. This is because the former does not confirm whether the permitted building was constructed or modified, and the latter faces issues related to survey-based data collection, where residents may provide incorrect or incomplete information about the building systems they inhabit.

The data provided by the real estate record and their uses in this work consist of eight files (from 2017 to 2024) that address urban areas and focus on parcels and building information. The details of the contents of the files can be seen in Table 1.

Table 1. Data available from the Chilean real state registry.

Column	Data in Real State Registry File	Use in This Work
1	SII Code of municipality	Identify Santiago's municipalities
2	Block number	No use in this work
3	Parcel number	No use in this work
4	Number of construction line	Identify building information
5	Structural material of construction line	Identify wood buildings (predominant material in bearing walls)
6	Quality of construction line	No use in this work
7	Year of construction line	Identify the age of wood buildings
8	Surface area of construction line	Identify the gross surface area of wood buildings
9	Use of construction line	No use in this work
10	Special condition of construction line	No use in this work

The file format remains the same for the eight years of data used in this study (2017–2024).

The Chilean real estate record also provides detailed information at the municipality, block, and property levels, with over 8 million entries nationwide. In this case study, 32 municipalities that make up the province of Santiago or city of Santiago are analyzed (the country's most densely populated urban area). For the purposes of this study, analyses are conducted at both the city of Santiago and municipality levels, summing the square meters accordingly, to understand how wood and its associated biogenic carbon are distributed within the urban environment. For the same reason, the results and analyses are limited exclusively to buildings that list wood as their primary construction system regardless of whether it represents only a small part of a building complex that may include a greater proportion of square meters constructed with other materials, either before or after.

It is important to note that while the real estate record is the most reliable source of this type of information for this case study, it contains some inconsistencies and lacks detail. An example of this is that it might include properties that have not been properly regularized or have undergone undeclared modifications not identified by inspectors. Additionally, the registry does not specify the construction methods used for wood, making it impossible to distinguish between frame systems or mass timber, which could affect the quantification of the material.

2.3. Historical, Traditional, and Contemporary Construction Systems

Since the record of wooden square meters in the Chilean real estate database refers to the gross built surface area and not the volume or specific mass of the material, a methodology is developed to quantify the material intensity (MI) of wood per square meter. The MI is an indicator that represents the mass of various materials per surface area or volume, typically expressed in kg/m^2 or kg/m^3. It is calculated by dividing the mass of a given construction material by the gross surface area or internal volume of the project that is being measured, which subsequently generates a record for the set of materials used [17].

In this way, by calculating the material in a project (whether based on design specifications, the material used during construction, or the material registered during demolition), it is possible to establish the MI for a specific material. In the case of wood, the mass depends on the equilibrium moisture content, so the MI of wood can be calculated either at the equilibrium moisture point or in a completely dry state.

$$MI_{wood} = \frac{V_{\omega} \times \rho_{\omega}}{GBA} \tag{1}$$

where

MI_{wood}: the material intensity of wood per gross built surface area (in kg/m^2).
V_{ω}: the volume of the wood product at the specified moisture content (in m^3).
ρ_{ω}: the density of the wood at the specified moisture content (in kg/m^3).
GBA: the gross built surface area (in m^2).
Alternatively,

$$MI_{wood-dry} = \frac{V_{\omega} \times \rho_{\omega}}{GBA \times \left(1 + \frac{\omega}{100}\right)} \tag{2}$$

where

$MI_{wood\text{-}dry}$: the material intensity of dry wood per gross built surface area (in kg/m^2).
ω: the moisture content of the wood, expressed as a percentage (in %).

The MI is commonly used by researchers to track the flow of construction materials within buildings, assess their lifespan, create inventories, study material recovery potential, analyze temporal–spatial identification, and evaluate environmental impacts, among other applications [15,17,23]. Materials associated with products that are prone to renovation, such as interior finishes or coverings, are more difficult to assess over time and to keep a consistent record of. Conversely, structural materials tend to exhibit fewer changes over time and remain throughout the entire lifespan of a building [24].

Since it is impractical to conduct an MI analysis of wood for every building in the case study, an assumed MI is proposed for different types of wooden construction systems. This is achieved by defining typical wooden walls and roofs linked to historical periods of construction in the country. Maximum and minimum MI indicators are proposed for the wood used in these structural elements. Considering that structural elements are the least likely to change over time and that the Chilean real estate registry primarily records load-bearing wall structures and their modifications, the focus remains on these aspects.

This study centers exclusively on typical construction systems based on sawn wood, acknowledging that this is the primary material used for the construction of such buildings in Chile. Engineered wood products, such as laminated timber, only gained significant market presence during the last decade of the 20th century—the first laminated structure is presumed to date from 1969, OSB products emerged in the early 2000s, and mass timber products like CLT only began to be used in the 2020s. While these products and construction systems may have higher MI values than frame-based systems, they are much less common and not representative of the historical wood construction practices in the country.

Thus, three historical periods are defined with the objective of evaluating the MI of wood in the case study: pre-1900 systems, those from the 20th century, and those from after the year 2000. It is important to note that while the real estate registry includes heritage structures dating back to the 17th century, most of the recorded built surface area is concentrated in the second half of the 20th century and the first decades of the 21st century. Additionally, an unknown number of historical wooden constructions in Santiago have been affected by fires, earthquakes, or have simply been replaced by masonry and reinforced concrete structures.

To facilitate the analysis of a large number of wooden architectures represented in the real estate record and to generate representative wood volumes for construction systems and historical periods, two typical wooden walls and two typical roofs are defined for analysis. These typologies assume building heights not exceeding three stories, in line with data collected by the INE in its building permit records [25]. This approach is also consistent with research conducted in Finland and other Northern European countries, which attribute nearly the entirety of wood MI in low-rise framed buildings to the walls and roof structures [15].

For the walls, a small wall of 2400 mm in length and a larger one of 7200 mm is defined, with a height set at 2400 mm. Although historical architecture often features taller ceilings than those found in constructions from the second half of the 20th century and the 21st century, due to variability depending on the project, a uniform and more conservative height of 2400 mm is used for all wall types regardless of the historical period.

Furthermore, to determine the volume of wood in walls per square meter of built structure, the wall density data from the floor plan are utilized. According to the literature, wooden structures should have a wall density in the floor plan between 3% and 6% [26–29], values that are higher than those found in equivalent concrete structures.

$$Wall\ Density = \frac{A_{wall}}{GBA} \times 100 \tag{3}$$

where

Wall Density: the floor plan area of structural walls expressed as a percentage of the GBA (in %).

A_{wall}: the floor area covered by structural walls (in m^2).

GBA: the gross built surface area (in m^2).

Regarding the defined roof types, both are gable roofs with a 40% pitch. One case considers a smaller volume of wood covering a span of 3000 mm and a length of 2400 mm, while the second case involves a higher volume of wood, spanning 6000 mm with a length of 4800 mm. Horizontal structural elements, such as ventilated floors or intermediate floors, are not considered in this analysis because they cannot be identified in the real estate database. It is assumed that these elements are partially represented within the wood volume addressed by the roof cases.

2.3.1. Historical Systems

Regarding historical construction systems, the most common type corresponds to heavy timber framing and its variations. Examples of these systems include the "Quincha" of Latin American origin and some reinterpretations of European and North American systems, which were mainly imported during the 18th century [30–33]. For this type of construction, walls are considered to include studs, beams, top and bottom plates, and diagonals with cross sections ranging from 100 mm × 100 mm to 150 mm × 150 mm, while smaller elements like noggings are considered to have cross sections in the range of 50 mm × 50 mm to 70 mm × 70 mm (Figure 2). Additionally, the shorter wall of 2400 mm in length has studs placed every 1200 mm, and the longer wall of 7200 mm in length features studs spaced every 800 mm.

As for the roofs, simple and traditional trusses from the period that utilize the minimum and maximum amounts of wood for such constructions are considered. The "King-post" truss is used for cases with the least amount of wood, while the "Queen-post" truss is applied to those with the highest material intensity (Figure 3). Regarding the cross sections of the legs, ties, posts, and diagonals, sizes of 41 mm × 138 mm and 41 mm × 185 mm are used. For purlins, sections between 19 mm × 90 mm and 41 mm × 80 mm are defined.

These trusses are evaluated with a spacing of 1200 mm for the roof with a lower volume of wood and 400 mm for the case with a higher volume.

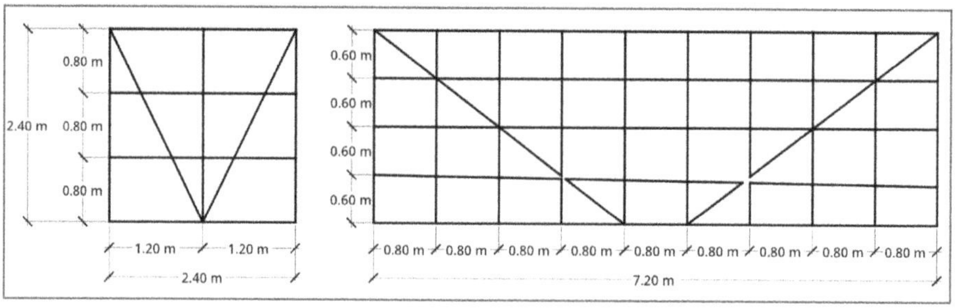

Figure 2. Diagrams of structural wood for short and long historical walls.

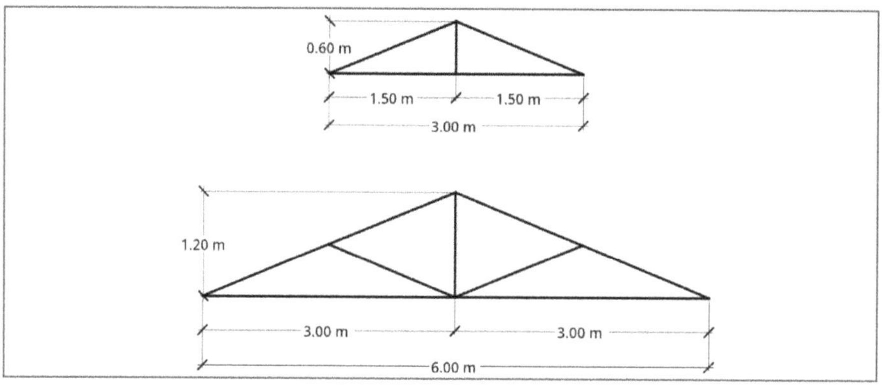

Figure 3. Diagrams of structural wood for short and long historical roof spans.

2.3.2. Traditional Systems

For traditional construction systems built between 1901 and 2000, the walls are considered to consist of a light-frame construction with cross sections of 41 mm × 65 mm and 41 mm × 90 mm, with noggings spaced every 800 mm and 600 mm, and diagonals with cross sections ranging from 19 mm × 90 mm to 41 mm × 90 mm (Figure 4). The potential sheathing of the wall is not included in the volume calculation as it does not serve a structural function, since the rigidity is provided by the diagonals, and it is subject to possible renovations, especially in the case of exterior sidings. Studs are spaced every 600 mm for the smaller wall and every 400 mm for the larger case.

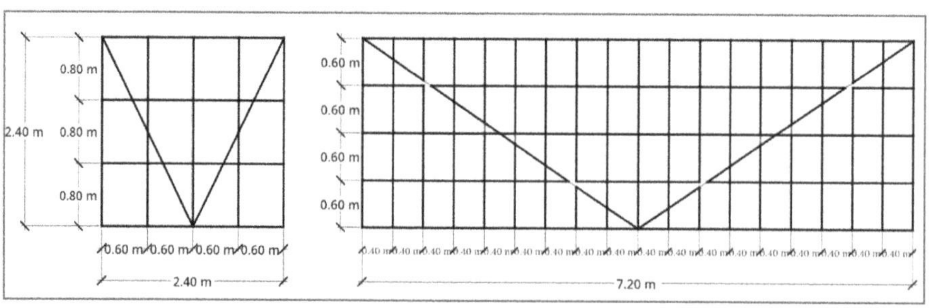

Figure 4. Diagrams of structural wood for short and long traditional walls.

As for the roof systems used during the 20th century, two systems are considered to represent the minimum and maximum volumes of wood required for a gable roof (Figure 5). These scenarios are intended to reflect the various types of trusses found within the context of the case study. In the minimum case, spanning a three-meter width, a simple truss is used, where the beams are placed at a slope on the structural walls and rest on a ridge beam with purlins. In contrast, for the scenario with a higher volume of wood, a "Pratt" truss with purlins and roof sheathing is proposed. The cross sections for the top chords, ridge beams, ties, and posts, as applicable, range from 41 mm × 90 mm to 41 mm × 114 mm, while the purlins are considered to have sections of 41 mm × 41 mm, and the sheathing is 19 mm thick in the maximum case.

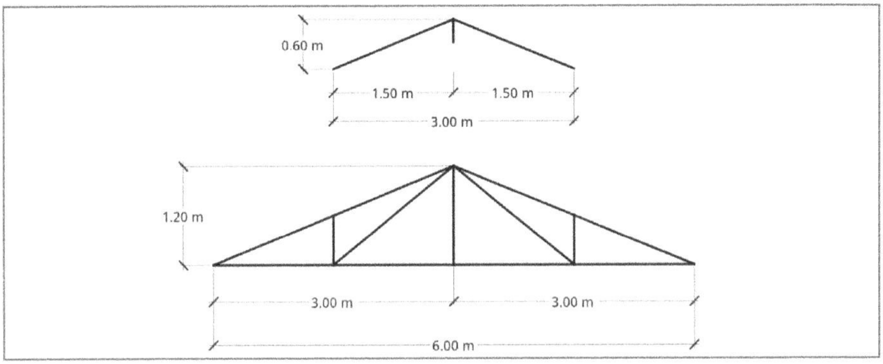

Figure 5. Diagrams of structural wood for short and long traditional roof spans.

2.3.3. Contemporary Systems

For walls representative of structures built after the year 2000, the main variation is the introduction of OSB (Oriented Strand Board) panels into the market, which led to their structural use for wall bracing. This results in the reduction or elimination of diagonals and a decreased use of noggings (Figure 6). While the cross sections for studs, top and bottom plates, and blocking pieces remain between 41 mm × 65 mm and 41 mm × 90 mm, the diagonals are replaced by OSB panels, which can vary in thickness between 9.5 mm and 11.1 mm. In the case of buildings up to two stories high, these panels are typically installed on the exterior side of the wall, with the interior finish being provided by a gypsum board panel.

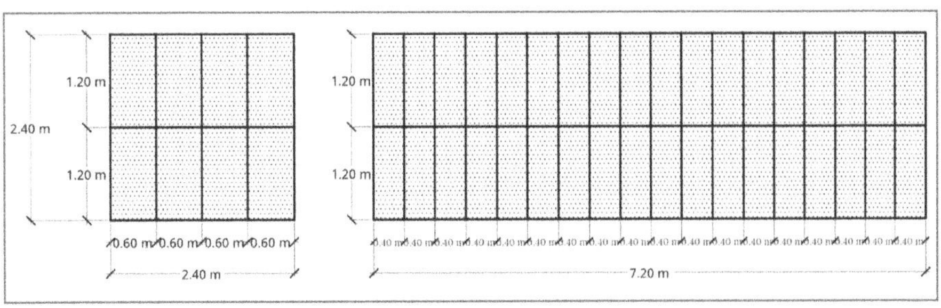

Figure 6. Diagrams of structural wood for short and long contemporary walls with OSB panels.

Regarding roofing systems, the truss structure remains similar to that of 20th-century cases, with the primary difference being that the roof sheathing can be replaced with OSB panels. In this regard, the sheathing thickness is reduced from 19 mm to 11.1 mm, which falls within the minimum and maximum volume of wood calculated for traditional cases.

Thus, the calculated minimum and maximum volumes of wood are found to be between 0.06 m^3/m^2 and 0.14 m^3/m^2 for "historical" cases before the 20th century, between 0.025 m^3/m^2 and 0.086 m^3/m^2 for "traditional" cases between 1900 and 2000, and within the range of 0.03 m^3/m^2 and 0.085 m^3/m^2 for constructions from the early decades of the 21st century.

2.4. Quantification of Biogenic Carbon

The methodology used to determine the biogenic carbon content in wood follows the guidelines established in the standard "UNE-EN 16449: Wood and wood-based products. Calculation of the biogenic carbon content of wood and conversion to carbon dioxide" [4]. The calculated volume of wood for each building and year recorded in the real estate database is used to estimate the biogenic carbon using the following formula:

$$C_{bio} = cf \times \frac{\rho_\omega \times V_\omega}{1 + \frac{\omega}{100}} \quad (4)$$

where

C_{bio}: the biogenic carbon stored in wood (in kg).

V: the volume of the wood product (in m^3).

ρ_ω: the density of the wood at the specified moisture content (in kg/m^3).

cf: the carbon fraction of wood biomass, where a standard value of 0.5 (50%) is used.

ω: the moisture content of the wood, expressed as a percentage.

Alternatively, for the purpose of this work, it can be expressed as

$$C_{bio} = f_c \times MI_{wood-dry} \times A \quad (5)$$

where

A: the built surface area of wood.

Meanwhile, regarding CO_2 emissions associated with biogenic carbon storing, UNE-EN 16449 presents the following equation:

$$P_{CO_2} = \frac{44}{12} \times C_{bio} \quad (6)$$

where

P_{CO_2}: the amount of carbon dioxide emitted by the oxidation of biogenic carbon at the end of the product's life cycle (in kg of CO_2).

44/12: the conversion factor from the molar mass of carbon (12 g/mol) to carbon dioxide (44 g/mol).

Additionally, since wood density can vary by species, the analysis considers two species that establish representative minimum and maximum values. From the second half of the 20th century onward, the period with the highest number of wooden constructions according to the real estate record, Pinus radiata (radiata pine) has been the most widely used wood species. Meanwhile, Nothofagus obliqua (Chilean oak) is one of the most common woods in the central–southern region of the country and represents the oldest constructions in the case study. Radiata pine, with a density of 450 kg/m^3 (oven-dry), is one of the lightest local woods used in construction, while Chilean oak, with a density of 558 kg/m^3 (oven-dry), is among the densest common construction woods [34]. The moisture content for both species is considered to be around 14%, in line with the equilibrium moisture content established for the climate in the case study [34], allowing for the determination of the carbon content per unit mass of wood.

While there is some debate regarding the carbon fraction associated with the dry mass of wood [5], this analysis follows the recommendations outlined by the UNE-EN 16449 standard. Therefore, the carbon fraction in the oven-dry state is established as 0.5, consistent with other similar studies [35,36]. This enables the quantification of both the biogenic carbon present in the mass of wood used in construction in the case study and its equivalent in carbon dioxide captured from the atmosphere.

2.5. Methodological Considerations

Although the methodology employed in this study offers an alternative approach for quantifying biogenic carbon stored in wooden structures within the built environment of Santiago, it has certain limitations. The reliance on real estate tax records, while based on a robust data acquisition system, can still introduce inconsistencies due to gaps in coverage. This is particularly relevant for properties with unregistered modifications, as the registry is updated semiannually, or for informal constructions/demolitions that fall outside the scope of data collection, especially in rural areas. However, given the sample size and its urban nature, it is expected that these unregistered cases follow the trends observed in the registry and represent a minor percentage of the study case.

The methodology used to identify the height of buildings and the presence of wood in them also depends on complementary analyses with other databases, such as the building permits registry. This is because the real estate registry cannot determine whether the square meters constructed with a specific material correspond to an entire building or just a portion of it (e.g., the top two floors). Consequently, the different methodologies used in each registry may lead to some taller wooden constructions being underrepresented in the analysis. Nonetheless, given the rarity of such buildings in the Chilean context, this underrepresentation is expected to be minimal.

Moreover, while the real state registry identifies the square meters constructed with a specific material, it considers only the predominant material used in bearing walls and does not provide the exact quantity of material present in the entire structure. Additionally, the registry does not include direct information about the specific construction system (e.g., post-and-beam or platform frame), making it difficult to estimate material quantities accurately. Similarly, it does not indicate the type of wood species used or whether engineered wood products are used, such as laminated timber, logs, or CLT.

Regarding the use of historical averages and generalized material intensity values to estimate the wood volume and biogenic carbon content, these may not fully capture variations in construction practices over time or across different geographic locations. Thus, the estimated wood quantity does not directly represent the mass of wood per square meter of structure, potentially leading to underestimation or overestimation. However, the proposed methodology also excludes secondary structural elements or non-structural components in its calculations, making the analyzed scenarios more conservative.

Additionally, for this methodology to be applicable in other case studies, it requires, at a minimum, adequate records of constructed areas by material for a given case, as well as relatively homogeneous wooden construction typologies, as presented in the analyzed case of construction systems. For this reason, urban contexts lacking proper records (either due to the absence of a data collection system or the age of the buildings) could hinder its implementation or result in significant errors.

3. Results

The results are presented by first quantifying the wood usage per square meter in buildings from the case study and generating wood MI data, categorized by historical period. Second, these results are cross-referenced with the real estate database to identify average values for the entire case study and to determine their variation by the studied municipality. Finally, the results for biogenic carbon storage and the equivalent CO_2 amount are presented, both aggregated for the entire city and by municipality.

3.1. Wood Material Intensity

The maximum and minimum wood volumes per gross built area, calculated by the studied time period, range from 0.78 m^3/m^2 to 1.04 m^3/m^2 for "historical" cases prior to the 20th century, between 0.42 m^3/m^2 and 0.52 m^3/m^2 for "traditional" cases from 1901 to 2000, and in the range of 0.59 m^3/m^2 to 0.64 m^3/m^2 for "contemporary" constructions from the early decades of the 21st century (Figure 7). These figures yield wood material intensity values with a maximum of 91.56 kg/m^2 and a minimum of 30.88 kg/m^2 for buildings from the 19th century or earlier; a maximum of 54.75 kg/m^2 and a minimum of 12.67 kg/m^2 for 20th-century constructions; and a maximum of 54.05 kg/m^2 and a minimum of 15.29 kg/m^2 for 21st-century structures (Figure 8). These values are in accordance with the densities presented in the methodological section and the ambient humidity.

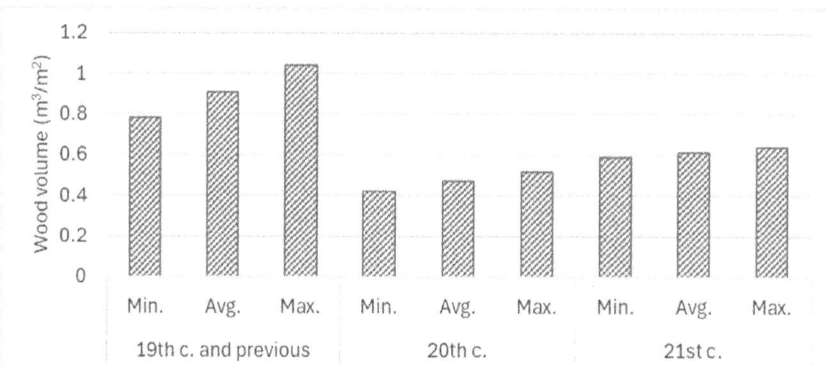

Figure 7. Wood volume per built square meter according to historical period and according to minimum, average, and maximum scenarios (city of Santiago).

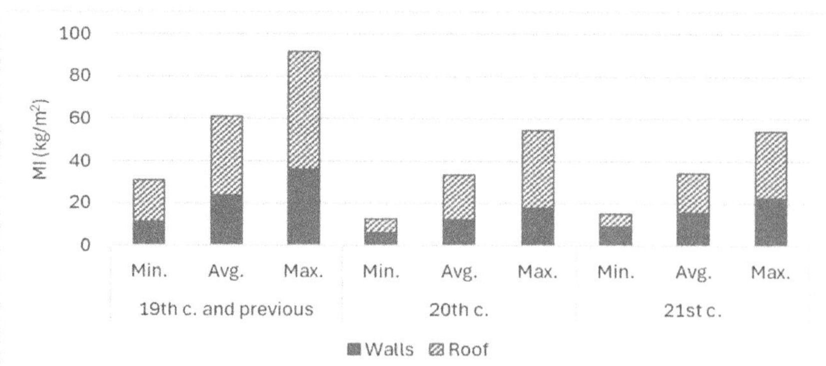

Figure 8. Material intensity of wood according to historical period and according to minimum, average, and maximum scenarios (city of Santiago).

Similarly, by correlating material intensities by time period with the results from the filtered database, it becomes possible to identify that the weighted wood material intensity ranges between 54.69 kg/m^2 and 13.07 kg/m^2, with an average of 33.88 kg/m^2 across all declared periods (Figure 9). Additionally, the ratio between the wood material intensity of the roof and walls is 2:1 in the most material-intensive case and nearly 1:1 in the least material-intensive case. This observation takes into account the differences in the amount of wood established for each case according to the historical period.

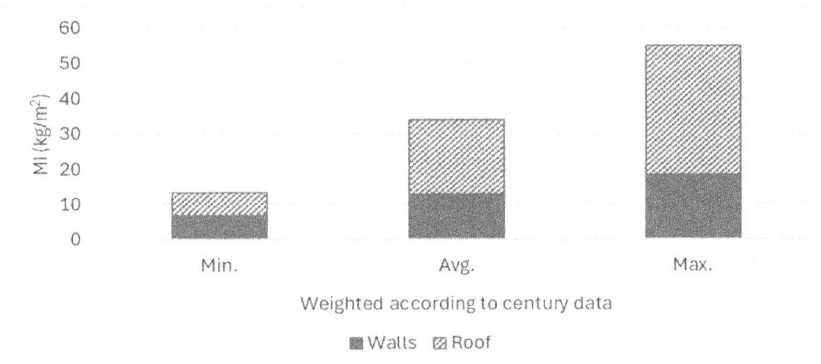

Figure 9. Weighted material intensity of wood over entire studied period, according to minimum, average, and maximum scenarios (city of Santiago).

3.2. Wood Temporal Distribution

From the processing of the real estate database, it is possible to isolate the square meters of wood construction per year, which increased from 13.4 million in 2017 to 13.6 million in 2024, with a growth of 192,831 square meters during the period (Figure 10). Additionally, these figures show an average growth rate of wood construction area of around 0.2% compared to the previous year. Moreover, a significant decline in new wood-built areas is seen in 2021, followed by a stabilization in the growth percentage between 2023 and 2024.

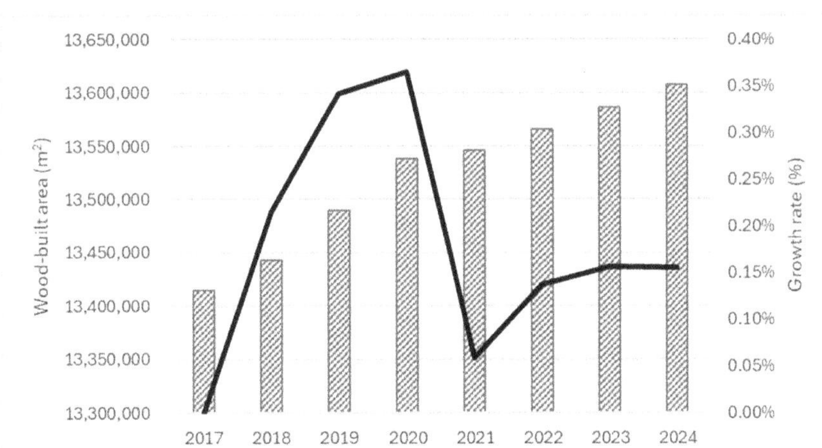

Figure 10. The annual wood-built surface area (bars) and growth rate (line) according to the average scenario during the study period (city of Santiago).

The variation in growth rate of the wood-built surface area in the case study (Figure 10) can be attributed to an expansion in construction at the beginning of the recorded period, which was abruptly interrupted by the effects of the COVID-19 pandemic in 2020 and followed by a limited recovery due to subsequent economic impacts. Similar effects have been observed in other economies worldwide, with recovery periods varying depending on the specific context.

Similarly, considering a minimum and maximum range according to the methodology presented, the volume of wood built in the case study varies between 0.3 and 1.2 million cubic meters of wood depending on the scenario (minimum, average, or maximum). This shows a ratio difference of approximately four times, with the average value varying between 0.75 and 0.76 million cubic meters of wood depending on the year. It is important to highlight that this volume of wood considers a 14% moisture content in accordance with the equilibrium moisture level of the case study and the methodology used for this work.

When considering the evolution of dry wood mass, the scenario in 2017 presents a minimum of 153,883 tons and a maximum of 645,219 tons, reaching 156,930 and 654,000 tons of dry wood, respectively, in 2024. Similarly, the average values increase from 399,552 tons to 405,468 tons of dry wood, representing a rise of 5914 tons over the 8-year recording period (Figure 11).

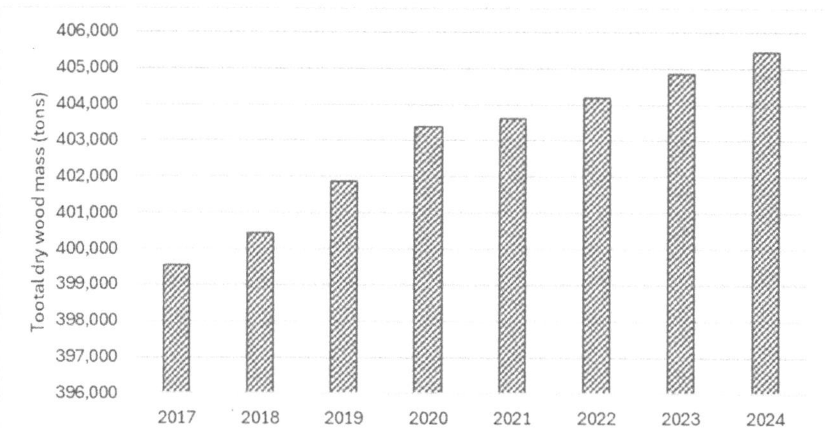

Figure 11. The annual wood mass per year according to the average scenario (city of Santiago).

Upon reviewing the age of wooden structures, it is possible to confirm that the largest stock of dry wood was produced during the second half of the 20th century, with most of the wood's age being concentrated in constructions from the 1960s to the 1990s (Figure 12). Thus, older municipalities located in the city center contain wood that has remained in the built environment for a longer time. In contrast, younger wood is represented by a large volume of buildings originating in municipalities that are part of the urban expansion in the late 20th and early 21st centuries.

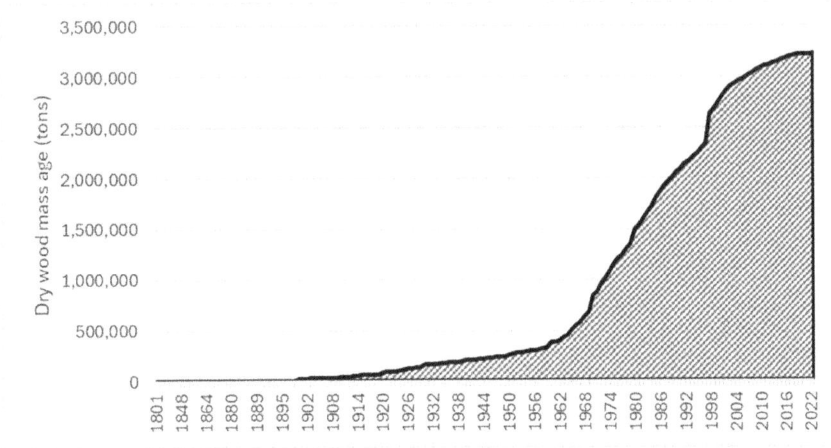

Figure 12. The age of dry wood mass in the average scenario (city of Santiago).

3.3. Biogenic Carbon and CO_2 Capture

If we examine the biogenic carbon associated with the mass of dry wood, it can be observed that it has grown steadily on average from 199.78 kt to 202.73 kt, with a cumulative increase of 1.5% in the total mass over the 8-year measurement period (Figure 13). Similarly, the scenario with the minimum biogenic carbon stored in 2024 accounts for 78.47 kt, while the maximum reaches 327.0 kt.

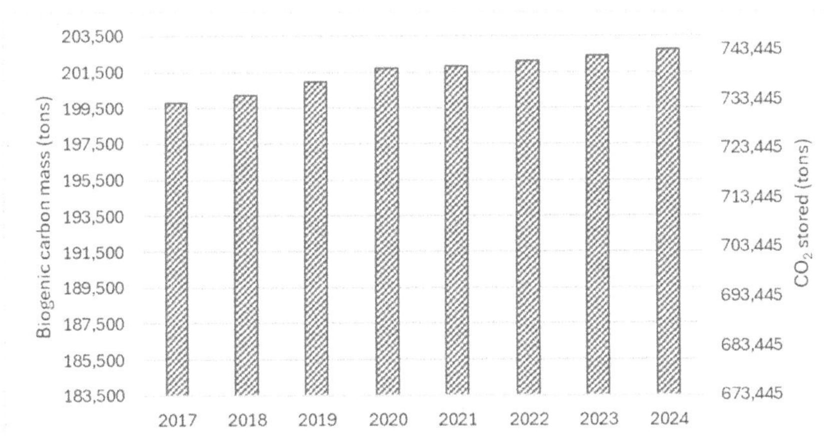

Figure 13. Annual biogenic carbon and CO_2 storage per year according to average scenario.

As expected, we observe the same trend with the CO_2 captured in wooden buildings, varying from 732.51 kt in 2017 to 743.35 kt in 2024 (Figure 13). Similarly, in 2024, there is a variation between a minimum scenario of 287.71 kt of CO_2 and a maximum of 1199.00 kt of CO_2 captured by the built wood biomass. The oldest CO_2 is concentrated in the central districts of the city, while the newest CO_2 is found in the urban periphery, with intermediate municipalities showing lower CO_2 capture in comparison (Figure 14).

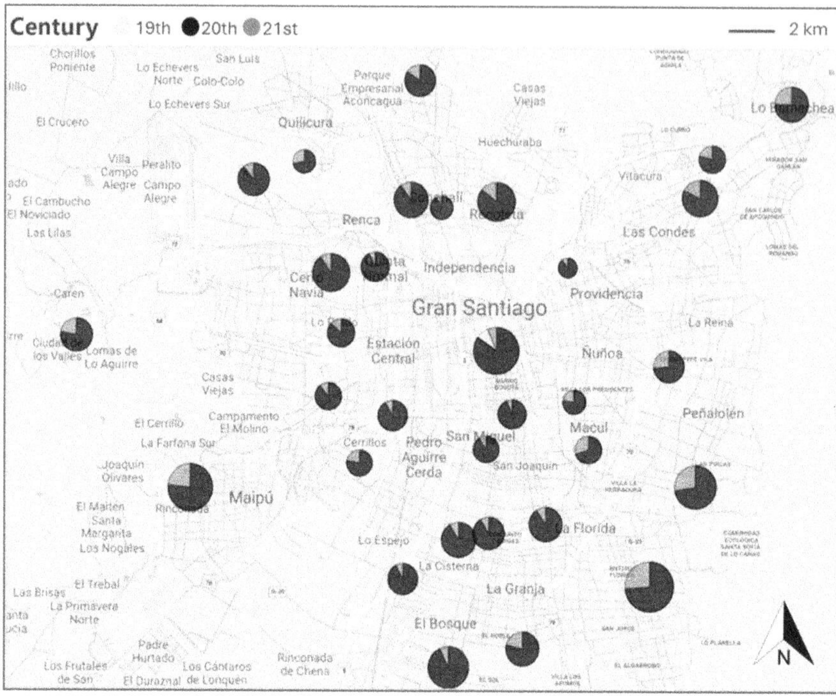

Figure 14. Temporal distribution of CO_2 storage per municipality (city of Santiago), made with Microsoft Power Bi Desktop version 2.138.1452.0 64-bit (November 2024).

4. Discussion

Regarding the wood MI results, which range from 54.69 kg/m^2 to 13.07 kg/m^2 with an average of 33.88 kg/m^2, previous studies have established values for housing, considering variations in wall and roof structures, between 15 kg/m^2 and 50 kg/m^2 [15]. Also, these studies show similar wall/roof wood usage ratios. Meanwhile, studies addressing the full set of wood-based materials have indicated a range of around 50 kg/m^2 of wood being used for framed buildings, while maintaining the roof-to-wall ratio identified in the Chilean case [17]. In this sense, the upward variation in some case studies is due to the fact that these analyses do not only include structural elements. However, an average wood MI of 37 kg/m^2 has been identified in the context of European and Asian studies [17] in accordance with the results obtained in this work.

On the other hand, it has been identified that historical dwellings based on other construction methods with high wood usage, such as stacked log systems, could have a wood intensity of up to 135 kg/m^2 on average for Swedish homes and 88 kg/m^2 on average for traditional Japanese homes [17]. However, these types of buildings are rare in the Chilean context as they were not recorded during the data collection period and should not represent a significant correction factor for the results presented. It is recognized, however, that in the case of Chilean historical buildings, wood usage may be slightly underestimated due to the height used in the quantification of material in the defined wall types. This is because a review of projects from that era suggests that the heights of historical buildings are greater than those of contemporary buildings [32]. Nevertheless, due to the high variability in this aspect, a conservative height was chosen in alignment with modern structures.

Regarding the quantification of wood buildings in urban environments, as expressed in the methodology, several assumptions had to be made to obtain the presented results. This may lead to underestimations or overestimations, which are addressed by the sensitivity analysis that considers minimum and maximum ranges for wood mass estimation. In this case, the representative structural elements and the wall density factor present an alternative to previous research, which used representative building types or random samples [15,17,23,37–39]. However, since the methodology considers only basic structural elements and not the entire wood construction, and it also underestimates the wood usage in historical structures, overall, these results tend to be conservative in relation to the total wood carbon storage.

With regard to the specific results of the case study, it is also noteworthy that there was a decline in timber construction in 2021, following continuous growth in previous years. This effect has been seen in other countries across the entire construction sector and can be easily explained by the COVID-19 pandemic and associated restrictions [40]. Furthermore, although a recovery in wood construction growth is evident between 2022 and 2024, this recovery is quite modest due to a local building sector crisis that the country has been experiencing for the past four years, related to investment issues and bureaucratic delays in permits [41]. This clearly demonstrates how global economic events and local conditions can affect wood construction growth and its capacity to store carbon.

Also, regarding the case study, it is interesting to note that historical parts of the city retain older wooden buildings, while urban expansion in the periphery through small wooden houses also contributes to increasing the wood stock. However, the ring surrounding the downtown area, where the modern city is more consolidated, shows less wood usage. This might imply that, in this case, urban densification through taller buildings has displaced wood in favor of more common construction systems for tall buildings, such as concrete or steel. Nevertheless, further study is needed to draw conclusions on this matter, and new trends in tall wooden buildings may also alter this tendency in the future.

It is also interesting to note that although the case study includes buildings that are centuries old, less than 0.4% of the wood-built surface area is older than the 19th century. In this regard, 15.5% of the wood-built surface area dates to the 21st century, while 84.1% corresponds to the 20th century, mostly from the second half of it. This is particularly notable considering that the case study area has never been affected by a major city fire like in London or Chicago, or by significant war events. Moreover, earthquakes that commonly affect the city tend to impact rigid structures like masonry more directly than wood. Nevertheless, old wooden structures are rare as they have been replaced by newer and taller structures made from other materials.

To understand the potential of wood buildings for carbon and CO_2 storage in the built environment, it is important to compare them with other carbon sinks, such as forests. In this regard, Chilean forest land covers 17.9 million hectares, of which 14.8 million are native forests and 3.1 million are plantations [20]. According to the Chilean national carbon inventory, based on the IPCC methodology, native forests have sequestered approximately 67,474 kt of CO_2 per year, while plantations have captured around 74,935 kt of CO_2 per year, averaged over the most recent eight-year reporting period from 2013 to 2021 [42]. This means that, on average, native forests have captured 5077 kg CO_2/ha per year, and plantations 21,440 kg CO_2/ha per year. However, most of the carbon captured by plantations is assumed to be released later during harvest. In contrast, over the eight years of study, the wooden surface area in the case study has stored 1002 kg CO_2/ha per year. However, this figure could increase in some peripheral municipalities to up to 7607 kg CO_2/ha per year, while municipalities undergoing a densification process could release up to 3794 kg CO_2/ha per year due to the replacement of low-height wood buildings by taller concrete structures. This indicates that wood constructions in the built environment have the potential to retain annual CO_2 amounts comparable to a native forest in the case study scenario, although this potential may be significantly influenced by building trends.

That said, through its wooden constructions and the timber products used in its structures, the city of Santiago has managed to retain a significant part of the biogenic carbon originating from forests. The results have shown that this carbon may have been stored for a few decades in some younger municipalities and for up to hundreds of years in the case of downtown Santiago. Furthermore, the rate of wood construction in some expanding municipalities indicates that the annual CO_2 storage from new buildings may even surpass the per-hectare capture of native forests in the Chilean case. This highlights a temporal and quantitative potential for annual CO_2 capture, suggesting, to some extent, the city's potential as a "Second Forest" in relation to its biogenic carbon storage capacity.

In a different context, while this work introduces an innovative approach to quantifying biogenic carbon in wood structures within Santiago's built environment, it is not without limitations. These are primarily due to the type of information recorded in the database utilized and the lack of data specifically tailored for studies of this nature. Despite these constraints, the sample size and its predominantly urban character help to minimize the impact of such discrepancies. However, less homogeneous areas, such as rural locations, may exhibit a higher margin of error. Similarly, urban areas where mass timber buildings with higher material intensity are a predominant typology may be difficult to differentiate from lightweight timber frame buildings with lower material mass.

Additionally, considering the limitations of the proposed methodology and its extrapolation to other case studies, Santiago, being a relatively young city with an urban expansion primarily concentrated in the second half of the 20th century and a detailed record of its built environment, may present a more favorable case compared to older cities in other regions. This advantage arises from the ages of buildings in those areas, where structures can be hundreds or even thousands of years old with little or no data available, the variability of construction systems influenced by diverse cultural exchanges, and partial or interrupted records caused by historical events such as major fires or wars.

Moreover, although this study attempts to address these uncertainties through sensitivity analyses, the results remain conservative and may not fully capture the potential for carbon storage in urban wooden structures. Future research should incorporate more detailed data and advanced modeling techniques to improve accuracy and applicability across diverse urban contexts. This is particularly relevant given the global development of public policies aimed at promoting the use of wood in construction to reduce emissions and enhance carbon capture within the sector. In this context, information on biogenic carbon storage will become increasingly significant, as policies like France's, which mandate that at least 50% of materials in state buildings must be bio-based, are expected to become more widespread. Similar initiatives have already been adopted in developed countries such as Croatia, Slovenia, and regions of Spain [43], and they are likely to emerge in developing nations like Chile in the coming decades.

5. Conclusions

This study presents a methodology for quantifying biogenic carbon and stored CO_2 in timber structures within the built environment. It has effectively assessed the city of Santiago's potential to function as a wood carbon sink or "Second Forest". Through the analysis of real estate tax databases, construction permits, timber building typologies, and carbon accounting models, this study quantifies the carbon stored in the city's timber structures, revealing storage volumes comparable to those of native forest ecosystems in some urban areas. Additionally, the analysis shows that urban expansion zones tend to increase the use of timber in their constructions, while areas undergoing vertical densification tend to reduce the presence of timber. This finding is significant in terms of CO_2 storage in the built environment and the impacts of vertical densification using traditional materials other than wood.

Moreover, maintaining a record of CO_2 stored in the built environment not only improves the understanding of buildings' carbon mitigation potential but also lays the groundwork for developing tools and public policies aimed at promoting low-emission, CO_2-storing construction practices. Data generated by studies like this are essential for creating regulatory frameworks that encourage timber construction, especially in taller buildings, thereby strengthening urban carbon storage strategies. Such methodologies also allow for extrapolation to other urban cases in countries lacking comprehensive material intensity records or timber use data within the built environment.

Finally, it is recommended that future studies focus on developing more accurate methodologies to quantify and characterize biogenic carbon storage in the built environment, particularly in regions where this methodology may not be suitable due to the stated limitations. Additionally, it would be valuable for future research to explore how urban planning trends influence the permanence of biogenic carbon in timber within the built environment, as well as to examine future scenarios for high-rise timber buildings and their potential impacts on urban biogenic carbon stock.

Author Contributions: Conceptualization, F.V.; methodology, F.V.; investigation, F.V. and W.B.; data curation, F.V.; writing—original draft preparation, F.V. and W.B.; writing—review and editing, F.V. and W.B.; supervision, W.B. All authors have read and agreed to the published version of the manuscript.

Funding: The authors want to acknowledge the support from Centro Nacional de Excelencia para la Industria de la Madera (CENAMAD, ANID BASAL FB210015). The authors also gratefully acknowledge the research support provided by CEDEUS under the research grant ANID FONDAP 1523A0004.

Institutional Review Board Statement: Not applicable.

Informed Consent Statement: Not applicable.

Data Availability Statement: Real state taxation data can be found at https://zeusr.sii.cl/AUT2000/InicioAutenticacion/IngresoRutClave.html?https://www4.sii.cl/sismunInternet6/?caller=DETALLE_CAT_Y_ROL_COBRO (accessed on 8 August 2024). Building permit data can be found at https://www.ine.gob.cl/estadisticas/economia/edificacion-y-construccion/permisos-de-edificacion (accessed on 8 August 2024). Chilean forest carbon inventory data can be found at https://snichile.mma.gob.cl/sector-uso-de-la-tierra-cambio-de-uso-de-la-tierra-y-silvicultura/ (accessed on 8 August 2024). Any additional data presented in this study are available upon request from the corresponding author.

Conflicts of Interest: The authors declare no conflicts of interest.

References

1. Churkina, G.; Organschi, A.; Reyer, C.P.O.; Ruff, A.; Vinke, K.; Liu, Z.; Reck, B.K.; Graedel, T.E.; Schellnhuber, H.J. Buildings as a Global Carbon Sink. *Nat. Sustain.* **2020**, *3*, 269–276. [CrossRef]
2. Hildebrandt, J.; Hagemann, N.; Thrän, D. The Contribution of Wood-Based Construction Materials for Leveraging a Low Carbon Building Sector in Europe. *Sustain. Cities Soc.* **2017**, *34*, 405–418. [CrossRef]
3. Zhang, L.; Yu, C.; Cheng, B.; Yang, C.; Chang, Y. Mitigating Climate Change by Global Timber Carbon Stock: Accounting, Flow and Allocation. *Renew. Sustain. Energy Rev.* **2020**, *131*, 109996. [CrossRef]
4. *UNE-EN 16449*; Wood and Wood-Based Products. Calculation of the Biogenic Carbon Content of Wood and Conversion to Carbon Dioxide. AENOR: Madrid, Spain, 2014.
5. Van Shaik, T.; Doraisami, M.; Martin, A.R. Carbon Fractions in Wood for Estimating Embodied Carbon in the Built Environment. *Sci. Total Environ.* **2024**, *921*, 171095. [CrossRef]

6. Ramage, M.H.; Burridge, H.; Busse-Wicher, M.; Fereday, G.; Reynolds, T.; Shah, D.U.; Wu, G.; Yu, L.; Fleming, P.; Densley-Tingley, D.; et al. The Wood from the Trees: The Use of Timber in Construction. *Renew. Sustain. Energy Rev.* **2017**, *68*, 333–359. [CrossRef]
7. *2023 Global Status Report for Buildings and Construction: Beyond Foundations—Mainstreaming Sustainable Solutions to Cut Emissions from the Buildings Sector*; United Nations Environment Programme: Nairobi, Kenya, 2024; ISBN 9789280741315.
8. Elhacham, E.; Ben-Uri, L.; Grozovski, J.; Bar-On, Y.M.; Milo, R. Global Human-Made Mass Exceeds All Living Biomass. *Nature* **2020**, *588*, 442–444. [CrossRef]
9. Watari, T.; Cao, Z.; Serrenho, A.C.; Cullen, J. Growing Role of Concrete in Sand and Climate Crises. *iScience* **2023**, *26*, 106782. [CrossRef]
10. Griscom, B.W.; Adams, J.; Ellis, P.W.; Houghton, R.A.; Lomax, G.; Miteva, D.A.; Schlesinger, W.H.; Shoch, D.; Siikamäki, J.V.; Smith, P.; et al. Natural Climate Solutions. *Proc. Natl. Acad. Sci. USA* **2017**, *114*, 11645–11650. [CrossRef]
11. Pugh, T.A.M.; Lindeskog, M.; Smith, B.; Poulter, B.; Arneth, A.; Haverd, V.; Calle, L. Role of Forest Regrowth in Global Carbon Sink Dynamics. *Proc. Natl. Acad. Sci. USA* **2019**, *116*, 4382–4387. [CrossRef]
12. Stephenson, N.L.; Das, A.J.; Condit, R.; Russo, S.E.; Baker, P.J.; Beckman, N.G.; Coomes, D.A.; Lines, E.R.; Morris, W.K.; Rüger, N.; et al. Rate of Tree Carbon Accumulation Increases Continuously with Tree Size. *Nature* **2014**, *507*, 90–93. [CrossRef]
13. Churkina, G.; Brown, D.G.; Keoleian, G. Carbon Stored in Human Settlements: The Conterminous United States. *Glob. Chang. Biol.* **2010**, *16*, 135–143. [CrossRef]
14. Reyes, N.; Rodríguez, B.; Wiegand, E.; Zilic, F.; Ramage, M.; Bukauskas, A.; Debnath, R.; Shah, D.U.; Colman, T.; De Wolf, C.; et al. *Achieving Zero Carbon Emissions in the Construction Sector*; Cambridge University Press: Cambridge, UK, 2021.
15. Nasiri, B.; Kaasalainen, T.; Hughes, M. Estimating the Material Intensity of Wooden Residential Houses in Finland. *Resour. Conserv. Recycl.* **2023**, *198*, 107142. [CrossRef]
16. Kalt, G. Carbon Dynamics and GHG Implications of Increasing Wood Construction: Long-Term Scenarios for Residential Buildings in Austria. *Carbon. Manag.* **2018**, *9*, 265–275. [CrossRef] [PubMed]
17. Gontia, P.; Nägeli, C.; Rosado, L.; Kalmykova, Y.; Österbring, M. Material-Intensity Database of Residential Buildings: A Case-Study of Sweden in the International Context. *Resour. Conserv. Recycl.* **2018**, *130*, 228–239. [CrossRef]
18. Arehart, J.H.; Hart, J.; Pomponi, F.; D'Amico, B. Carbon Sequestration and Storage in the Built Environment. *Sustain. Prod. Consum.* **2021**, *27*, 1047–1063. [CrossRef]
19. Gobierno de Chile. *Estrategia Climática de Largo Plazo de Chile*; Gobierno de Chile: Santiago, Chile, 2021.
20. Poblete Hernández, P. *Anuario Forestal 2023*; INFOR: New York, NY, USA, 2023.
21. del Río Arteaga, M.; Contreras Biekhert, J.L.; Rojas Pinochet, E.; Tokman Ramos, C.; Valenzuela Benavides, A.; Romero Cocuy, C.; Miranda Toledo, R.; Stipo Antiao, D.; Krell Loy, R. *Comisión Nacional de Productividad: Productividad En El Sector de La Construcción*; CNEP: Santiago, Chile, 2021; Available online: https://cnep.cl/estudios-finalizados/productividad-en-elsectorde-la-construccion/ (accessed on 23 October 2024).
22. Servicio de Impuestos Internos Detalle Catastral y Rol de Cobro. Available online: https://zeusr.sii.cl/AUT2000/InicioAutenticacion/IngresoRutClave.html?https://www4.sii.cl/sismunInternet6/?caller=DETALLE_CAT_Y_ROL_COBRO (accessed on 21 October 2024).
23. Schandl, H.; Marcos-Martinez, R.; Baynes, T.; Yu, Z.; Miatto, A.; Tanikawa, H. A Spatiotemporal Urban Metabolism Model for the Canberra Suburb of Braddon in Australia. *J. Clean. Prod.* **2020**, *265*, 121770. [CrossRef]
24. Kalcher, J.; Praxmarer, G.; Teischinger, A. Quantification of Future Availabilities of Recovered Wood from Austrian Residential Buildings. *Resour. Conserv. Recycl.* **2017**, *123*, 143–152. [CrossRef]
25. Instituto Nacional de Estadísticas Permisos de Edificación. Available online: https://www.ine.gob.cl/estadisticas/economia/edificacion-y-construccion/permisos-de-edificacion (accessed on 23 October 2024).
26. Estrella, X.; Guindos, P.; Almazán, J.L.; Malek, S.; Santa María, H.; Montaño, J.; Berwart, S. Seismic Performance Factors for Timber Buildings with Woodframe Shear Walls. *Eng. Struct.* **2021**, *248*, 113185. [CrossRef]
27. Carcamo, S.; Valdivieso, D. Morphologic Study of Hybrid Tall Building Towards an Interdisciplinary Design. In Proceedings of the World Conference on Timber Engineering 2023 (WCTE2023), Oslo, Norway, 19–22 June 2023.
28. Cárcamo Chávez, S.I.; Caamaño González, J.L.; Cid Olivares, A.A. Interdisiplinary Design for the Light Timber Frame System: Recommendation for a Good Structuration. In Proceedings of the World Conference on Timber Engineering, Vienna, Austria, 22–25 August 2016.
29. Berwart, S.; Estrella, X.; Montaño, J.; Santa-María, H.; Almazán, J.L.; Guindos, P. A Simplified Approach to Assess the Technical Prefeasibility of Multistory Wood-Frame Buildings in High Seismic Zones. *Eng. Struct.* **2022**, *257*, 114035. [CrossRef]
30. Chile, M.E.; Schmitt, C. *Lecciones de La Tradición Constructiva En Estudio de Obras Patrimoniales Del Siglo XX*; Pontificia Universidad Católica de Chile, Escuela de Arquitectura: Santiago, Chile, 2020.
31. Acevedo Oliva, R.M.; Broughton Weine, J.; Carrillo Zuñiga, O. *Construcción En Quincha Liviana Constructivos Sustentables de Reinterpretación Patrimonial*; Ministerio de Vivienda y Urbanismo: Santiago, Chile, 2019. Available online: https://csustentable.

minvu.gob.cl/wp-content/uploads/2020/03/CONSTRUCCION_CON_QUINCHA_LIVIANA_1a_edicion.pdf (accessed on 21 October 2024).
32. *Patrimonio Construido En Madera: Manual Para Su Valorización e Intervención*, 1st ed.; Schmitt, C., Ed.; Ediciones UC: Santiago, Chile, 2023.
33. Tomasi, J.; Bellmann, L. La quincha y los sistemas de entramados. *Estructuras* **2018**, *1*, 28–34. Available online: https://revistas.unc.edu.ar/index.php/estructuras/article/view/24741/24011 (accessed on 8 August 2024).
34. *NCh1198:2024*; Timber—Timber Constructions—Calculation. Instituto Nacional de Normalización—INN: Santiago, Chile, 2024. Available online: https://ecommerce.inn.cl/nch1198202488155 (accessed on 8 August 2024).
35. Hoxha, E.; Passer, A.; Saade, M.R.M.; Trigaux, D.; Shuttleworth, A.; Pittau, F.; Allacker, K.; Habert, G. Biogenic Carbon in Buildings: A Critical Overview of LCA Methods. *Build. Cities* **2020**, *1*, 504–524. [CrossRef]
36. Pittau, F.; Krause, F.; Lumia, G.; Habert, G. Fast-Growing Bio-Based Materials as an Opportunity for Storing Carbon in Exterior Walls. *Build. Environ.* **2018**, *129*, 117–129. [CrossRef]
37. Han, J.; Xiang, W.N. Analysis of Material Stock Accumulation in China's Infrastructure and Its Regional Disparity. *Sustain. Sci.* **2013**, *8*, 553–564. [CrossRef]
38. Tanikawa, H.; Fishman, T.; Okuoka, K.; Sugimoto, K. The Weight of Society over Time and Space: A Comprehensive Account of the Construction Material Stock of Japan, 1945–2010. *J. Ind. Ecol.* **2015**, *19*, 778–791. [CrossRef]
39. Tanikawa, H.; Hashimoto, S. Urban Stock over Time: Spatial Material Stock Analysis Using 4d-GIS. *Build. Res. Inf.* **2009**, *37*, 483–502. [CrossRef]
40. ILO Sectoral Brief. *Impact of COVID-19 on the Construction Sector*; ILO: Geneva, Switzerland, 2021. Available online: https://researchrepository.ilo.org/esploro/outputs/encyclopediaEntry/Impact-of-COVID-19-on-the-construction/995219336302676 (accessed on 8 August 2024).
41. Forbes Staff Iconstruye: Industria de La Construcción En Chile Atraviesa La Crisis Más Aguda En 30 Años. Available online: https://forbes.cl/actualidad/2024-08-12/iconstruye-industria-de-la-construccion-en-chile-atraviesa-la-crisis-mas-aguda-en-30-anos (accessed on 24 October 2024).
42. Ministerio del Medio Ambiente Sector UTCUTS. Available online: https://snichile.mma.gob.cl/sector-uso-de-la-tierra-cambio-de-uso-de-la-tierra-y-silvicultura/ (accessed on 24 October 2024).
43. Chalupovitsch, M.; Buhary, N.; Gesley, J.; Figueroa, D.; Umeda, S.; Hofverberg, E.; Feikert-Ahalt, C. *Mass Timber Construction*; The Law Library of Congress, Global Legal Research Directorate: Washington, DC, USA, 2022. Available online: https://www.loc.gov/item/2022666107/ (accessed on 24 August 2024).

sustainability

MDPI

Article

A New Approach to Improving the Seismic Performance of Existing Reinforced Concrete Buildings Using Laminated Timber

Yusuf Yıldız and Fethi Şermet *

Department of Civil Engineering, Igdir University, Igdir 76000, Türkiye; ce.yusufyildiz@gmail.com
* Correspondence: fethi.sermet@igdir.edu.tr

Abstract

Following recent earthquakes in Van, Istanbul, Izmir, and Kahramanmaraş, concerns have once again been raised regarding whether existing buildings possess adequate seismic performance and the necessity of strengthening those that do not. A common theme in all related discussions is how to improve the seismic resilience of the existing building stock most efficiently and cost-effectively. In particular, seismic retrofitting efforts should be accelerated for residential buildings in areas where urban transformation has not been feasible due to low added value, as well as for public buildings in smaller settlements and school or dormitory structures in rural towns and villages. In this study, the seismic performance of a reinforced concrete (RC) frame was evaluated using the nonlinear single-mode pushover analysis method in accordance with the Turkish Building Earthquake Code (TBEC). For frames with inadequate performance, a retrofitting method was proposed using glued laminated timber (glulam), a renewable and sustainable material, as diagonal bracing. This intervention aimed to improve the structural performance to an acceptable level specified by the code. The results indicated that glulam braces can effectively enhance the seismic performance of RC buildings and may be considered a viable solution for this purpose.

Keywords: earthquake; reinforced concrete buildings; performance analysis; glued laminated timber (glulam); sustainability

1. Introduction

In countries with high seismic risks, such as Turkey, a significant portion of the existing building stock consists of RC buildings with weak structural systems that have not received engineering services. Therefore, strengthening solutions are vital to increase the seismic safety of these structures. To date, traditional strengthening techniques, such as reinforced concrete jacketing, steel bracing, and shear wall addition methods, have been widely used.

A fundamental vulnerability of frame systems and one of the primary reasons for building collapse during earthquakes is inter-story drift. In braced frame systems, these inclined elements are subjected to tension and compression, depending on the design and construction details, thereby helping to reduce the relative story drifts [1]. Although braced frame systems generally offer lower strength than shear wall systems, they tend to provide superior ductility [2]. Strengthening RC buildings with bracing elements is a common practice. Since it is subsequently added to an existing structure, the steel bracing method is generally preferred. In some cases, RC buildings are also strengthened using precast RC bracing elements [3–5]. Another method, the application of reinforced concrete

https://doi.org/10.3390/su17177690

jacketing to strengthen existing columns, can significantly improve the compressive behavior of these elements [6–10]. However, this method has some disadvantages, such as a significant increase in structural weight and occupied area, and it requires a complex construction process. In situations where the load-bearing capacity of damaged columns must be significantly increased in a short time, traditional strengthening techniques may be insufficient to meet these requirements. Another traditional strengthening method involves the use of RC shear walls, which play a fundamental role in the structural system design of buildings, especially in seismic regions [11,12]. These walls are critically important for resisting horizontal loads and enhancing the structural integrity and safety of buildings during earthquakes [13,14]. Although traditional strengthening methods, such as steel braces, fiber-reinforced polymers, and ultra-high-performance concrete, have been applied to improve the seismic performance of these structures, these techniques have limitations, such as durability, application difficulties, and cost [15–20].

Traditional strengthening methods often involve long application times, add extra weight to the structure, and are costly and labor-intensive. Due to these limitations, modern strengthening techniques have been developed. Energy dissipators (liquid viscous dampers), buckling-restrained braces (BRBs), and fiber-reinforced polymer (FRP) composites are notable examples in this regard. With advances in research on controlling seismic vibrations, the addition of external control systems has emerged as a prominent solution for reducing the impact of earthquake vibrations on structures [21–23]. Passive energy dissipation devices are an effective solution that provides additional damping to structures without the need for an external energy source. The addition of passive energy dissipation devices reduces damage to the main structural system by decreasing the need for inelastic energy dissipation, rather than increasing the strength of the structure [24–27]. Significant developments, including viscoelastic, liquid viscous, frictional, and metallic dampers, have occurred in the application of passive devices to mitigate severe earthquake and wind effects [28]. In addition to passive damping systems, alternative strengthening methods that act as part of the structural system also exist. One such method, BRBs, not only increases the energy dissipation capacity of the structural system but also offers ductile behavior by eliminating the risk of buckling. Frame systems strengthened with BRBs can significantly reduce seismic structural demands and residual displacements compared with conventional moment frames [29–35]. Furthermore, many studies have shown that the residual displacements occurring in BRB-strengthened frames are well below the limit values that would allow the repair of the structure [36–40]. In essence, the collapse resistance capacity and seismic reliability of traditional moment frame systems can be significantly enhanced using BRBs.

However, increasing sustainability goals have encouraged a shift towards environmentally friendly materials in the construction sector. In this context, engineered wood, such as glued laminated timber, stands out due to its environmental and structural advantages. The prominent characteristics of wood are its renewability, low carbon footprint, lightness, and workability. Wood has a lower embodied impact than concrete during production and construction processes. Replacing concrete with wood can reduce a building's carbon footprint by 25–75% [41,42]. Shin B. and Kim S. stated that wood is more advantageous than concrete structures in reducing environmental impacts and suitability for building reuse, utilizing the building circularity method (BCI) [43]. Existing research has highlighted the sustainability of using wood. It has been determined that converting the original structural system to a cross-laminated timber (CLT) structural system can reduce lifecycle sustainability impacts by over 51% [42]. In addition to these benefits, wood can be sourced sustainably from nature and is easier to process than other materials, making it an inherently sustainable construction material. Structurally, wood possesses a higher strength-to-weight ratio than concrete and has a density that is about one-fifth that of concrete [44]. In recent years, the development of engineered wood (EW), also known as mass timber, such as glulam or CLT, with high structural rigidity, has enabled its diverse use in mid- and high-rise buildings [44,45]. Jayalath et al. [46] demonstrated that CLT mid-rise residential buildings in Australia could reduce life cycle greenhouse gas emissions by 30% compared to RC structures. Pierobon et al. [41] reported that in commercial buildings, a CLT hybrid building reduced the Global Warming Potential (GWP) by an average of 26.5% when compared to RC structures.

Various experimental and numerical studies have been conducted on timber elements and timber connections. Chen et al. [47] presented a novel approach for using laminated wood as a filler material within the square steel tube of the core of BRBs, replacing mortar or concrete. Their study shaped this new BRB design based on various parameters and evaluated its cyclic behavior (mechanical properties, fatigue performance, and energy dissipation capacity) through experimental tests. They suggested that while glulam may not have the same high strength as concrete, it could offer a suitable, economical, and sustainable alternative, especially for laminated timber frame structures or low-rise, earthquake-resistant structures subjected to vertical loads. Dorn et al. [48] conducted experiments on 64 steel-wood single-dowel connections. Connections using high-density wood showed a higher load-bearing capacity but also exhibited more brittle behavior. Meng et al. [49] revealed the significant role of gravitational potential energy in the seismic performance of traditional timber structures, stating that this energy conversion mechanism enables traditional timber structures to resist large earthquakes. Chen and Popovski [50] mathematically demonstrated the effect of connection properties on the system performance in braced timber frames, highlighting that factors such as connection design and column stiffness play a crucial role in system ductility. In a study by Sustersic and Dujic, the seismic performance of a previously full-scale modeled and dynamically analyzed RC structure was improved by applying CLT panels to its exterior walls. The study revealed that these panels significantly increased the structural rigidity and strength of the building [51]. An experimental study by Juhász et al. on the damage behavior of laminated timber-concrete composite beams showed that such composite systems allow for spanning long spans while maintaining sustainability and improving structural efficiency [52]. Selçukoglu and Zwicky emphasized that composite slabs with a concrete top layer and laminated timber bottom layer exhibit better structural performance and vibration behavior than traditional timber slabs, specifically highlighting the significant ductility of laminated timber under compression [53]. A study published by Öztürk et al. stated that structural system elements produced with laminated timber technology are components of a modern and advanced building system, further emphasizing that these elements can be produced in Turkey using local timber and adhesives [54]. Similarly, Sandoli et al.'s work stated that CLT structures offer a strong alternative to heavier structural systems in terms of seismic resistance and environmental sustainability [55]. Furthermore, Furukawa and his team evaluated whether the flexural stiffness and load-bearing capacity of composite beams formed from H-section steel beams, using CLT floor panels with high-strength friction-type bolted connections, could be simply predicted from a design perspective based on the properties of individual components and connection details [56].

In recent years, various studies have explored the use of glued laminated timber elements to strengthen existing RC buildings. However, the literature in this field is still limited, and there is a shortage of practical application examples. This study aims to develop an alternative strengthening method to improve the seismic performance of existing RC buildings with inadequate earthquake resistance. Additionally, it seeks to present a sample approach that may pave the way for future research exploring the use of timber elements in this field, offering both aesthetic value and ease of application. Within this scope, the seismic performance level of a two-dimensional RC frame without shear walls was first assessed using a nonlinear pushover analysis method. Subsequently, the system was retrofitted with pinned glued-laminated timber braces in various configurations. The seismic performance of the strengthened system was re-evaluated, and the contribution of the glulam braces to overall structural performance was examined. Although numerous studies in the literature focus on the comparative performance analyses of RC structures and the use of materials such as fiber-reinforced polymers or steel braces for seismic strengthening, research addressing the use of laminated timber elements to enhance the seismic performance of RC buildings remains limited.

2. Materials

2.1. Timber

Timber primarily consists of cellulose (≈60%), lignin (≈30%), and other sugar-based compounds. Its load-bearing capacity stems from longitudinally aligned cellulose fibers bonded by lignin [57]. Glued laminated timber, one of the earliest engineered wood products, is produced by gluing timber laminations in the same direction. Glulam can be classified as homogeneous ("h") if all laminations have similar strength, or combined ("c") if laminations of different strength classes are used.

2.1.1. Strength and Stiffness of Timber

Timber is an orthotropic material with distinct mechanical properties in three principal directions: longitudinal (L), radial (R), and tangential (T) directions. Its compressive strength is significantly higher parallel to the grain (L) than perpendicular to it, making loading direction a critical factor in design. While radial and tangential differences are often negligible in structural applications, accurate modeling requires the elastic moduli (E_L, E_R, E_T), shear moduli (G_{LR}, G_{LT}, G_{RT}), and Poisson's ratios (ν_{LR}, ν_{RL}, etc.) for these directions [58].

Considering the similarity between radial and tangential directions and assuming equal Poisson's ratios, the 12 orthotropic parameters can be reduced to six: $E_{//}, E_{\perp}, G_{//}, G_{\perp}$, $\upsilon_{//}, \upsilon_{\perp}$. In this study, the full orthotropic properties of glulam were defined for completeness; however, since the braces were modeled as axial link elements, only the axial behavior governed by the fiber direction properties was relevant. Orthotropic parameters were used solely for capacity checks under axial loading.

2.1.2. Timber Material Model

Since timber is an anisotropic material, it is challenging to accurately model its mechanical behavior. In this study, the glulam braces to be used will be modeled to carry only axial forces. Therefore, the material model shown in Figure 1 (Elastic Perfectly Plastic) will be used, with the mechanical properties of timber in the direction parallel to the fibers, specifically the modulus of elasticity and characteristic strengths.

2.2. Concrete and Reinforcement Steel

For the material models of concrete and reinforcing steel, the material models provided in Appendix 5A of the "Principles for the Design of Buildings Under Earthquake Effects" annexed to the Turkish Building Earthquake Code (TBEC 2018) are used [59].

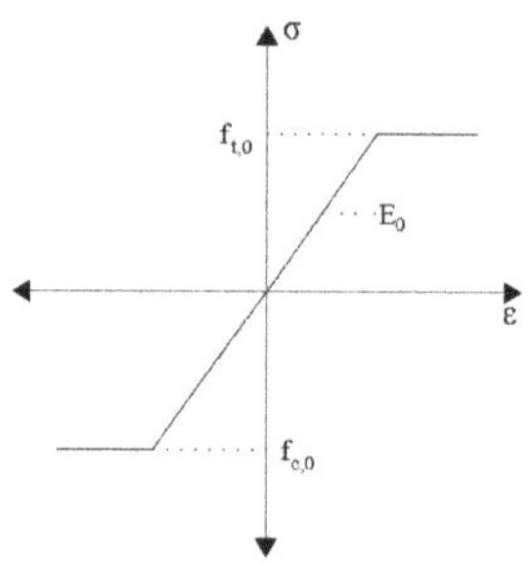

Figure 1. Stress-strain curve of the uniaxial material model for timber.

2.2.1. Confined and Unconfined Concrete Material Models

The Poisson's ratio for concrete, υ_c, is assumed to be 0.2. The uniaxial compressive strength of unconfined concrete and its corresponding strain are denoted as f_c and ε_c, respectively. The value of ε_c typically, it ranges from 0.002 to 0.003 [60]. When concrete is subjected to lateral confining pressure, its uniaxial compressive strength, f_{cc}, and the corresponding strain, ε_{cc}, become significantly higher compared to unconfined concrete. The compressive strengths of confined and unconfined concrete, f_c and f_{cc}, can be calculated using the formulas provided in Equation (5)A-1 and Equation (5)A-2 of TBEC-2018 [59].

2.2.2. Reinforcement Steel Material Model

The stress-strain curve of reinforcing steel has been assumed to be elastic-perfectly plastic. The elastic modulus of reinforcing steel, E_S, is taken as 200 GPa. To be used in performance assessment under nonlinear analysis based on displacement demands, the stress–strain relationships for reinforcing steel are defined according to Equation (1) [59].

$$
\begin{aligned}
f_s &= E_s.\varepsilon_s && (\varepsilon_s \leq \varepsilon_{sy}) \\
f_s &= f_{sy} && (\varepsilon_s < \varepsilon_{sy} \leq \varepsilon_{sh}) \\
f_s &= f_{su} - (f_{su} - f_{sy})\frac{(\varepsilon_{su} - \varepsilon_s)^2}{(\varepsilon_{su} - \varepsilon_{sh})^2} && (\varepsilon_{sh} < \varepsilon_s \leq \varepsilon_{su})
\end{aligned}
\quad (1)
$$

3. Methods

3.1. Determination of Analysis Models

In the analyses, a 2D existing RC frame consisting of a ground floor + 4 normal floors with a 3 m story height and a 5 m beam span, and reinforced models created by adding glulam braces with a central diagonal and inverted V arrangement to the edge spans of the RC frame (without earthquake shear walls) were used.

In the reinforced models, the effect of using diagonally placed braces and braces with an inverted V arrangement on the increase in rigidity and the resulting changes in the system was investigated. An example of an inverted V-glulam brace arrangement is shown in Figure 2.

3.2. Determination of Earthquake Performance by Nonlinear Analysis

Using the nonlinear analysis method, specifically single-mode pushover analysis, the 2D existing frame was analyzed to determine the damage zones, earthquake performance demand, and earthquake performance level based on this demand.

Subsequently, glulam braces were added to the edge spans of the frame system, which were subjected only to axial loads, and the nonlinear single-mode pushover analysis was repeated. As a result, the damage zones, earthquake performance demands, and the earthquake performance level based on this demand were determined for the reinforced frame.

For the performance analysis, a plastic hinge model based on accumulated plastic behavior was used for the RC frame elements, while for the glulam braces, brittle (force-based) hinges were defined to check the axial compression/tension capacity.

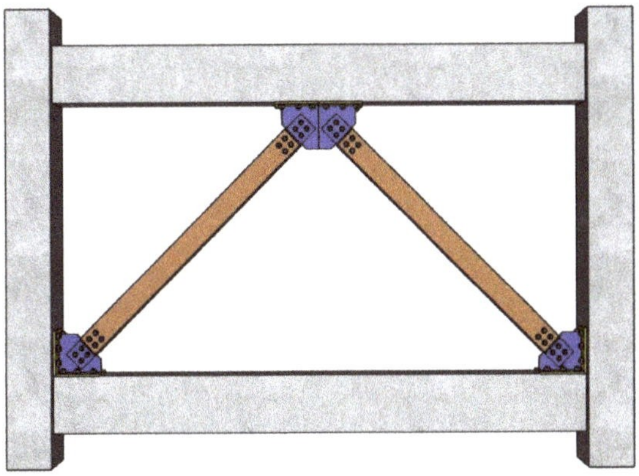

Figure 2. Example of Inverted V Glulam Brace Arrangement.

3.3. *Evaluation of System Performance*

The calculation principles to be applied in the evaluation of the performance of existing buildings and building-type structures under earthquake effects are defined in Chapter 15 of TBEC-2018. Within the scope of this study, the principles in this chapter were used for the frame models, whose performance under earthquake effects was evaluated. In cases where no specific information was available, the closest applicable conditions were approximated using generally accepted approaches or methods, in accordance with the principles of this chapter.

3.3.1. Knowledge Level

Depending on the scope of information obtained from the assessment of the existing condition of buildings, knowledge level coefficients defined in Section 15.2.12 of TBEC-2018 are used [59].

Within the scope of this study, considering disadvantaged structures such as residential buildings in regions where urban transformation projects cannot be implemented due to low added value, public buildings in small settlements, and schools and dormitories in rural or town areas, the limited knowledge level coefficient (0.75) was used. This coefficient was applied to the material strengths and incorporated into the capacity calculations of the structural-system elements.

3.3.2. Damage Boundaries and Damage Zones

For ductile elements, the section damage state and damage limits are defined as Limited Damage (LD), Controlled Damage (CD), and Pre-Collapse Damage (PCD), along with their corresponding threshold values. However, these classification states and limit values are not applicable to elements that fail in a brittle manner. The damage limits and damage zones are illustrated in Figure 3.

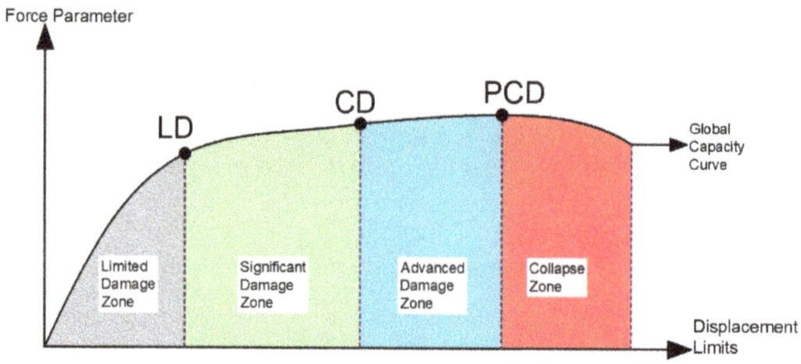

Figure 3. Damage limits and damage zones.

3.3.3. General Principles and Rules Regarding Earthquake Calculations

In TBEC-2018, both linear and nonlinear analysis methods can be used to evaluate the seismic performance of existing structures. Although it is stated that both analysis methods are applicable, it should not be expected that these fundamentally different approaches will yield similar results.

The principles and rules to be applied for both linear and nonlinear methods are outlined in Section 15.4 of TBEC-2018 [59]. In this study, the Nonlinear Static Single-Mode Pushover Method was used.

3.4. *Design of Structural Timber Elements*

As only glulam braces subjected solely to axial tension or compression loads are used in this study, only the effects of axial loads are addressed in this section. Calculations will be carried out in accordance with Eurocode 5: Design of Timber Structures [61].

3.4.1. Force Limits for Laminated Timber Braces

For the glulam braces under axial loading, force-controlled brittle hinges will be defined. These hinges will be used to control both the axial tension and axial compression capacities under seismic effects by setting force control limits at 50%, 75%, and 100% of their capacities.

These defined force limits allow the capacity states of the glulam braces to be checked at the performance point of the system.

3.4.2. Ultimate Modulus of Elasticity

In structural analysis, when stiffness plays a critical role, the influence of load duration and moisture content on deformations must be considered. To account for these effects, the final (adjusted) modulus of elasticity is used in both the serviceability and ultimate limit states [61].

In this study, a reduction factor was applied to the moment of inertia instead of directly modifying the modulus of elasticity (E). This approach preserves the inherent properties of the material while realistically reducing the section stiffness for an accurate analysis.

$$E_{mean,fin} = \frac{E_{mean}}{1 + \Psi_2 \cdot k_{def}} \tag{2}$$

Here;

$E_{mean,fin}$: Ultimate modulus of elasticity,

E_{mean}: Average modulus of elasticity,

Ψ_2: Semi-permanent load combination factor,

k_{def}: It is the adjustment factor to take into account the effect of humidity on deformation.

4. Analysis

Within the scope of this study, SAP2000 V.23, a widely used finite element-based analysis and design program in structural engineering, was utilized. It allows the modeling of RC, steel, and composite systems, as well as the performance of linear/nonlinear and static/dynamic analyses. The structural behavior under loads, such as earthquakes, wind, and temperature, can be examined in detail [62].

Analyses were conducted on a 2D RC frame with a floor height of 3 m and a beam span of 5 m, consisting of a ground floor plus 4 normal floors without seismic shear walls. To enhance the system, glulam braces with a central diagonal and reverse V arrangement were added to the edge spans of the frame. Glulam of GL24H strength class with dimensions of 20 × 20 cm was used for the braces.

As a result of all the analyses, the changes induced by the glulam braces and the different bracing configurations were examined and reported.

4.1. Analysis Models

The names of the analysis models are listed in Table 1. The framework models are shown in Figures 4–6.

4.2. Material Properties

The material properties of the concrete (C16/20) and reinforcement steel (S420) used in the frames are shown in Tables 2 and 3. The strength values and properties of the existing concrete and reinforcement steel have been reduced by considering the information level factor (0.75). This ensures that the capacities of the elements are determined according to the information level.

Table 1. Naming of analysis models.

Number of Floors	Unbraced Frame Model	Diagonal Braced Frame Model	Inverted V Braced Frame Model
Ground + 4 Normal Floors	Model 01	Model 01D	Model 01V

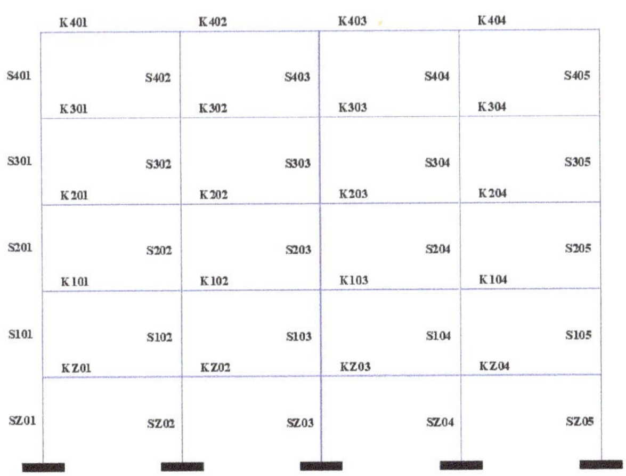

Figure 4. Appearance of Model 01.

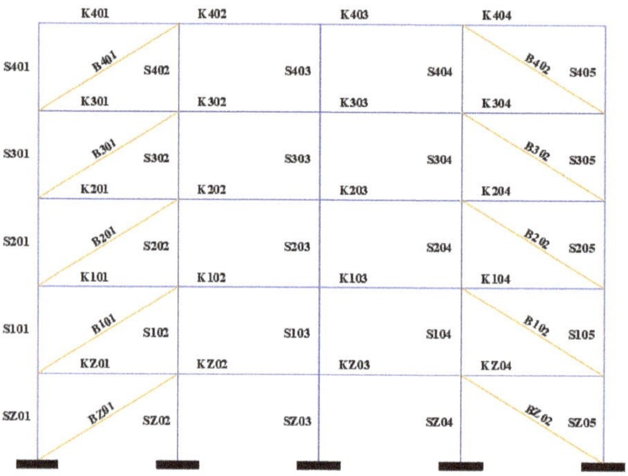

Figure 5. Appearance of Model 01D.

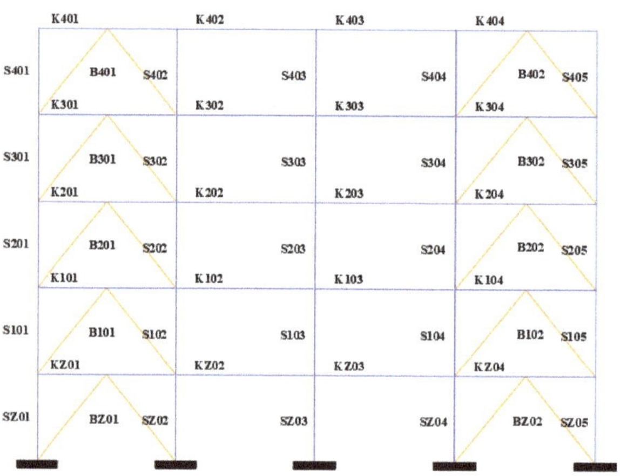

Figure 6. Appearance of Model 01V.

Table 2. Material properties of concrete.

Property	Value
$E(kN/m^2)$	27×10^6
υ	0.2
$f_c(kN/m^2)$	16×10^3
ε_{co}	2×10^{-3}
ε_{cu}	5×10^{-3}

Table 3. Material properties of the reinforced steel.

Property	Value
$E(kN/m^2)$	2×10^8
$f_y(kN/m^2)$	420×10^3
$f_u(kN/m^2)$	550×10^3
ε_{sh}	8×10^{-3}
ε_{su}	0.08

The unit deformations for compression and tension capacities of GL24H have been calculated according to the material model in Section 2.1.2, and the properties of GL24H are shown in Table 4.

Table 4. Material properties of GL24H laminated timber.

Property	Value
$f_{t,0,k}\,(N/mm^2)$	19.20
$f_{c,0,k}\,(N/mm^2)$	24.00
$E_{0.05}\,(N/mm^2)$	9600.00
$G_{0.05}\,(N/mm^2)$	540.00
$E_{0,mean}\,(N/mm^2)$	11,500.00
$E_{90,mean}\,(N/mm^2)$	300.00
$G_{0,mean}\,(N/mm^2)$	650.00
$G_{90,mean}\,(N/mm^2)$	65.00
$\rho_{mean}\,(kg/m^3)$	420.00
$\varepsilon_{t,0}$	1.7×10^{-3}
$\varepsilon_{c,0}$	2.1×10^{-3}

4.3. Section Properties

The section definitions for the concrete and reinforcement steel have been created using the Section Designer tool in SAP2000 V.23. Thus, all stress-strain relationships, moments, curvatures, and rotation values for the relevant materials have been obtained using the program.

4.3.1. Reinforced Concrete Column and Beam Sections

All frame columns have the same reinforcement arrangement and dimensions (40 × 40 cm). The column section is modeled with a 2.5 cm unwrapped shell concrete outside and a 35 × 35 cm confined core concrete inside, with 12Φ14 reinforcement. The column confinement reinforcement has a diameter of Φ8, and the spacing between the confinement reinforcement is 20 cm.

All frame beams also have the same reinforcement arrangement and dimensions (25 × 50 cm). The beam section is modeled with a 2.5 cm unwrapped shell concrete outside and a 20 × 45 cm confined core concrete inside, with 6Φ12 reinforcement. The beam confinement reinforcement has a diameter of Φ8, and the spacing between the confinement reinforcement is 20 cm. The cross sections of the columns and beams are shown in Figure 7.

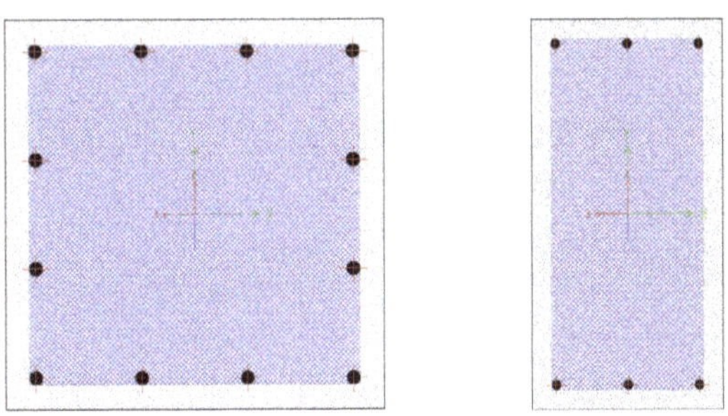

Figure 7. 40 × 40 cm column and 25 × 50 cm beam cross sections (The blue shaded area shows the confined part, the red shaded area shows the unconfined part of column and beam).

4.3.2. Glulam Cross Sections

Timber cross sections are defined as 20 × 20 cm in size according to the properties of laminated timber of GL24H strength class. The effective cross-sectional area value multiplier mentioned in Section 3.4.2 is calculated as follows:

$$E_{mean,fin} = \frac{E_{mean}}{1 + \Psi_2.k_{def}} = \frac{11,500.00}{1 + 0.3 \times 0.8} = 9274.19\text{N/mm}^2$$

$$Effective\ section\ area\ multiplier = \frac{E_{mean,fin}}{E_{mean}} = \frac{9274.19}{11,500.00} \cong 0.81$$

4.4. Load and Mass Source

The fixed and live distributed loads from the floors and walls applied to the frame beams are shown in Table 5. The material unit weights were included in the calculations with a unit weight multiplier of 1.0 when defining fixed loads in SAP2000 V.23.

Table 5. Dead and live loads act on the beams.

Type of Load	Value
G(kN/m)	25.00
Q(kN/m)	5.00

The mass source is defined with the G + 0.3Q combination, where the dynamic load mass participation factor is taken as 0.3.

4.5. Earthquake Load Parameters

The earthquake load parameters used in the analyses are shown in Table 6.

Table 6. Earthquake load parameters.

Soil Class	S_S	S_1	S_{DS}	S_{D1}	PGA	PGV	T_L
ZC	1.957	0.517	2.348	0.767	0.065	0.327	6.00

4.6. Definition of Force Limits for Glulam Cross-Braces

The calculated axial tension and compression capacity values for the 20 × 20 cm glulam cross-braces under axial load are shown in Table 7. To verify the calculated axial tension and compression capacities, the acceptance criteria were defined at 50%, 75%, and 100% of their capacities, and force-controlled (force-controlled) hinges were defined for the glulam cross-braces in SAP2000 V.23.

Table 7. Axial tension and compression capacities of glulam cross-braces.

Values	Diagonal Brace (GL24H)	Inverted V Brace (GL24H)
$f_{t,0,d}\,(\text{N/mm}^2)$	16.90	
$f_{c,0,d}\,(\text{N/mm}^2)$	21.12	
ε_t	0.0017	
ε_c	0.0021	
$N_{t,0,Rd}\,(\text{N/mm}^2)$	675.84	675.84
$N_{c,0,Rd}\,(\text{N/mm}^2)$	305.84	595.22

4.7. *Analysis with Constant Single Mode Pushover Method*

According to TBEC-2018, Section 15, the application principles for Nonlinear Analysis Methods, specifically the Single Mode Pushover Method, used in the evaluation of existing buildings, are defined in Section 5.6. Analyses were conducted following these principles.

The displacements and internal force demands obtained from the analyses were compared with the displacement and internal force capacities to perform a structural performance assessment at the section and frame levels. Based on this, the performance level was determined.

4.8. *Validation*

Obtaining the Performance Point and Capacity Curve of the System

Using the methods defined in Annex 5B of the TBEC 2018 for the fixed single-mode pushover analysis, the performance point and capacity curve of Model 01 were calculated, and the results are presented below. The linear earthquake spectrum for the system was obtained based on the spectral values provided in Table 6. According to the modal analysis performed in SAP2000 V.23, the period of the first mode of vibration is 1.168 s.

The maximum displacement of the modal single-degree-of-freedom system (nonlinear spectral displacement) was obtained by superimposing the modal capacity curve and the linear earthquake spectrum, as shown in Figure 8.

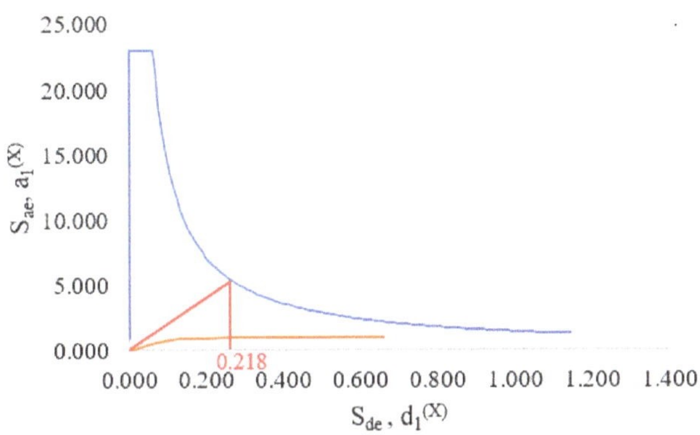

Figure 8. Obtaining the maximum displacement of the system (The blue color shows the linear earthquake spectrum, the orange color shows the modal capacity curve).

Using the maximum displacement, the system's peak displacement (the peak displacement demanded by the earthquake) was determined to be 0.278 m.

$$u_{Nx1}^{(p)} = \phi_{Nx1}^{(1)} \cdot d_{1,max}^{(X)} \cdot \Gamma_1^{(X,1)} = 0.278 \text{ m}$$

The peak displacement of the system can be obtained by plotting the linear earthquake spectrum and modal capacity diagram together, or it can be directly calculated using the following equations:

$$d_{1,\,max}^{(X)} = S_{di}(T_1) = C_R \cdot S_{de}(T_1)$$

$$T_1 > T_B \text{ since it is } C_R = 1$$

In this case, the maximum displacement is equal to the elastic design spectral displacement.

$$d_{1,\,max}^{(X)} = S_{di}(T_1) = 1 \times S_{de}(T_1) = \frac{T^2}{4\pi^2} \cdot g \cdot S_{ae}(T)$$

Horizontal elastic design spectral acceleration,

$$T_B \leq T \leq T_L \text{ since it is } S_{ae}(T) = \frac{S_{D1}}{T_1}$$

is calculated as.

$$d_{1,\ max}^{(X)} = \frac{T^2}{4\pi^2} \cdot g \cdot S_{ae}(T) = \frac{1.142^2}{4\pi^2} \times 9.80665 \times \frac{0.767}{1.142} \cong 0.218\ m$$

And as a result, the top displacement of the system is,

$$u_{Nx1}^{(p)} = \phi_{Nx1}^{(1)} \cdot d_{1,max}^{(X)} \cdot \Gamma_1^{(X,1)} = 0.0772 \times 0.218 \times 16.51 = 0.278\ m$$

is found as.

The system was pushed again to the obtained peak displacement value, and a capacity curve consisting of displacement–base shear force coordinates was obtained. The performance point of the system was determined as the intersection point of the peak displacement and base shear force.

According to the "ATC-40: Seismic Evaluation and Retrofit of Concrete Buildings" guideline [63] available in SAP2000 V.23, it is possible to obtain performance point and capacity curve values close to those of TBEC-2018 using the capacity spectrum method.

For a peak displacement of 0.278 m, the base shear force obtained using TBEC-2018 is 272.29 kN. According to the adapted ATC-40 guideline, the base shear force corresponding to a peak displacement of 0.289 m is calculated by the program as 271.60 kN. The base shear forces are 99.75% close to each other. In the performance comparisons of the models, a high correlation was observed between the performance points and capacity curve values obtained from SAP2000 V.23 and those calculated using the TBEC-2018 design principles.

5. Findings

As a result of the analyses and calculations, the peak displacement and base shear force values at the performance point for Models 01, 01D, and 01V are provided in Table 8.

The damage zones at the performance point are shown in Figures 9–11.

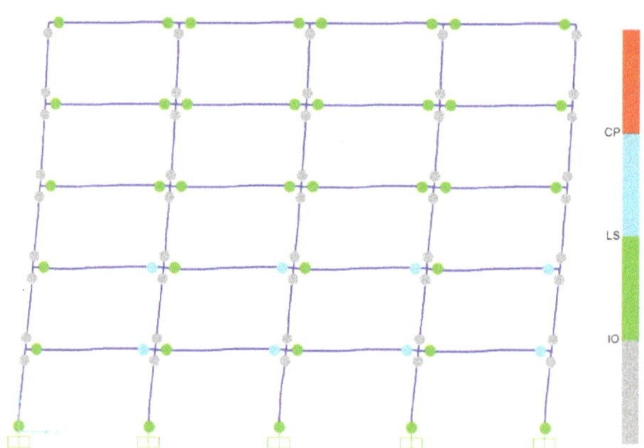

Figure 9. Section damage conditions at the performance point for Model 01.

Table 8. Peak displacement—base shear force values at the performance point.

Model Nu.	Peak Displacement (m)	Base Shear Force (kN)
Model 01	0.289	272.37
Model 01D	0.106	1017.98
Model 01V	0.087	913.76

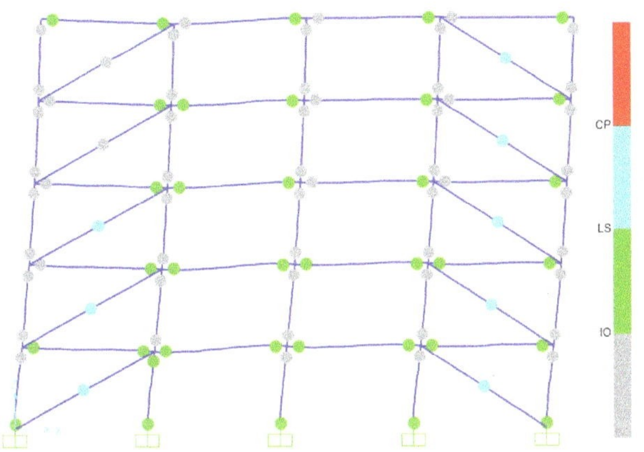

Figure 10. Section damage conditions at the performance point for Model 01D.

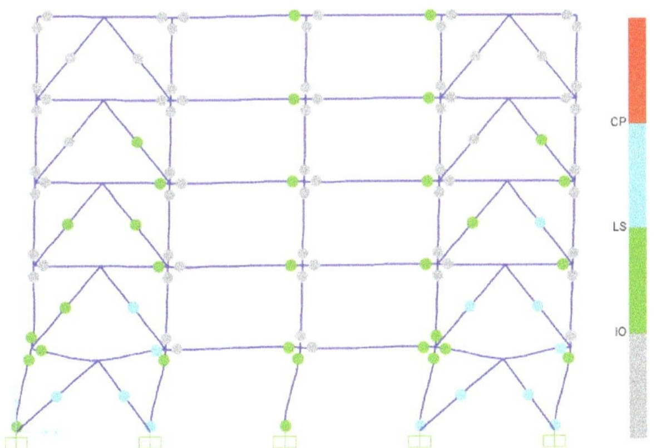

Figure 11. Damage conditions of the section at the performance point for Model 01V.

The numerical distributions of the column and beam damage regions are presented in Tables 9 and 10.

Table 9. Numerical distribution of column-damage regions at the performance point.

Model Nu.	The Number of Columns in the Damage Regions (Number)			
	LDZ (Gray)	SDZ (Green)	ADZ (Blue)	CDZ (Red)
Model 01	20	5	0	0
Model 01D	20	5	0	0
Model 01V	18	4	3	0

Table 10. Numerical distribution of beam-damage regions at the performance point.

Model Nu.	The Number of Beams in the Damage Regions (Number)			
	LDZ (Gray)	SDZ (Green)	ADZ (Blue)	CDZ (Red)
Model 01	0	8	12	0
Model 01D	0	20	0	0
Model 01V	4	14	2	0

In the Figure 9, the tilting and displacement of the structure towards the upper floors demonstrate the effect of lateral displacement. Looking at the color distribution, the vast majority of the system remains within the IO (Immediate Occupancy) and LS (Life Safety) performance levels. This indicates that the structure behaves safely under the influence of an earthquake and does not carry a serious risk of collapse. None of the laminated timber braces, as shown in Figures 10 and 11, reached their full capacity at the performance point. Therefore, the strengths and dimensions of the selected laminated braces were sufficient. The numerical distribution of the laminated timber braces according to their capacity regions is shown in Table 11.

Table 11. Numerical distribution of laminated timber braces according to axial load capacity ratios.

Model Nu.	Number of Braces According to Capacity Regions			
	Number of Braces	%0–50	%50–75	<%100
Model 01D	10	2	0	8
Model 01V	20	6	6	8

According to the principles specified in Section 15.8 of TBEC-2018, performance level checks were carried out by progressing from the cross-section to the structural elements, and from the structural elements to the entire structural system. The performance levels of the systems are presented in Table 12.

Table 12. Performance levels of frame models.

Model Nu.	Performance Level
Model 01	Collapse Prevention
Model 01D	Controlled Damage
Model 01V	Collapse Prevention

The performance level of the existing structural system (Model 01) is Collapse Prevention. According to TBEC–2018, at this performance level, the current condition and use of a building pose safety concerns for life and property.

As specified in TBEC–2018 Section 15.8.1, under the DD-2 earthquake ground motion level, the Building Performance Target for the existing structural system is the Controlled Damage performance level, as shown in Table 13.

Table 13. Performance targets for the existing structural system (Model 01).

Building Use Category (TBEC-2018 Table 3.1)	Earthquake Design Class (TBEC-2018 Table 3.2)	Building Height Class (TBEC-2018 Table 3.3)	Normal Performance Target (TBEC-2018 Table 3.4.c)	Evaluation Approach
BUC = 3	EDC = 1	BHC = 6	LD	DDM (Deformation Design Method)

The strengthening approach using laminated timber cross braces in Model 01D successfully raised the system performance level to the controlled damage performance level.

However, in Model 01V, the use of reverse V braces, which are both more numerous and have a higher axial compression capacity, made the structure more rigid. As a result, during the push to the peak displacement, the base shear force reached 1592.88 kN by the 6th step, and due to the high plastic rotations in the ground floor columns, the load-carrying capacities of the columns significantly decreased. Along with these sudden capacity losses, by the 11th step of the push, the base shear force dropped to 913.76 kN at the peak displacement value.

This situation is illustrated by the system capacity curve in Figure 12. In such a case, if the push steps were continued, the system capacity would rapidly deplete, and after reaching the maximum peak displacement, collapse would occur. The system then collapses entirely while still carrying a small amount of load through forces.

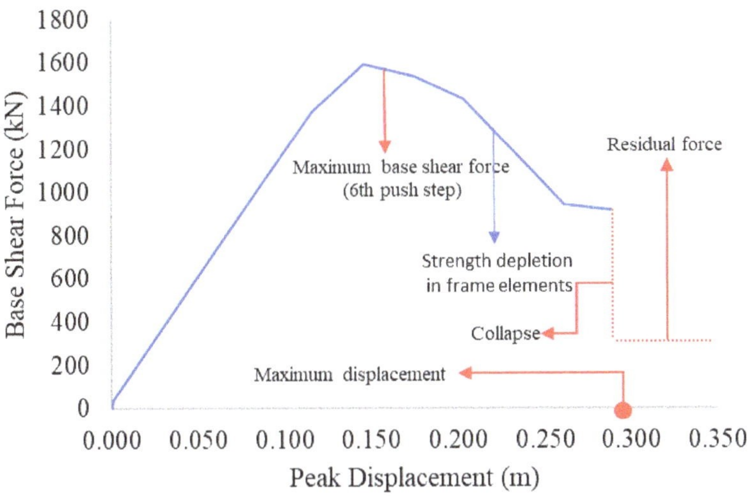

Figure 12. Capacity curve of Model 01V and probable collapse scenario.

The capacity curves of Models 01, 01D, and 01V are shown together in Figure 13. While the frame is quite ductile in its initial state, adding laminated timber cross-braces results in a significant reduction in ductility, accompanied by a considerable increase in the base shear force. In Model 01D with diagonal cross-braces, which successfully achieves the targeted performance level, the optimal performance point is reached under the current conditions with minimal section damage.

The story displacements of the frames are shown in Figure 14. Without laminated timber braces (Model 01), the peak displacement of the system is approximately 2% of the total height. Models 01D and 01V exhibit similar displacements, with the peak displacements for the two models being approximately 0.7% and 0.6% of the total height, respectively.

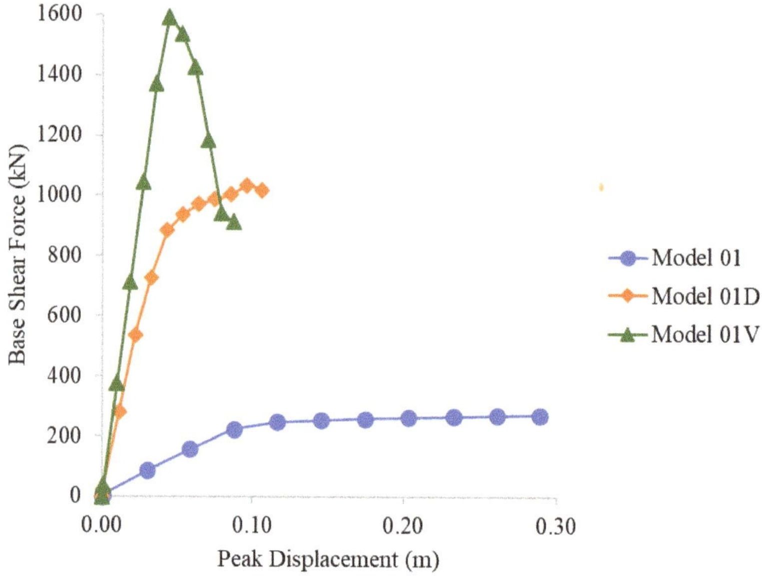

Figure 13. Comparison of the capacity curves of frames.

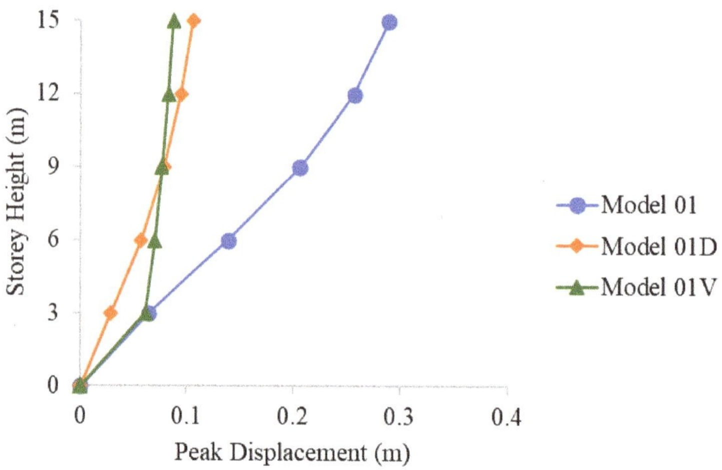

Figure 14. Comparison of the story displacements.

6. Conclusions and Recommendations

In this study, a 2D frame with a height of 3 m per story and a 5 m beam span, lacking earthquake shear walls, with insufficient seismic performance, consisting of a Ground + 4 Normal Floors, was analyzed. To improve the seismic performance, diagonal and reverse V-arrangement laminated glued-timber braces were added at the edge openings. The performance of the created models was examined according to TBEC-2018, and the controlled damage performance level requested by the regulation for the existing load-bearing system at the DD-2 earthquake ground motion level was achieved using Model 01D.

As a result of the analyses, the following observations were made:

a. To improve the seismic performance of structures in disadvantaged areas, laminated glued timber braces can be used due to their speed and ease of application.

b. Glued laminated timber has a high potential to restrict the displacements and increase the strength of reinforced concrete systems.
c. For buildings with inadequate seismic performance, the rigidity of the glued-laminated timber braces used in strengthening works should be compatible with the system rigidity. Therefore, repeated analyses may be required.
d. If glued laminated timber braces or a brace arrangement with higher rigidity than the existing system's rigidity is applied, the base shear force is resisted through the columns, which could lead to the system reaching a collapse mechanism, particularly in the ground floor columns, after a sudden loss of strength, without achieving the desired performance target. Therefore, in addition to adding laminated braces to the system, column strengthening may also be necessary.
e. In the future, further integration of glulam systems with performance criteria such as energy efficiency, thermal comfort, and structural strength holds great potential for sustainable architectural and engineering practices.
f. Glued laminated timber is a significant structural material in sustainable building design due to its high load-bearing capacity, low density, and suitability for prefabrication. Compared to concrete and steel, glulam has a significantly lower embodied energy, which substantially reduces its carbon footprint.

Furthermore, its environmental impact is minimized because its production process consumes less energy and is derived from renewable resources. Glulam's ability to enhance the energy efficiency of buildings throughout their lifecycle makes it an advantageous choice, especially for projects developed with low-carbon targets in mind. Considering all these aspects, glulam offers a strong and viable alternative for designing sustainable structural systems and excels in both environmental and structural performance.

Additionally, studies on the connections of glued laminated timber braces with reinforced concrete frames would be beneficial for the development of this type of strengthening pproach.

Author Contributions: Conceptualization, Y.Y.; Methodology, Y.Y.; Software, Y.Y.; Resources, F.Ş.; Writing—original draft, Y.Y.; Writing—review & editing, F.Ş.; Visualization, Y.Y.; Supervision, F.Ş.; Project administration, F.Ş. All authors have read and agreed to the published version of the manuscript.

Funding: This research received no external funding.

Institutional Review Board Statement: Not applicable.

Informed Consent Statement: Not applicable.

Data Availability Statement: The original contributions presented in this study are included in the article. Further inquiries can be directed to the corresponding author.

Conflicts of Interest: The authors declare no conflict of interest.

References

1. Jadallah, M.; Almustafa, M.K.; Dogangün, A.; Nehdi, M.L. Performance of X and inverted V bracing systems in controlling progressive collapse of reinforced concrete buildings. *Results Eng.* **2005**, *27*, 105812. [CrossRef]
2. Doğangün, A. *Earthquake-Soil and Earthquake Resistant Building Design*; Birsen Publishing: Istanbul, Turkey, 2021; p. 545. ISBN 978-975-511-710-2. (In Turkish)
3. Watanabe, F.; Miyazaki, S.; Tani, M.; Kono, S. Seismic strengthening using precast prestressed concrete braces. In Proceedings of the 13th World Conference on Earthquake Engineering, Vancouver, BC, Canada, 1–6 August 2004.
4. Alahi, F.N.; Vatandoost, M.M. Single diagonal precast prestressed concrete bracing for strengthening existing concrete frames. *Int. J. Adv. Struct. Eng.* **2018**, *10*, 339–347. [CrossRef]
5. Kono, S.; Katayama, T. Seismic retrofit of reinforced concrete building structures with prestressed braces. *J. Adv. Concr. Technol.* **2009**, *7*, 337–345. [CrossRef]

6. Vandoros, K.G.; Dritsos, S.E. Concrete jacket construction detail effectiveness when strengthening RC columns. *Constr. Build. Mater.* **2008**, *22*, 264–276. [CrossRef]
7. Minafo, G. A practical approach for the strength evaluation of RC columns reinforced with RC jackets. *Eng. Struct.* **2015**, *85*, 162–169. [CrossRef]
8. Meda, A.; Mostosi, S.; Rinaldi, Z.; Riva, P. Corroded RC columns repair and strengthening with high performance fiber reinforced concrete jacket. *Mater. Struct.* **2016**, *49*, 1967–1978. [CrossRef]
9. Yang, X.; Zhang, B.; Zhou, A.; Wei, H.N.; Liu, T.J. Axial compressive behaviour of corroded steel reinforced concrete columns retrofitted with a basalt fibre reinforced polymer-ultrahigh performance concrete jacket. *Compos. Struct.* **2023**, *304*, 116447. [CrossRef]
10. Shayanfar, J.; Omidalizadeh, M.; Nematzadeh, M. Analysis-oriented model for seismic assessment of RC jacket retrofitted columns. *Steel Compos. Struct.* **2020**, *37*, 371–390.
11. Epackachi, S.; Sharma, N.; Whitaker, A.; Hortacsu, A. A cyclic backbone curve for squat reinforced concrete shear walls. In Proceedings of the 11th National Conference on Earthquake Engineering, Los Angeles, CA, USA, 25–29 June 2018.
12. Guan, M.; Sha, M.; Wang, J.; Hang, X.; Jin, G. Cyclic behavior of self-slitting squat composite shear walls with concrete-filled steel tubes: Experiment. *J. Constr. Steel Res.* **2023**, *210*, 108054. [CrossRef]
13. Arafa, A.; Farghaly, A.S.; Benmokrane, B. Prediction of flexural and shear strength of concrete squat walls reinforced with GFRP bars. *J. Compos. Constr.* **2018**, *22*, 854. [CrossRef]
14. Han, W.; Zhao, Z.; Qian, J.; Zhang, Y.; Ma, T. Experimental seismic behavior of squat shear walls with precast concrete hollow molds. *Earthq. Eng. Eng. Vib.* **2019**, *18*, 540–547. [CrossRef]
15. Nagib, M.T.; Sakr, M.A.; El-khoriby, S.R.; Khalifa, T.M. Cyclic behavior of squat reinforced concrete shear walls strengthened with ultra-high-performance fiber-reinforced concrete. *Eng. Struct.* **2021**, *246*, 112999. [CrossRef]
16. Tariq, M.; Khan, A.; Ullah, A.; Zamin, B.; Kashyzadeh, K.R.; Ahmad, M. Gene expression programming for estimating shear strength of RC squat wall. *Buildings* **2022**, *12*, 918. [CrossRef]
17. Sivaguru, V.; Rao, G.A. Strength and behavior of reinforced concrete squat shear walls with openings under cyclic loading. *ACI Struct. J.* **2021**, *118*, 51732832. [CrossRef]
18. Peng, Y.; Wu, H.; Zhuge, Y. Strength and drift capacity of squat recycled concrete shear walls under cyclic loading. *Eng. Struct.* **2015**, *100*, 356–368. [CrossRef]
19. Zheng, S.; Li, Q.; Qin, Q.; Gan, C.; Liu, W.; Dong, L. Restoring force model of freezing-thawing damaged squat reinforced concrete shear walls. *J. Build. Struct.* **2018**, *39*, 111–119.
20. Yang, K.H.; Mun, J.H.; Hwang, Y.H.; Song, J.K. Cyclic tests on slip resistance of squat heavyweight concrete shear walls with construction joints. *Eng. Struct.* **2017**, *141*, 334–344. [CrossRef]
21. Kaleybar, R.S.; Tehrani, P. Effects of using different arrangements and types of viscous dampers on seismic performance of intermediate steel moment frames in comparison with different passive dampers. *Structures* **2021**, *33*, 3382–3396. [CrossRef]
22. Lavan, O.; Levy, R. Optimal Peripheral Drift Control of 3d Irregular Framed Structures Using Supplemental Viscous Dampers. 2006. Available online: www.worldscientific.com (accessed on 11 April 2025).
23. De Domenico, D.; Ricciardi, G.; Takewaki, I. Design strategies of viscous dampers for seismic protection of building structures: A review. *Soil Dyn. Earthq. Eng.* **2019**, *118*, 144–165. [CrossRef]
24. Pekcan, G.; Mander, J.B.; Chen, S.S. The seismic response of a 1:3 scale model R.C. structure with Elastomeric Spring Dampers. *Earthq. Spectra* **1995**, *11*, 249–267. [CrossRef]
25. Ryan, K.L.; Kelly, J.M.; Chopra, A.K. Nonlinear model for lead–rubber bearings including axial-load effects. *J. Eng. Mech.* **2005**, *131*, 1270–1278. [CrossRef]
26. Ozdemir, G.; Bayhan, B. Response of an isolated structure with deteriorating hysteretic isolator model. *Res. Eng. Struct. Mater.* **2015**, *1*, 1–9. [CrossRef]
27. Taylor Devices Inc. *Fluid Viscous Dampers General Guidelines for Engineers Including a Brief History*; Taylor Devices Inc.: North Tonawanda, NY, USA, 2020.
28. Almajhali, K.Y.M. Review on passive energy dissipation devices and techniques of installation for high rise building structures. *Structures* **2023**, *51*, 1019–1029. [CrossRef]
29. Tremblay, R.; Dehghani, M.; Fahnestock, L.; Herrera, R.; Canales, M.; Clifton, C.; Hamid, Z. Comparison of seismic design provisions for buckling restrained braced frames in Canada, United States, Chile, and New Zealand. *Structures* **2016**, *8*, 183–196. [CrossRef]
30. Deylami, A.; Mahdavipour, M.A. Probabilistic seismic demand assessment of residual drift for Buckling-Restrained Braced Frames as a dual system. *Struct. Saf.* **2016**, *58*, 31–39. [CrossRef]
31. Khorami, M.; Alvansazyazdi, M.; Shariati, M.; Zandi, Y.; Jalali, A.; Tahir, M. Seismic performance evaluation of buckling restrained braced frames (BRBF) using incremental nonlinear dynamic analysis method (IDA). *Earthq. Struct.* **2017**, *13*, 531. [CrossRef]

32. Fang, C.; Zhong, Q.; Wang, W.; Hu, S.; Qiu, C. Peak and residual responses of steel moment-resisting and braced frames under pulse-like near-fault earthquakes. *Eng. Struct.* **2018**, *177*, 579–597. [CrossRef]
33. Veismoradi, S.; Darvishan, E. Probabilistic seismic assessment of mega bucklingrestrained braced frames under near-fault ground motions. *Earthq. Struct.* **2018**, *15*, 487–498.
34. Li, H.N.; Liu, Y.; Li, C.; Zheng, X.W. Multihazard fragility assessment of steel-concrete composite frame structures with buckling-restrained braces subjected to combined earthquake and wind. *Struct. Des. Tall Spec. Build.* **2020**, *29*, e1746. [CrossRef]
35. Beiraghi, H. Seismic response of buckling restrained braced frame with strong back system. *Ing. Sismica* **2021**, *37*, 65–81.
36. Pandikkadavath, M.S.; Sahoo, D.R. Analytical investigation on cyclic response of buckling-restrained braces with short yielding core segments. *Int. J. Steel Struct.* **2016**, *16*, 1273–1285. [CrossRef]
37. Pandikkadavath, M.S.; Sahoo, D.R. Mitigation of seismic drift response of braced frames using short yielding-core BRBs. *Steel Compos. Struct.* **2017**, *23*, 285–302. [CrossRef]
38. Hoveidae, N. Numerical investigation of seismic response of hybrid buckling restrained braced frames. *Period. Polytech. Civ. Eng.* **2019**, *63*, 130–140. [CrossRef]
39. Hoveidae, N.; Radpour, S. A novel all-steel buckling restrained brace for seismic drift mitigation of steel frames. *B Earthq. Eng.* **2021**, *19*, 1537–1567. [CrossRef]
40. Beiraghi, H. Near-fault ground motion effects on the responses of tall reinforced concrete walls with buckling-restrained brace outriggers. *Sci. Iran.* **2018**, *25*, 1987–1999.
41. Pierobon, F.; Huang, M.; Simonen, K.; Ganguly, I. Environmental benefits of using hybrid CLT structure in midrise non-residential construction: An LCA based comparative case study in the U.S. Pacific Northwest. *J. Build. Eng.* **2019**, *26*, 100862. [CrossRef]
42. Tokede, O.O.; Rodgers, G.; Waschl, B.; Salter, J.; Ashraf, M. Harmonising life cycle sustainability thinking in material substitution for buildings, Resour. *Conserv. Recycl.* **2022**, *185*, 106468. [CrossRef]
43. Shin, B.; Kim, S. Advancing the circular economy and environmental sustainability with timber hybrid construction in South Korean public building. *Build. Environ.* **2024**, *257*, 111543. [CrossRef]
44. Asdrubali, F.; Ferracuti, B.; Lombardi, L.; Guattari, C.; Evangelisti, L.; Grazieschi, G. A review of structural, thermo-physical, acoustical, and environmental properties of wooden materials for building applications. *Build. Environ.* **2017**, *114*, 307–332. [CrossRef]
45. Ramage, M.H.; Burridge, H.; Busse-Wicher, M.; Fereday, G.; Reynolds, T.; Shah, D.U.; Wu, G.; Yu, L.; Fleming, P.; Densley-Tingley, D.; et al. The wood from the trees: The use of timber in construction. *Renew. Sustain. Energy Rev.* **2017**, *68*, 333–359. [CrossRef]
46. Jayalath, A.; Navaratnam, S.; Ngo, T.; Mendis, P.; Hewson, N.; Aye, L. Life cycle performance of Cross Laminated Timber mid-rise residential buildings in Australia. *Energy Build.* **2020**, *223*, 110091. [CrossRef]
47. Chen, C.H.; Firdausy, A.I.; Lin, Y.W. Experimental investigation of buckling-restrained braces with glulam-filled steel tube. *J. Constr. Steel Res.* **2025**, *234*, 109688. [CrossRef]
48. Dorn, M.; De Borst, K.; Eberhardsteiner, J. Experiments on dowel-type timber connections. *Eng. Struct.* **2013**, *47*, 67–80. [CrossRef]
49. Meng, X.; Lia, T.; Yang, Q. Experimental study on the seismic mechanism of a full-scale traditional Chinese timber structure. *Eng. Struct.* **2019**, *180*, 484–493. [CrossRef]
50. Chen, Z.; Popovski, M. Connection and System Ductility Relationship for Braced Timber Frames. *J. Struct. Eng.* **2020**, *146*, 4020257. [CrossRef]
51. Susteric, I.; Dujic, B. Seismic strengthening of existing buildings with cross laminated timber panels. *World* **2012**, *15*, 19.
52. Juhász, T.; Lee, Y.; Holtzman, R.; Balogh, J. Experimental study of failure of glulam-concrete composite beams. *Pollack Period.* **2024**, *19*, 13–18. [CrossRef]
53. Selçukoğlu, E.; Zwicky, D. Towards the Plastic Design of Glulam Concrete Structures. In Proceedings of the IABSE Symposium, Bangkok, Thailand, 9–11 September 2009; pp. 20–29.
54. Beceren, Ö.R.; Arıoğlu, N. Mechanical properties of laminated timber beams from Turkish Scots pine. *ITU J-FET* **2006**, *5*, 25–36.
55. Sandoli, A.; D'Ambra, C.; Ceraldi, C.; Calderoni, B.; Prota, A. Sustainable Cross-Laminated Timber Structures in a Seismic Area: Overview and Future Trends. *Appl. Sci.* **2021**, *11*, 2078. [CrossRef]
56. Furukawa, S.; Iwami, R.; Kimura, Y. Development of an Environmentally Friendly Steel Structural Framework: Evaluation of Bending Stiffness and Yield Bending Moment of Cross-Laminated Timber Slab–H-Shaped Steel Composite Beams for Component Reuse. *Sustainability* **2025**, *17*, 2073. [CrossRef]
57. Aghayere, A.; Vigil, J. *Structural Wood Design: A Practice-Oriented Approach Using the ASD Method*; John Wiley & Sons, Inc.: Hoboken, NJ, USA, 2007.
58. Swedish Forest Industries Federation. *Design of Timber Structures—Structural Aspects of Timber Construction*; Swedish Forest Industries Federation: Stockholm, Sweden, 2016; Volume 1.
59. Disaster and Emergency Management Authority (AFAD). *Turkish Building Earthquake Code: Design Principles for Buildings Under Seismic Action*; Disaster and Emergency Management Authority: Ankara, Turkey, 2018.

60. *ACI 318-99*; Building Code Requirements for Structural Concrete and Commentary. American Concrete Institute (ACI): Detroit, MI, USA, 1999.
61. The European Union Per Regulation. *Design of Timber Structures—Part 1-1: General—Common Rules and Rules for Buildings*; The European Union Per Regulation: Brussels, Belgium, 2004.
62. Computers and Structures Inc. *SAP2000 V.23 v.23.3.1: Structural Analysis and Design*; Computers and Structures Inc.: New York, NY, USA, 2022.
63. Applied Technology Council. *Seismic Evaluation and Retrofit of Concrete Buildings, ATC-40*; Applied Technology Council: Redwood, CA, USA, 1996; Volumes 1 and 2.

MDPI AG
Grosspeteranlage 5
4052 Basel
Switzerland
Tel.: +41 61 683 77 34

Sustainability Editorial Office
E-mail: sustainability@mdpi.com
www.mdpi.com/journal/sustainability

www.ingramcontent.com/pod-product-compliance
Lightning Source LLC
Chambersburg PA
CBHW040823250926
51850CB00007BA/1249
* 9 7 8 3 7 2 5 8 7 3 8 6 9 *